# Bionanotechnology Towards Sustainable Management of Environmental Pollution

This book highlights the characteristics, aims, and applications of bionanotechnology as a possible solution for sustainable management and bioremediation of environmental pollutants. It covers remediation of toxic pollutants, removal of emerging contaminants from industrial wastewater, eco-design, and modification study of bionanoparticles and life cycle assessment, nano-filtration, bionanomaterial-based sensors for monitoring air and water pollution, resource recovery from wastewater, and highlights Internet of things based green nanotechnology.

- Provides a comprehensive solution of environmental problems in sustainable and cost-effective mode.
- Reviews bionanotechnological applications in nanomaterial design, modification, and treatment of emerging contaminants from industrial wastewater
- Covers eco-design study of bionanomaterials, bio-nano filters, and assessment for the treatment of emerging pollutants.
- Includes IoT-based bionanotechnology.
- Explores future research needs on bionanotechnology and scientific challenges in the mitigation of environmental pollutants.

This book is aimed at researchers, professionals, and graduate students in nanobiotechnology, environmental engineering, and biotechnology.

# Advances in Bionanotechnology

***Series Editors:***
***Ravindra Pratap Singh***
*Department of Biotechnology, Indira Gandhi National Tribal University, Anuppur, Madhya Pradesh, India*

***Jay Singh***
*Department of Chemistry, Institute of Science, Banaras Hindu University, Varanasi, Uttar Pradesh, India*

***Charles Oluwaseun Adetunji***
*Department of Microbiology, Edo State University Uzairue, Iyamho, Edo State, Nigeria*

**Series Description**

Bionanotechnology is a multidisciplinary field that shows immense applicability in different domains, namely chemistry, physics, material sciences, biomedical, agriculture, environment, robotics, aeronautics, energy, electronics, and so forth. This book series will explore the enormous utility of bionanotechnology for biomedical, agricultural, environmental, food technology, space industry, and many other fields. It aims to highlight all the spheres of bionanotechnological applications and its safety and regulations for using biogenic nanomaterials that are a key focus of researchers globally.

**Bionanotechnology Towards Sustainable Management of Environmental Pollution**
*Edited by Naveen Dwivedi & Shubha Dwivedi*

# Bionanotechnology Towards Sustainable Management of Environmental Pollution

Edited by
Naveen Dwivedi
Shubha Dwivedi

CRC Press is an imprint of the
Taylor & Francis Group, an **informa** business

First edition published 2023
by CRC Press
6000 Broken Sound Parkway NW, Suite 300, Boca Raton, FL 33487-2742

and by CRC Press
4 Park Square, Milton Park, Abingdon, Oxon, OX14 4RN

*CRC Press is an imprint of Taylor & Francis Group, LLC*



**Library of Congress Cataloging-in-Publication Data**

Names: Dwivedi, Naveen, editor. | Dwivedi, Shubha (Professor of biotechnology), editor.
Title: Bionanotechnology towards sustainable management of environmental pollution / edited by Naveen Dwivedi, Shubha Dwivedi.
Description: First edition. | Boca Raton : CRC Press, 2023. | Series: Advances in bionanotechnology | Includes bibliographical references and index.
Identifiers: LCCN 2022012483 (print) | LCCN 2022012484 (ebook) | ISBN 9781032220383 (hardback) | ISBN 9781032220390 (paperback) | ISBN 9781003270959 (ebook)
Subjects: LCSH: Bioremediation. | Nanobiotechnology. | Pollution prevention. | Environmental engineering.
Classification: LCC TD192.5 .B555 2023 (print) | LCC TD192.5 (ebook) | DDC 628.5--dc23/eng/20220725
LC record available at https://lccn.loc.gov/2022012483
LC ebook record available at https://lccn.loc.gov/2022012484

ISBN: 9781032220383 (hbk)
ISBN: 9781032220390 (pbk)
ISBN: 9781003270959 (ebk)

DOI: 10.1201/9781003270959

Typeset in Times
by Deanta Global Publishing Services, Chennai, India

# Contents

# Preface

The high rate of industrialization and urbanization has led to extreme ecological disturbances and deprivation of natural resources. The orthodox practices in industrial and agricultural sectors have produced large amounts of hazardous wastes which cause water, air, and soil pollution, devastating human health and the environment. Most of the pollutants have been identified as major sources of environment and health hazards due to their high levels of toxicity. Existing conventional treatment techniques employ expensive chemical methods; however, these methods are disagreeable from both economic and environmental viewpoints. Therefore, there is an exigent demand for the development of sustainable, effective, eco-friendly, cost-effective, and reliable technology to monitor and prevent all the striking environmental challenges. Hence, bionanotechnology can have a substantial impact on developing "cleaner" and "greener" technologies with significant health and environmental benefits. The applications of bionanotechnology are being explored for their potential to provide solutions to manage, mitigate, and clean up air, water, and soil pollution, as well as to improve the performance of conventional technologies used in environmental clean-up and sustainability. Therefore, this book represents a study of bionanotechnology concepts and their role in revolutionizing conventional treatment methods accompanied with eliminating or minimizing negative influence of hazardous contaminants on human health and the environment. The realization of smart and sustainable treatment technology will substantially provide significant economic and social benefits.

In this book, an interdisciplinary team of researchers comprises a sustainable approach of green/bionanotechnology in the field of environment with their advancement and highlighting the technical, scientific, regulatory, safety, and societal impacts. This book provides a solid understanding of the subject knowledge with current ongoing sustainable solutions and contemporary challenges of bioremediation of environmental pollutants using bionanotechnology-based interventions. Readers will learn all about the recent and sustainable progress in both theoretical and practical aspects and future potential applications of bionanotechnology in the environment. We are thankful to the publisher (Engineering) team, CRC Press, and all our contributors whose great efforts made this book possible.

**Naveen Dwivedi**
*Muzaffarnagar, Uttar Pradesh, India*

**Shubha Dwivedi**
*Meerut, Uttar Pradesh, India*

# Editors

**Dr Naveen Dwivedi** is currently working as Professor and Head in the Department of Biotechnology at the S. D. College of Engineering and Technology, Muzaffarnagar, Uttar Pradesh, India. He earned his M. Tech. degree in Biotechnology from the Institute of Engineering & Technology, Lucknow, Uttar Pradesh, India. Dr Dwivedi completed his M. Tech. dissertation work from the Fermentation Division of Central Drug Research Institute (CSIR-CDRI), Lucknow, Uttar Pradesh, India, and Ph.D. research work from the Bioenergy & Environmental Engineering Research Lab, Indian Institute of Technology Roorkee, Uttarakhand, India. Dr Dwivedi is broadly interested in the fields of environmental biotechnology, nanobiotechnology, biological remediation of wastewater, bioprocess engineering, and fermentation biotechnology. Dr Dwivedi has more than 17 years of teaching experience. He has about 50 research publications in peer-reviewed journals and conferences. He has also contributed 20 chapters in books published by Elsevier, CRC Press, AAP, Springer, and Wiley. He is an author of the book *Introduction to Biotechnology*, published by University Science Press, New Delhi. He is also engaged in editing various book projects of international repute with Wiley and CRC Press. Dr Dwivedi has received many research awards and grants. Dr Dwivedi has delivered several talks and guest lectures on various topics of biotechnology, nanobiotechnology, and bioenergy at different places in India and abroad. He is a member of editorial boards and is a reviewer of several international journals. He is the senior member of the Universal Association of Civil, Structural, and Environmental Engineers (IRED), New York, USA, and the Society of Chemical Industry (SCI).

**Dr Shubha Dwivedi**, a researcher by profession, is an Associate Professor in the Department of Biotechnology, School of Life Science and Technology, IIMT University, Meerut, India. She received her M.Sc. degree in Biochemistry from Jiwaji University, Gwalior, Madhya Pradesh, India, and MBA degree from Punjab Technical University, Kapurthala, India. Dr Dwivedi completed her M. Sc. dissertation work from the Institute of Genomics and Integrative Biology (CSIR-IGIB), Delhi, India, and Ph.D. research work from the Sustainable Processing and Water Treatment Research Lab, Indian Institute of Technology Roorkee, Uttarakhand, India. She has more than 16 years of teaching experience and research work in the area of biochemical engineering and environmental biotechnology. Her area of work focuses on eco-friendly technology development for the removal of hazardous contaminants from wastewater. She has published more than 47 research papers in international journals and conference proceedings. Moreover, she has published more than a dozen book chapters with CRC Press, AAP, Wiley, Elsevier, and Springer. Dr Dwivedi has published one book, titled *Introduction to Biotechnology* (University Science Press, New Delhi). She is also engaged in three international book projects from Wiley and CRC Press. Recently, she was appointed session lead on an IPR workshop organized in RAIB-2021 at Precious Cornerstone University, Nigeria. She has chaired sessions at many national and international conferences. She has been invited as keynote speaker in various institutes of national and international repute. She is engaged in many international projects with various universities in terms of research exchange programs and organizing workshops and conferences, etc.

# List of Contributors

**Md Shahid Alam**
Department of Biosciences and Bioengineering
Indian Institute of Technology Roorkee
Roorkee, Uttarakhand, India

**Shruti Awasthi**
Garden City University
Bangalore, India

**Divyesh Bhisikar**
Symbiosis Institute of Technology
Symbiosis International (deemed a university)
Lavale, Pune, India

**Animesh Chatterjee**
School of Life Science and Technology
IIMT University
Meerut, India

**Subhakanta Dash**
Department of Chemistry
Synergy Institute of Engineering and Technology
Dhenkanal, Odisha, India

**Doli**
Faculty of Pharmacy
RBS Engineering Technical Campus
Bichpuri, Agra, Uttar Pradesh, India

**Naveen Dwivedi**
Department of Biotechnology
S. D. College of Engineering and Technology
Muzaffarnagar, India

**Shubha Dwivedi**
Department of Biotechnology
IIMT University
Meerut, India

**Deena Nath Gupta**
Department of Biosciences and Bioengineering
Indian Institute of Technology Roorkee
Roorkee, Uttarakhand, India

**Namrata Gupta**
RBS Engineering Technical Campus
Bichpuri, Agra, Uttar Pradesh, India

**Piyush Gupta**
Department of Chemistry
Faculty of Engineering and Technology
SRM Institute of Science and Technology
Delhi-NCR Campus
Ghaziabad, Uttar Pradesh, India

**Harry Kaur**
Department of Biosciences and Bioengineering
Indian Institute of Technology Roorkee
Roorkee, Uttarakhand, India

**Sumit Kaushik**
Faculty of Pharmacy
RBS Engineering Technical Campus
Uttar Pradesh, India

**Salman Khan**
Faculty of Pharmacy
IIMT University, Meerut, India

**M. Suresh Kumar**
CSIR-National Environmental Engineering Research Institute (CSIR-NEERI) and AcSIR
Ghaziabad, India

**Sunil Kumar**
CSIR-National Environmental Engineering Research Institute (CSIR-NEERI) and AcSIR
Ghaziabad, India

**Ram Kumar Lakshminarayana**
Department of Information Technology
University of Technology and Applied Sciences-Sur, Sultanate of Oman

**Sapna Lonare**
Department of Biosciences and Bioengineering
Indian Institute of Technology Roorkee
Roorkee, Uttarakhand, India

**Biljana S. Maluckov**
Technical Faculty in Bor
University of Belgrade,
Serbia

**Ashootosh Mandpe**
Department of Civil Engineering
Indian Institute of Technology Indore
Simrol, Madhya Pradesh, India

**Ved Kumar Mishra**
Department of Computational Biology & Bioinformatics
JIBB, SHUATS
Prayagraj

**Sonam Paliya**
CSIR-National Environmental Engineering Research Institute (CSIR-NEERI) and AcSIR
Ghaziabad, India

**Mehak Puri**
CSIR-National Environmental Engineering Research Institute (CSIR-NEERI) and AcSIR
Ghaziabad, India

**Preethi Rajesh**
Garden City University
Bangalore

**Neha Rana**
Department of Pharmacy
SRM Institute of Science and Technology, Delhi-NCR Campus
Delhi-Meerut Road, Modinagar, Ghaziabad, Uttar Pradesh, India

**Surabhi Rode**
Department of Biosciences and Bioengineering
Indian Institute of Technology Roorkee
Roorkee, Uttarakhand, India

**Deepa Sharma**
Department of Chemistry
IIMT University Meerut, Uttar Pradesh, India

**Rachita Sharma**
Department of Biotechnology
S. D. College of Engineering and Technology
Muzaffarnagar, India

**Gyanendra Singh**
Anand College of Pharmacy
Agra, Uttar Pradesh, India

**Monika Singh**
Faculty of Pharmacy
RBS Engineering Technical Campus
Bichpuri, Agra, Uttar Pradesh, India

**Priya Singh**
Department of Biotechnology
S. D. College of Engineering and Technology
Muzaffarnagar, India

**Nikita Singhal**
IIMT University Meerut, Uttar Pradesh, India

**Sanjukta Vidyant**
School of Life Science and Technology
IIMT University
Meerut, India

**Amit Yadav**
Faculty of Pharmacy, RBS Engineering Technical Campus
Bichpuri, Agra, Uttar Pradesh, India

**Mohd. Zafar**
Department of Applied Biotechnology
University of Technology and Applied Sciences-Sur
Sultanate of Oman

# 1 Advances and Challenges of Bionanotechnology in the Sustainable Management of the Environment and Green Economy

*Shubha Dwivedi and Naveen Dwivedi*

## CONTENTS



DOI: 10.1201/9781003270959-1

## 1.1 INTRODUCTION

The environment which was preserved by our ancestors is degrading rapidly. Global warming, carbon dioxide emissions, waste production, and uncontrolled exploitation of natural resources create a catastrophic situation and lead to climate change. "Clean India, Green India, and Healthy India" is not just a slogan—in fact it means a lot. A healthy mind lives in a healthy body and healthy bodies live in a healthy environment. Several issues arise due to the lack of proper management of various available technologies. Bionanotechnology can have a considerable influence on developing 'cleaner' and 'greener' technologies with important health and environmental benefits. The applications of bionanotechnology are being investigated for their potential to afford solutions to manage, mitigate, and clean up air, water, and land pollution, as well as to improve the performance of conventional technologies used in environmental clean-up. With the upsurge of the global human population, resources are dwindling. The development of pollution-free technologies for environmental clean-up and clean energy provisions for the sustainable growth of human society are the needs of the hour. Bionanotechnology has been developed as an applied technology in various arenas in recent times. It is a convergence of various sciences, providing the way to work at the nano level and create new structures. It includes the creation of nano-sized constituents and devices. In view of this description, it is found that bionanotechnology has numerous applications in various areas of life sciences, engineering, and medicine. The environment is one of the areas where bionanotechnology can be used extensively.

So many issues arise in terms of solid waste management, remediation, clean water supply, energy, etc. Bionanotechnology provides solutions to problems in a sustainable way. This chapter deals with a prominent application of bionanotechnology to resolve major environmental problems, such as municipal solid waste (MSW) management, air pollution, water scarcity, nanomaterial safety, and developing trash to treasure.

## 1.2 THE ROLE OF BIONANOTECHNOLOGY IN WASTEWATER TREATMENT

The shortage of water has become a global issue. It is an essential natural resource for sustaining life and environment that we have always thought would be freely available in abundance—a free gift of nature. Due to various ecological factors, both natural and anthropogenic, groundwater is getting polluted due to the presence of various hazardous contaminants, like fluoride, arsenic, nitrates, sulfates, heavy metals, etc. The contamination of surface and inland waters caused by the release of toxic chemicals from different industries can be dangerous to all classes of living organisms if discharged without proper treatment. Poisonous chemicals in industrial wastewater include inorganic and organic chemicals, both of which, even in extremely low concentrations, may be toxic to aquatic and terrestrial life. Therefore, treatment of these toxic materials is required before their discharge. To overcome the issue of water shortage, it is essential to reuse and recycle wastewater at all consumption levels, domestic as well as industrial.

There are several possible methods for treatment of contaminated wastewater. Existing conventional wastewater treatments for removal of contaminants employ expensive chemical methods, such as ion exchange, electrochemical, coagulation–precipitation, etc. However, these methods are not very friendly from both economic and environmental viewpoints because they require the use of chemical compounds and do not degrade the complete range of contaminants. The first one, i.e., adsorption is better than other wastewater treatment techniques in terms of efficiency, ease of operation, and low cost. The second—the biological approach—is better than either the physical or chemical methods. Bionanotechnology is an environmentally friendly and quite inexpensive treatment method. It is based on synergistic interaction of two technologies, and the method of preparation, bionanoparticles, is the most appropriate and effective technique for wastewater treatment. In subsequent sections of this chapter, the intriguing pathways and processes that researchers are developing, which are also naturally safer and more economical, will be described. These have led to new trends that involve bio-inspired nanoscale bioadsorbents for the remediation of a wide range of water contaminants. Various biomolecules *viz* carbohydrates, proteins, polymers, flavonoids, alkaloids, and several antioxidants obtained from various sources of live materials, like plants, bacteria, fungi, and algae have demonstrated their usefulness as capping and stabilizing agents during the process of manufacturing the bionanomaterials. It is a relatively newer and greener approach in the area of wastewater treatment that directly or indirectly boosts the green economy.

Nano-engineered particles, nano-membranes, photocatalysts, nano-adsorbents, and nano-metals offer prominent options for wastewater technologies which are capable of fulfilling customer requirements. State-of-the-art engineered nanomaterials are very promising for remediation of hazardous contaminants, as they have high surface areas due to their small size and remarkable reactivity. Subsequently, there is a pressing need to formulate these effective, high activity/efficiency, eco-friendly, and ease-to-handle green nanomaterials from live organisms. In view of this, green-structured biological nanomaterials could be an excellent option for the photocatalysis application in industrial water treatment systems.

The classic physicochemical approach for the formation of nanomaterials involves hazardous and volatile materials. This motivated researchers to discover biogenic and greener approaches which are eco-friendly, harmless, and cost-effective for the development of innovative and efficient nanoscale adsorbents for removing and degrading various contaminants in water. Certainly, the various phenolic antioxidants present in plants and other microorganisms serve as covering and reducing agents for the production of nanostructures of varied shapes, namely, flowers, wires, rods, and tubes.

Diarrhea is the third most common cause of death in children under five years, due to improper sanitation and unsafe drinking water, and it is responsible for 13% deaths in this age-group, killing an estimated 300,000 children in India every year (Bassani et al., 2010). Drinking water contamination is mainly categorized under physical contamination, chemical contamination, biological contamination, and radioactive contamination. Nanotechnology proves to be the favored technology for overcoming these challenges. Recently, the application of nanostructured materials for foraging and degrading toxic water contaminants is gaining attention due to the unique size-dependent properties of these particles, like large surface area, short intra-particle diffusion distance, and compressibility with negligible reduction in surface area. Excellent stability and adherence to the 5R (Reduce, Reuse, Recycle, Recovery, Removal) approach of sustainable development also contribute to the value of this method. For the formation of nanostructures with the preferred shape and size, there are two classic methods, based on the musters followed: top-down and bottom-up approach as shown in Figure 1.1. There are fundamental differences between these approaches for the synthesis of nanostructures. In the top-down approach, bulk materials are crushed into bits and pieces, leading to the production of fine nanostructures. Whereas, the bottom-up (or self-assembling) approach is characterized by amalgamation or assembling of atoms by atoms, molecules by molecules, cluster by cluster, to generate a diverse range of nanoparticles (NPs).

### 1.2.1 Advanced Bio-based Green Approach for the Synthesis of Bionanostructures

Biogenic nanostructures can be useful for remediation of pollutants in treatment plants; membrane bioreactors and sewage systems are the other state-of-the-art water purification strategies used to diminish or remove hazardous contaminants in water resources. These bio-based, biogenic nanostructures have tremendous potential in the removal of heavy metals as well as the degradation or adsorption of inorganic, organic, radioactive, pharmaceutical pollutants, nitro compounds, fluoride, cyanide, arsenic, etc. There are several eco-friendly and biological methods for the biogenic production of nanomaterials, using plants and microorganisms. Basically, there are four types of synthesis of nanostructures reported by researchers *viz* plant-mediated, bacteria-mediated, algae-mediated, and fungi-mediated. The presence of terpenoids, proteins, flavonoids, phenolic acid, vitamins, glycosides, polymers, alkaloids carbohydrates, and various antioxidants in such sources serve as covering/stabilizing and reducing agents for the production of sustainable bionanostructures, namely, nanowires, nanorods, nanoflowers, nanotubes, and nanoparticles. Various process parameters, like pH, temperature, incubation, time, aeration, mixing ratio, etc., are studied during the process (Zhang et al., 2018; Singh et al., 2016). Beveridge and Murray (1980) reported that utilization of microorganisms for the production of bionanostructures dates back to when their research group evaluated the synthesis

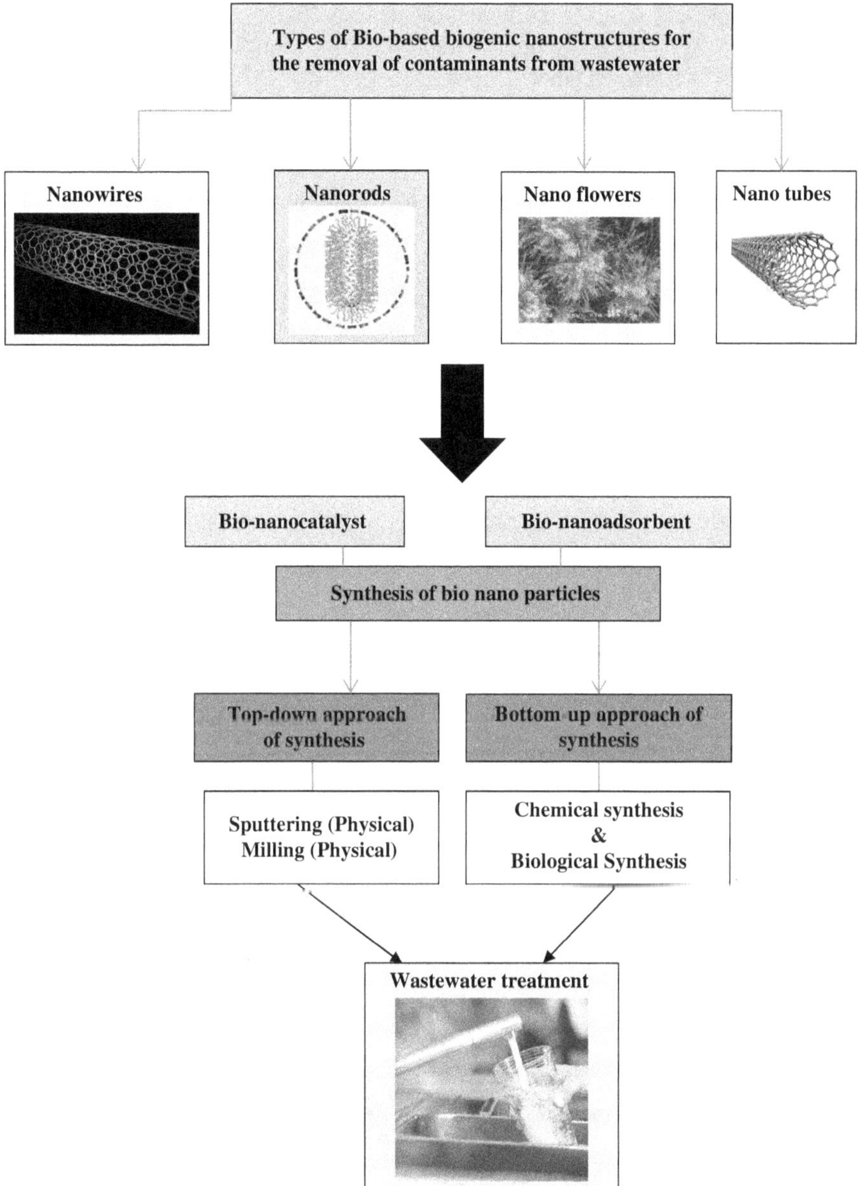


**FIGURE 1.1** Types of bio-based biogenic nanostructures for purposes in wastewater treatment.

of gold NPs by using the *Bacillus subtilis* as an aerobic, gram-positive bacterium. Undeniably, microorganisms have the capability to adsorb and accumulate metal ions, which can secrete a higher number of enzymes by cell activities, thereby increasing the reduction of metal ions to their elemental form (Beveridge and Murray, 1980; Sengani et al., 2017). Biomaterial can serve as a capable matrix and provide support and host for the bionanostructures (BNSs). Synthesis of bio-based biogenic nanostructures using phyto-nanotechnology is a sustainable technique for generating BNSs as shown below:

$$\textbf{Plant material} + \textbf{Salts of Metal} \rightarrow \textbf{BNS}\left(\textbf{Bio-nanostructures}\right)$$

$$+ \textbf{biodegradable by products}$$

Das et al., 2012 reported the mechanism of biosynthesis of silver nanoparticles by the fungi source *Rhizopus oryzae*, proving that intra or extracellular enzymes and proteins play an important role in the reduction of $AuCl_4^-$ ions to silver nanoparticles and their subsequent stabilization by the capping activity of the enzymes. Many researchers reported that some bacterial enzymes, like keratinase, nitrate reductase, alpha-amylase, and sulfite reductase, as well some plant enzymes, possess the ability to reduce and stabilize metal ions to nanostructures (Parandhaman et al., 2019; Duran et al., 2015). These enzymes, secreted from microorganisms and plants extracellularly or intracellularly, are absolutely suitable for the bulk synthesis of bionanostructures via a simplistic and eco-friendly approach (Thapa et al., 2017).

Figure 1.2 shows that when A reacts with B in the presence of various operational parameters, such as pH, heat, temperature, and rotation per minute, A reduces B into particular metal ions (P) and by-products. The rate of reduction and generation of BNS is affected by various factors, such as time, temperature, and pH.

In 2010, Coker et al. defined a new, gentle, and eco-friendly approach for the synthesis of biogenic magnetite nanoparticles ($Fe_3O_4$ NPs), and their subsequent embellishment with Pd NPs, by using bacterium *Geobacter sulfurreducens* to reduce $Fe_3^+$-oxyhydroxide and $Na_2PdCl_4$ ions without modifying the surface of the bio-mineral.

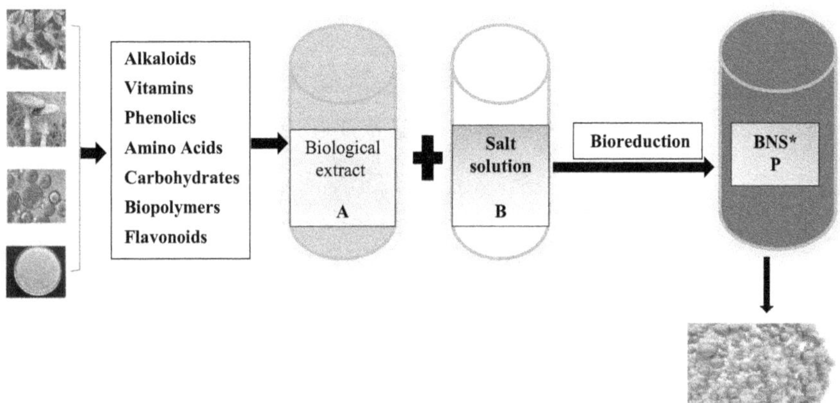


**FIGURE 1.2** Advance mechanism of synthesis of bionanostructures.

The overall conclusion is that biological/biogenic methods exploit microorganisms, algae, enzymes, plant polyphenols, and agricultural and industrial wastes. Among biosolids/biomolecules, enzymes and their metabolites (e.g., carbohydrates, proteins, nucleic acids, and peptides) have been employed as reducing/capping agents for the reduction of metal/metal oxide ions to generate various nanostructures. Further plant-mediated synthesis of bionanomaterials and bacterial-mediated synthesis of bionanostructures employing various species are presently under continued exploration. Various metal and metal oxide salts, including chlorides, acetates, and nitrates, possess high reduction potential because halogen also have an electron donation tendency, which can improve the electron density of metals on their conjugative salts, and/or metals were attached to acetate. In bacterial-mediated synthesis of bionanostructures, the bacterial cells perform numerous bio-processes, like biotransformation, bioleaching, bio-mineralization, and bioaccumulation, to solubilize the metal ions by altering their oxidation state through reduction and/or oxidation. Bacterial cells are capable of synthesizing metallic nanostructures by employing their defensive and protective strategies against soluble metal ions (Gautam et al., 2019).

Myco-nanotechnology is the term used for the synthesis of nanostructures from fungi. It is gaining attention due to the rich diversity of fungal species and their unique characteristics and metal-binding affinity (Dhillon et al., 2012). Similar to bacterial-mediated synthesis, fungi may also go under the process of biotransformation and biomineralization for the synthesis of myco-nanostructures (Das et al., 2012). Phyco nanotechnology is the term used for the synthesis of nanostructures from algal species. Algae has proven to be an efficient generator of nanostructures due to its diverse nature and with the help of a biotransformation mechanism.

### 1.2.2 Various Methods for the Removal of Contaminants from Polluted Water by Applying Nanostructures

There are various physical, chemical, and biological technologies available for treating wastewater, including electrochemical reduction, ion-exchange, adsorption, reverse osmosis, flocculation, sedimentation, membrane separation, ultrafiltration, and advanced oxidation processes (AOPs) as tabulated in Table 1.1. These are gradually implemented in the degradation of contaminants, due to their great competence, easy handling, easiness, and good reproducibility. AOP comprises the *in-situ* generation of highly reactive and nonselective chemical oxidants to degrade nonbiodegradable and resistant organic contaminants. Several pathways, such as UV photolysis/photocatalysis, adsorption, reduction, and photodegradation, are considered advanced approaches, includes bionanotechnology, and have been deployed for treating contaminants and removing organic/inorganic pollutants from groundwater, fresh water sediments, and wastewater (Jaafara et al., 2019).

A collection of biogenic nanostructures prepared for the remediation and degradation of various contaminants from wastewater are listed in Table 1.2 and Figure 1.3.

**TABLE 1.1**
**Various Methods of Removal of Contaminants from Polluted Water by Applying Nanostructures**

| S. No. | Removal method | Basic principle | Merits | Demerits | References |
|---|---|---|---|---|---|
| 1 | Electrochemical | Contaminants can be removed from analyte by electrode potential. The oxidation reduction reaction takes place. | The process was operated under a wide pH range (pH 5–9). The process was highly promising and reliable. It was a little expensive. | It was quite expensive. | Cui et al. (2012) |
| 2 | Ion exchange | Contaminants can be removed from water supplies with a strongly basic anion-exchange resin containing quaternary ammonium functional groups. The removal takes place according to the following reaction (Example): $\text{Matrix-NR3}^+\text{Cl}^- + \text{F}^- \rightarrow \text{Matrix-NR3}^+\text{F}^- + \text{Cl}^-$ | Removes up to 90–95%. Retains the taste and color of water intact | The technique is expensive because of the cost of resin, pretreatment required to maintain the pH, regeneration, and waste disposal. | Sundaram et al. (2008), Chubar et al. (2008) |
| 3 | Adsorption | Adsorption occurs quickly and may be monomolecular (unimolecular) layer or monolayer, or two, three or more layers thick (multi-molecular). Commonly used adsorbents are activated carbon fibers, alum impregnated activated alumina, activated carbon etc. | Treatment is cost-effective. The process can remove fluoride up to 90%. | Process is highly pH-dependent. The process has low adsorption capacity, poor integrity. Pretreatment is required. | Tor et al. (2009), Mohapatra et al. (2006) |
| 4 | Coagulation–precipitation | Coagulation is the destabilization of colloids by neutralizing the forces that keep them apart. Addition of lime leads to precipitation of contaminant as insoluble and raises the pH value of water up to 11–12. $Ca(OH)_2 + 2F^- \rightarrow CaF_2 + 2OH^-$ Lime and alum are the most commonly used coagulants. | - | It is not an automatic process, and the maintenance cost of the plant is very high. | Mohapatra et al. (2009) |

*(Continued)*

**TABLE 1.1 (CONTINUED)**
**Various Methods of Removal of Contaminants from Polluted Water by Applying Nanostructures**

| S. No. | Removal method | Basic principle | Merits | Demerits | References |
|---|---|---|---|---|---|
| 5 | Reverse osmosis | Molecules and ions from solutions by applying pressure to the solution when it is on one side of a selective membrane. | Process is highly effective for removal.<br>No chemicals are required and very little maintenance is needed.<br>It works under a wide pH range. | Remineralization is required after treatment.<br>The process is expensive in comparison to other options. The water becomes acidic and needs pH correction. | Sourirajan and Matsurra (1972); Simons (1993) |
| 6 | Membrane separation | Membrane separation involves the use of membrane for the separation of different chemicals. This method also uses some amount of force for removal as in reverse osmosis and electrodialysis. | Life of membrane is long and the process operates with minimal manpower.<br>The process is highly reliable.<br>The disinfection and treatment of water takes place in a single step. | Other common ions are also removed.<br>Remineralization of water was required after treatment so as to add the other essential components. | Meenakshi and Maheshwari (2006) |
| 7 | Electrodialysis | Electrodialysis is used to transport saltions from one solution through ion-exchange membranes to another solution under the influence of an applied electric potential difference. | It is used to purify small and medium-scale drinking water | Higher molecular weight, uncharged, and less mobile ionic species will not typically be significantly removed. | Kabay et al. (2008) |

*(Continued)*

**TABLE 1.1 (CONTINUED)**
**Various Methods of Removal of Contaminants from Polluted Water by Applying Nanostructures**

| S. No. | Removal method | Basic principle | Merits | Demerits | References |
|---|---|---|---|---|---|
| 8 | UV photolysis/ photocatalysis | Nanomaterial either adsorbs the contaminants or they degrade them by diverse catalytic methods | Potential utilizations in the green degradation of toxic organic contaminants | High-cost<br>Ozone may form toxic by-products<br>UV light penetration can be obstructed by turbidity | Chong et al. (2010) |
| 9 | Fenton's reaction | Able to degrade soluble and insoluble dyes from industrial wastewater | Novel approach | Low pH required,<br>pH adjustment increases the process cost | |
| 10 | Photodegradation | Compared with conventional water treatment processes (e.g., adsorption, conventional oxidation process, etc.), (photo) degradation and Fenton-like reaction have been broadly utilized in the pollutant treatment. | the potential applications of these individual approaches to economically dispose the toxic contaminants | Low efficiency/activity, low oxidation rate, low pH levels | Bremner et al. (2009) |
| 11 | Reduction | Among different reducing agents, $NaBH_4$ has been extensively considered a favored water-soluble reductant and preferred alternative to hydrogen sources in the reduction of toxic nitro compounds to significant and useful amino compounds in an aqueous medium. | Stimulated by the biosynthetic mineralization process | A metal substrate is required for the process | |

**TABLE 1.2**
**Biogenic Nanostructures in the Removal/Degradation of Contaminants from Wastewater**

| S. No | Biogenic nanostructures | Mechanism of removal | Prime contaminants chosen for removal | Bio-based sources used for synthesis | References |
|---|---|---|---|---|---|
| **1** | ZnO-Ag nano custard apple | Degradation | MB | Pomegranate peel | Kaviya and Prasad (2015) |
| **2** | Ag NPs | Degradation | RB-21, reactive Red-141 (RR-141) and Rhodamine-6G | Palm shell | Vanaamudan et al. (2016) |
| **3** | Ag-ZnO | Photo-degradation | MB | *Azadirachta indica* (Neem) leaf | Patil et al. (2016) |
| **4** | Ag NPs | Reduction | 4-NP | *Coleus forskohlii* root | Naraginti and Siva Kumar (2014) |
| **5** | Ag NPs | Photodegradation | Coomassie Brilliant Blue G-250 | *Coccinia grandis* leaf | Arunachalam et al. (2012) |
| **6** | Ag NPs | Degradation | MB | *Plectranthusamboinicus* leaf | Zheng et al. (2017) |
| **7** | Ag/$TiO_2$ NPs | Photodegradation | MB | Rambutan (*Nepheliumlappaceum* L.) peel | Kumar et al. (2016) |
| **8** | Ag NPs | Biosorptrion | Industrial effluents | *Morinda tinctoria* leaf | Vennila and Prabha (2015) |
| **9** | Ag NPs | Photo degradation | Methyl Red (MR) | *Piper pedicellatum* leaf | Tamuly et al. (2014) |
| **10** | Ag/CN-$TiO_2$ | Photodegradation | RhB | Bamboo leaf | Jiang et al. (2014) |
| **11** | Ag NPs | Photodegradation | MB | *Biebersteinia multifida* | Miri et al. (2018) |
| **12** | Au-Ag bimetallic nanocomposite | Reduction | 4-NP | *Silybum marianum* seed | Gopalakrishnan et al. (2015) |
| **13** | Ag NPs | Reduction | 4-NP | *Ficus hispida* Linn. f. leaf | Ramesh et al. (2018) |
| **14** | Ag NPs | Degradation | CR and MO | *Salvia microphylla* Kunth leaf | Lopez-Miranda et al. (2018) |
| **15** | Ag NPs | Reduction | 4-NP | *Allium ampeloprasum* L. leaf | Khoshnamv et al. (2019) |
| **16** | Pd NPs | Photodegradation | MB | *Andean blackberry* | Kumar et al. (2015) |

(*Continued*)

**TABLE 1.2 (CONTINUED)**
**Biogenic Nanostructures in the Removal/Degradation of Contaminants from Wastewater**

| S. No | Biogenic nanostructures | Mechanism of removal | Prime contaminants chosen for removal | Bio-based sources used for synthesis | References |
|---|---|---|---|---|---|
| 17 | Pd/walnut shell nanocomposite | Degradation | RhB, CR, and MB | *Equisetum arvense* L | Bordbar and Mortazavimanesh (2017) |
| 18 | Pd NPs | Reduction | organic dyes | *Terminalia arjuna* | Garai et al. (2018) |
| 19 | Cu/ZnO NPs | Degradation | MB and CR | *Euphorbia prolifera* leaf | Momeni et al. (2016) |
| 20 | Cu nanoflowers | Degradation | MB | *Ficus benghalensis* leaf | Robati et al. (2016) |
| 21 | Silver nanoparticles | Reduction | 4-NP | *Phoenix dactylifera L.* (date palm) leaf | Aitenneite et al. (2016) |
| 22 | Ag nanocomposite hydrogels based on sodium alginate | Bioremoval | MB | *Mukia maderaspatna* leaf | Karthiga et al. (2016) |
| 23 | CuO NPs/clinoptilolite | Degradation | 4-NP, RhB and MB | *Rheum palmatum* L. root | Bordbar et al. (2017) |
| 24 | Cu NPs | Removal | nitrate | Extract of *Hibiscus sabdariffa* flowers | Paixao et al. (2018) |
| 25 | Se | Removal | Hg | *Citrobacter freundii* Y9 | Wang et al. (2018) |
| 26 | Ag and Au NP | Reduction | 4-nitroaniline | *Citrus aurantifolia peel* | Dauthal and Mukhopadhyay (2015) |
| 27 | Ag NPs | Degradation | MB | *Trichodesma indicum leaf* | Kathiravan (2018) |
| 28 | Ag NPs | Reduction | Poisonous nitro compounds | *Extract of date palm* | Farhadi et al. (2017) |
| 29 | Ag NPs | Reduction | Eosin Blue (EB) and 4-NP | *Sapindus mukorossi* fruit | Dinda et al. (2017) |
| 30 | Pd NPs | Reduction | 4-NP | *Frimiana simplex* | Peng et al. (2019) |

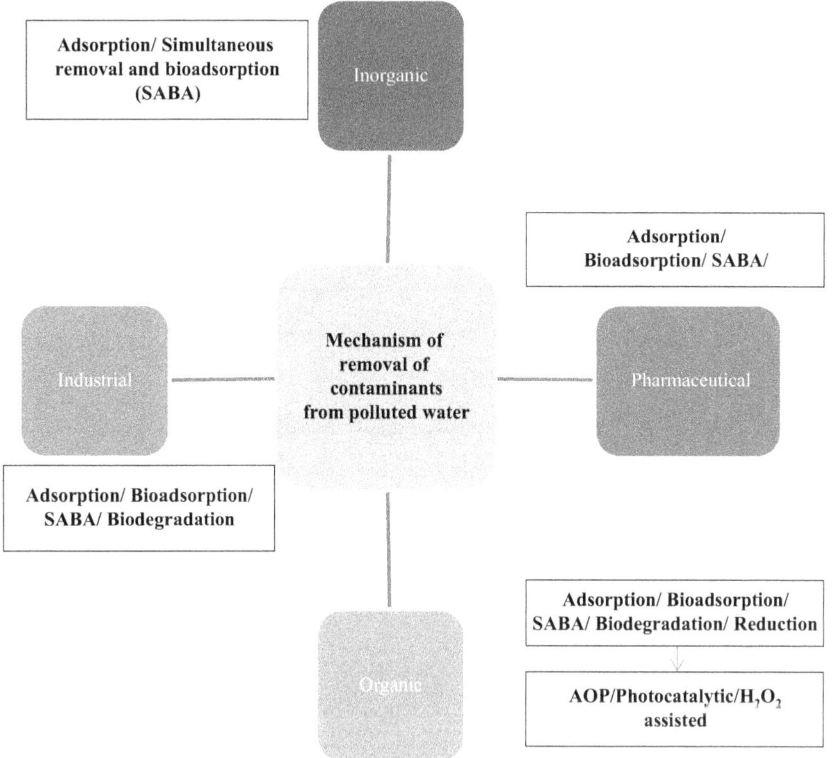


**FIGURE 1.3** Removal of contaminants by applying nanomaterials to various available technologies.

### 1.2.3 Challenges in the Development of Bio-based Biogenic Composites

There is an immediate requirement for revolutionary enhanced water technologies that can assure high-quality drinking water while also removing micropollutants. Water treatment technologies that are flexible and adaptive must be used to boost industrial production processes. When compared to traditional water technologies, one of the most significant benefits of nanomaterials is their ability to integrate diverse features, resulting in multifunctional systems such as nanocomposite membranes that enable particle retention as well as pollutant removal, as shown in Figure 1.4. Furthermore, because of their unique properties, such as a large surface area, nanomaterials offer improved process efficiency. However, certain significant disadvantages must be mentioned at this time. Aggregation, stability, size control, and sedimentation are still considered challenges for the commercial applications of biogenic nanostructures in the treatment of industrial effluents.

For instance, materials nanostructured with NPs integrated into or deposited on their surface pose a concern since the NPs may be discharged into the environment and accumulate over time. Several national and international rules and legislation are

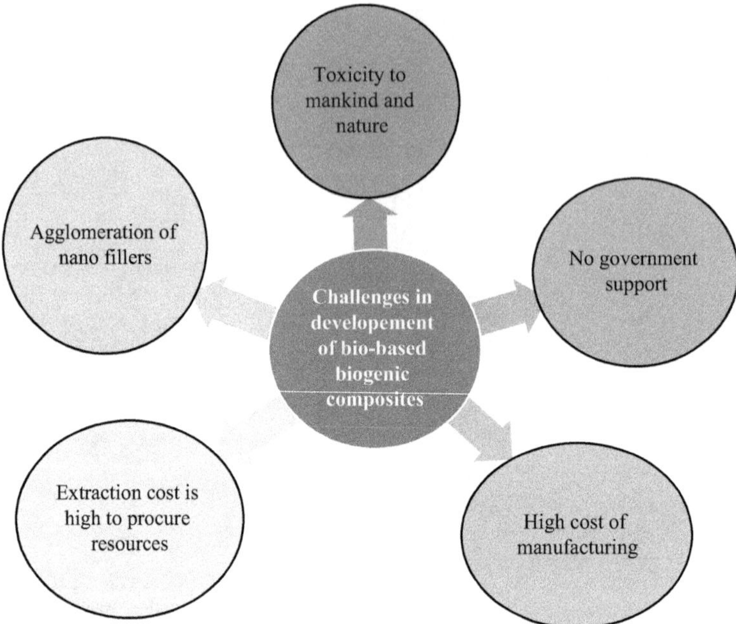


**FIGURE 1.4** Challenges in the development of bio-based biogenic composites.

being enacted in order to reduce the risk to public health. The fundamental technical restriction of nano-engineered water technologies is that they are seldom adaptable for large-scale procedures and are currently not competitive with traditional treatment methods in many circumstances. Nonetheless, safer and more plentiful nano-engineered materials hold considerable promise for future advances, notably in decentralized treatment systems, point-of-use devices, and highly biodegradable pollutants. Because of the intrinsic greenness and sustainability of the manufacturing techniques, as well as their high performance in the reduction of environmental toxins, biogenic NPs are promising materials (Gautam et al., 2019). The advancement of improved analytical and imaging technologies has opened up new avenues for the evaluation and measurement of nano-sized items, particularly for water treatment applications. Because of the applications of hazardous chemicals and materials for producing nano-objects, the chemical industry has been under pressure to substitute toxic reagents and hazardous solvents; the main push has been to deploy biomolecules from organisms as an alternative to damaging synthetic chemicals to produce biocompatible nano-objects.

Bio-prepared nanoparticles appear to be capable of adsorbing toxins from aqueous watercourses and catalyzing the breakdown of organic pollutants into harmless states. Because of their bio-renewable nature, biogenic nanomaterials are sustainable, relatively affordable, can be generated in an energy-efficient way, and are environmentally dependable. They might play important roles in decontamination regimens for drinking and industrial wastewaters (Gautam et al., 2019).

### 1.2.4 Future Recommendations

Some significant future views must be addressed while using biogenic nanomaterials for water treatment and purification:

- More research is needed to analyze the sustainability and toxicity problems, as well as to apply these green-synthesized nano-catalysts and nanomaterials on industrial and commercial levels. The use of nanomaterials may lead to secondary contamination, thus this essential problem must be handled and analyzed thoroughly.
- Although the production of these nanomaterials is convenient and environmentally friendly, some crucial and difficult aspects, such as the effects of reaction parameters and stability issues, should be analyzed and optimized because these factors can modify the behavior of nanomaterials, morphologies, and pollutant removal performance. Furthermore, the purification and extraction of the created biogenic nanomaterials for future applications is critical, and they must be separated with high purity, particularly in the case of water treatment.
- More research is needed to identify new nano-hybrids and multifunctional nanomaterials to improve their efficacy.
- Cost-effectiveness studies should be conducted to compare the manufacturing of green-synthesized nanomaterials with conventionally manufactured NPs.
- Efficacy problems and remedial performance evaluations are often created on laboratory scales, replicating the varying degrees of genuine exposure settings; nonetheless, it is critical to explore and assess the findings from realistic environmental situations.

## 1.3 THE ROLE OF BIONANOTECHNOLOGY IN WASTE MANAGEMENT

Swift worldwide industrial and population expansion has drastically increased the amount of daily municipal solid waste, and disposal of this waste is muddled and unscientific. Due to globalization, energy necessities also increase. Uncontrolled dumping of waste in the outer areas of cities and towns has created overflowing landfills, which are impossible to reclaim and present serious environmental issues. Solid waste is hazardous for the community as it leads to odors, spread of diseases, and other nuisances. These technologies decline waste volumes, environmental impact, extortions to public health and dependence on fossil fuels for power production. As per the United Nations Environment Program, nearly 11.2 billion tons of solid waste are generated every year, a significant source of environmental degradation and negative health impacts—particularly in developing countries, where more than 90% of waste is openly dumped or burned (UNEP, 2020).

Solid waste is the discarded solid material, generated from domestic areas, trade centers, commercial blocks, industries and agriculture, public services, mining activities, and institutions. India is the second-most heavily populated country in the world. The population India is approximately 1.3 billion, and each day, a single person generates 1.2 kilograms of waste. Massive piles of garbage are land-filled in an unsanitary manner. According to research projections, by 2025, that amount will have increased to around 1.42 kg waste per person. It is observed that there will be an increase of at least 4–6 times compared to the year 1999. These findings show the effects of population increase and waste generation. Out of the total population in the world, 68% lives in rural areas, while 32% lives in urban areas. Now India is in an industrialization phase, which directly affects the rate of urbanization. Uncontrolled urbanization leads to increasing rate of per capita waste generated. Around 127,486 tons municipal waste is generated per day due to commercial and institutional activities and household activities.

In India, composition and quality of municipal solid waste generated differs when compared to municipal waste generated by western countries. In India, it consists mainly of a large biodegradable (50%), recyclable (20%), inert (22%), and others (8%). Ahmadabad and Mumbai districts are ranked second and third in generating hazardous waste. The technical wing of the Ministry of Urban Development has classified solid waste in different categories as shown in Figure 1.5 based on their properties, origin and type of waste, and effect on human health and environment (which includes street sweeping, domestic waste, garbage, dead animals, rubbish, construction, ashes, demolition waste, bulky waste, industrial waste, hazardous waste).

### 1.3.1 Present Services for Waste Management and Their Drawbacks

There is no suitable and focused process for waste collection from door to door. Street sweeping is the only method of waste collection. Unexpectedly, it is also not a day-by-day process. Tools for waste collection and sweeping are outdated and in disrepair. Collected waste is transported for further processing and disposal. The most neglected part of the process is disposal of waste. This inattentiveness leads to wastewater, air pollution, and establishment of breeding grounds for flies and vermin, which are all severe environment threats.

### 1.3.2 Possible Solutions for Waste Management

Presently, researchers are looking at solid waste management as a twin-based approach system.

i) Managing solid waste.
ii) Converting waste into value-added products, which achieves the sustainability goal (Ferronato and Torretta, 2019; Sabah et al., 2020).

There is 5R technology to remove or convert waste into valuable products. Major objectives for the generation of energy follow:

- Achieving sustainable development.
- Decreasing generation of waste and power demand.
- Optimizing the energy potential of some solid waste components.
- Reduction of inert in MSW.

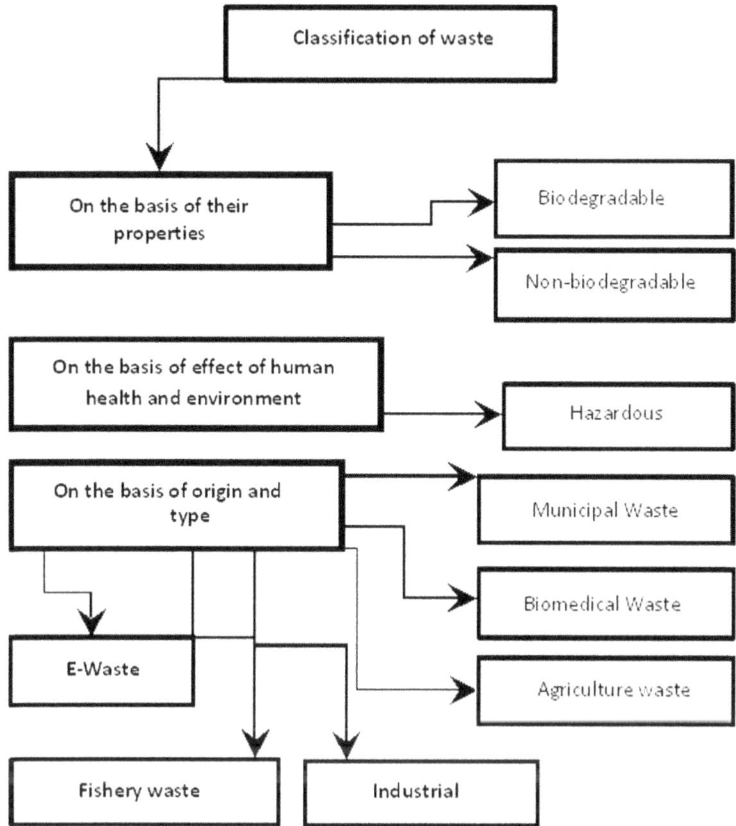


**FIGURE 1.5** Various types of waste.

#### 1.3.2.1 Technologies Available for the Treatment of Solid Waste

The waste to energy transformation processes generally working in India are the dispensation techniques, including thermal conversion (incineration, gasification, pyrolysis, refuse-derived fuel (RDF)); biochemical and biological conversion processes (vermicomposting, composting, bio-methanation, anaerobic digestion); physical conversion (mechanical extraction, briquetting of biomass, and distillation); and chemical conversion (solvent extraction, hydrolysis, transesterification). The bio-conversion process, as shown in Figure 1.6, works on organic waste and forms compost as a product, which is further used to generate biofuel, biogas, and resident sludge. These are used by power-generating plants to generate energy. Each one of the different technological options has its benefits and constraints.

Over the last few decades, the abundant waste-generated, -structured, and -engineered nanomaterials have emerged as potent agents that have the capability to deal with environmental issues such as the following (Sabah et al., 2020):

- Pollution monitoring.
- Membrane technology.

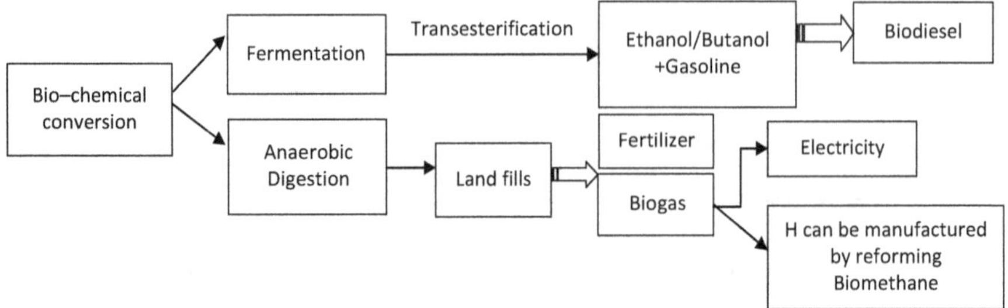


**FIGURE 1.6** Routes of bio-chemical conversion.

- Wastewater treatment.
- Energy transformation and storage.
- Precision agriculture and controlled delivery of food ingredients.
- Drug delivery and diagnostics.
- Tissue engineering.

In view of this, industrial, electric, electronic, and plastic wastes are proven to be suitable inputs for sustainable production of engineered nanoparticles, which can be further used in the development of advanced technologies for environmental applications (Dutta et al., 2018). From industry, rubber tires, batteries, biosolids, biosludges, and wastewater are prominent sources of carbon, palladium, zinc, and copper, and are studied as possible low-cost and pervasive preliminary materials for nanoparticle synthesis. In 2019, one researcher reported an innovative approach of metal recovery through the bioremediation and biomining process. The process uses metal-binding peptides to functionalize fungal mycelia. Recently published research from Malaysia showed the new path in which, with the help of bionanotechnology, fish and poultry waste was converted into gelatin-based film, which has application potential in the food packaging industry and achieved the objectives of clean, green, and sustainable. This film emerged as the efficient solution of a biodegradable packaging system that works as a moisture, gas, and microbial barrier, an eco-friendly approach that also achieves the stated function (Tuan et al., 2020).

In conclusion, the concept of converting waste into advanced technologies for environmental applications is gaining high momentum these days (Elsayed et al., 2020). At first glance, it looks like an attractive full-circle approach, creating and setting trends in the field of bionanotechnology in waste management. But still there are noteworthy knowledge gaps about structured nanomaterials that should be addressed before transitioning these ideas from the lab to the real world—particularly in cases of energy generation and to find out new routes for management of environmental issues.

### 1.3.3 Prerequisites of Energy Production from Waste for the Growth of a Sustainable Society

Energy is a necessity in our everyday life, a way of refining human development leading to economic growth and productivity. The "return to renewables" may help

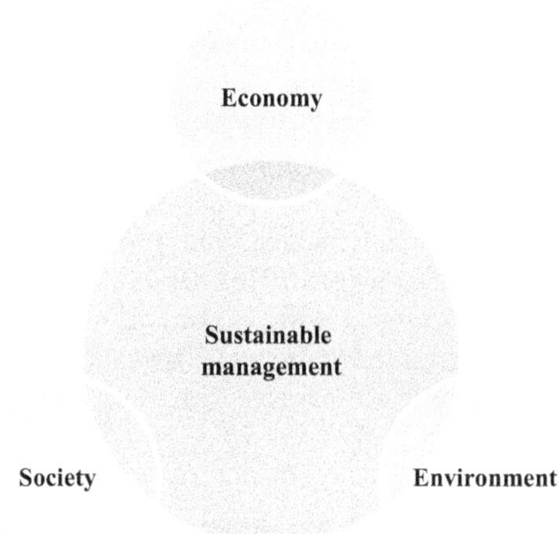


**FIGURE 1.7** Sustainable management of the environment.

allay climate change to a certain degree but, in order to ensure a sustainable future and bequeath to future generations the ability to meet their own energy needs, more needs to be done. Very little research is available regarding the interrelation between sustainable development and renewable energy in particular. There are three pillars of sustainability, i.e., social, environmental, and economic. The definition of sustainable development is "without compromising the present needs, it is to maintain the capability of future generations to meet their own needs." A Venn diagram of sustainable development shown in Figure 1.7 sheds light on the inter-relationships of social, environmental, and economic phases of sustainable development. Only by virtue of balancing these three pillars will we achieve true sustainability.

## 1.4 SUSTAINABLE MANAGEMENT OF THE ENVIRONMENT

Sustainable management is described as a set of methods that achieve a balance between three areas of human activity, namely, the economy, society, and the environment. As a result, sustainable management is defined as the use of sustainable techniques in various aspects of the economy that benefit both society and the environment as shown in Figure 1.7. The use of environmental resources in a manner and at a level that preserves and improves the adaptability of ecosystems and effectively preserves the quality of life on the planet is what sustainable environmental management is all about. Sustainable environmental management is the practice of serving the requirements of current generations of people without jeopardizing future generations' ability to satisfy stated or implied needs, while also contributing to the accomplishment of well-being goals. As coming generations confront the problem of increased population levels and growth of development activity, the demand for more ecological environmental assets grows. Despite significant human activity, environmental resources should be carefully conserved. This necessitates an agreement with the sustainable environment so that future generations are not harmed. Because man is a part of the ecosystem, environmental management should be ingrained in people all around the world. Human participation is necessary for long-term environmental management. When the environment is effectively managed by man's desire to satisfy his ongoing needs, the ecosystem is protected. Human actions that are reckless in their use of natural resources, on the other hand, will affect the ecosystem. Poverty, deforestation, pollution, resource depletion, and a general reduction in population well-being will be the eventual outcome. This needs a rethinking of man's methods of coexisting with the environment without causing natural conflict. Better environmental coping mechanisms are necessary. These include enhanced agricultural practices, forestation, and pollution management via the use of improved research and technology.

### 1.4.1 Practices for Sustainable Management of the Environment

Some human activities result in inadequate management of environmental resources, which eventually leads to surface runoff, woodland incursion, and pollution, all of which are together referred to as "environmental destruction." Environmental destruction is the consequence of multilateral activities including socioeconomic, institutional, and technical practices that have an impact on the environment. High essence for economic expansion, agricultural intensification, growing energy and transportation costs, and urbanization all result in mismanagement of environmental resources.

Here are several ways for every one of us to live more sustainably:

- Reorganizing living circumstances toward green infrastructure, such as sustainable villages, eco-municipalities, and sustainable cities.
- Reevaluating areas of the economy (sustainable living, green construction, green goods, and sustainable farming) or work methods (such as sustainable architecture).
- Technological innovation (green technologies, renewable energy, etc.).
- Changing one's lifestyle to save natural the environment.

### 1.4.2 The Role of Bionanotechnology in the Sustainable Management of the Environment

The supply of resources has been constrained as the degree of socioeconomic, administrative, and technical activity has increased. The invention of pollution-free technology for environmental cleanup and clean energy resources for human society's long-term progress are critical. Bionanotechnology has the potential to have a big effect on the development of 'cleaner' and 'greener' solutions with considerable health and environmental advantages. Bionanotechnology solutions are being investigated for their potential to give remedies to control, reduce, and clean up air, water, and land pollution, and thus to boost the effectiveness of current environmental clean-up methods. Bionanotechnology is a subfield of nanotechnology that aims to promote sustainability through a range of its uses. Bionanotechnology applications are utilized to alleviate environmental challenges by lowering total energy consumption throughout the synthesis and processing conditions, allowing items to be recycled after use and developing and using eco-friendly resources.

## 1.5 EXPANSION OF THE GREEN ECONOMY WITH THE HELP OF THE BIONANOTECHNOLOGY REVOLUTION

The green economy is envisioned as an economic state in which the expansion of growth is supported by economic activity that allows for lower carbon emissions and pollution, increased resource and energy efficiency, and the avoidance of biodiversity and ecosystem service degradation. Bionanotechnology has the potential to play a key role in the development of a 'cleaner' and 'greener' economy, with considerable health and environmental advantages. Bionanotechnology techniques comprise construction materials and chemicals, manufacturing goods and processes, and generating energy from renewable by-products, such as plant biomass and fisheries waste, that contribute to environmental sustainability and a greener economy as shown in Figure 1.8 (Quintero and Palencia, 2021).

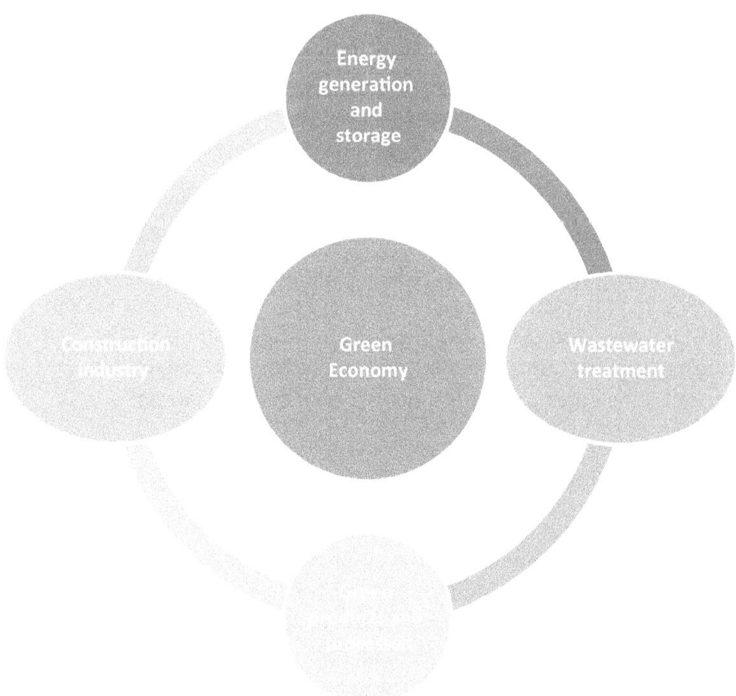


**FIGURE 1.8** Bionanotechnology revolution and green economy.

The potential bionanotechnology roles in a green economy are outlined in the following sections.

### 1.5.1 Bionanotechnology for Energy Generation

Renewable energy solutions that employ bionanomaterials to generate energy are one of the most intriguing uses of bionanotechnology. Several researchers have proposed that a green economy vision can be realized by utilizing various applications of bionanotechnology in energy generation, such as direct conversion of sunlight into electric power (Abdin et al., 2018), hydrogen generation from seawater, energy-efficient lightbulbs using a nano-engineered polymer matrix and plasmonic cavities, electricity generation by windmills, generating electricity from waste heat, storing hydrogen for fuel cell-powered cars, and so on.

### 1.5.2 Bionanotechnology for Water Clean-up Technologies

Nanotechnology-enabled water supply and sewage treatment promises to not only overcome key obstacles faced by present treatment systems but also to deliver novel treatment capabilities that might allow for the cost-effective use of unusual water sources to increase water supply (Gottardo et al., 2021). Integration of nanostructured materials, such as antimicrobial nanoparticles and photocatalytic nanoparticles into membrane surfaces to improve permeation, fouling resistance, biofilm control, mechanical and thermal stability, pollutant degeneration, and self-cleaning capacity could be significant applications. Furthermore, fullerene and metal-based nano-adsorbents may boost the adsorption properties of organic compounds, metal ions, and toxic substances significantly.

### 1.5.3 Bionanotechnology for the Construction Industry

Manufactured nanomaterials and nanostructured materials have several prospects in infrastructure projects. The durability, longevity, and weight of various materials, as well as heat-insulating, self-cleaning, fire-retardant, anti-fogging, and structural health-sensing qualities, may be enhanced or given. For wood, highly water-resistant coatings, including silica and hydrophobic polymers, are appropriate.

### 1.5.4 Green Products and Processes

The advantages of using nanomaterials in processes and products that lead to sustainable outcomes may carry with them environmental, health, and safety problems, ethical and social difficulties, market and consumer acceptability uncertainties, and intense rivalry with existing technology. Green nanotechnology is expected to play a critical role in bringing an additional capability across the value chain of a product, both through the beneficial properties of nanomaterials included as a small proportion in a final device and through nano-enabled processes that do not include any nanomaterials in the final product (Diallo et al., 2013). However, the majority of possible green nano-solutions are in the lab/start-up stages, and relatively few solutions have yet to come to market. More research is needed to evaluate the usability, efficacy, and sustainability of bionanotechnology under more realistic settings, as well as to verify nanomaterial-enabled systems in contrast to existing methods.

## 1.6 ECO-VILLAGE CONCEPT TO PROMOTE A GREEN ECONOMY

The "green economy" concept has been driven into the mainstream of policy debate by the global economic crisis, expected increase in global demand for energy by more than one-third between 2010 and 2035, rising commodity prices as well as the urgent need for addressing global challenges in domains such as energy, environment, and health. The most extensively used and trustworthy definition of "green economy" comes from the United Nations Environment Program which states that "a green economy is one that results in improved human well-being and social equity, while significantly reducing environmental risks and ecological scarcities. It is low carbon, resource competent, and socially comprehensive."

The green economy is the backbone of the sustainable concept of development of structural basis of the "eco-village." The concept of eco-village lies on "zero waste" discharge, i.e., proper exploitation of the waste generated. Generated waste should convert into value-added products. However, inappropriate urban management, often based on inaccurate perspicacity and information, can turn opportunity into catastrophe. Recently, most metropolises are designed and planned based on eco-village concepts, and those cities are designed to convey a high quality of life to their residents. The key 5R rules (Reduce, Reuse, Recycle, Recovery, Removal) are depicted in Figure 1.9.

**FIGURE 1.9** A rounded zero-waste city model, with the five inter-connected key principles that need to be applied simultaneously.

- To reduce our requirements and whatever we used, it should be properly consumed (comportment change and justifiable consumption).
- We have to be responsible toward usage (prolonged manufacturer and consumer responsibility).
- To adopt strategies which recycle the waste (100% reprocessing of municipal solid waste).
- Focused on recovery process (100% resource recapture from waste).
- Removal or dumping of the hazardous waste (established zero land fill and incineration).

Figure 1.9 presents the five parts that are fundamental in transforming global villages into zero-waste eco-villages. The utensils, capitals, or methods established for recycling or treatment of waste should be inexpensive in the socioeconomic framework, governing or controllable in the socio-political framework, relevant in the policy and technological framework, and effective or efficient in the background of economy and technology. In conclusion, all these phases should be directly related to environmental sustainability (Atiq et al., 2011). The eco-village ideologies are well explained by the four-sphere diagram in Figure 1.10.

## 1.7 CONCLUSION

"*Clean, green, and sustainable world*" is not just a slogan – in fact, it means a lot. A healthy mind lives in a healthy body, and a healthy body lives in a healthy and sustainable environment. The environment which was preserved by our ancestors is degrading rapidly. Several issues arise due to the lack of proper municipal solid waste

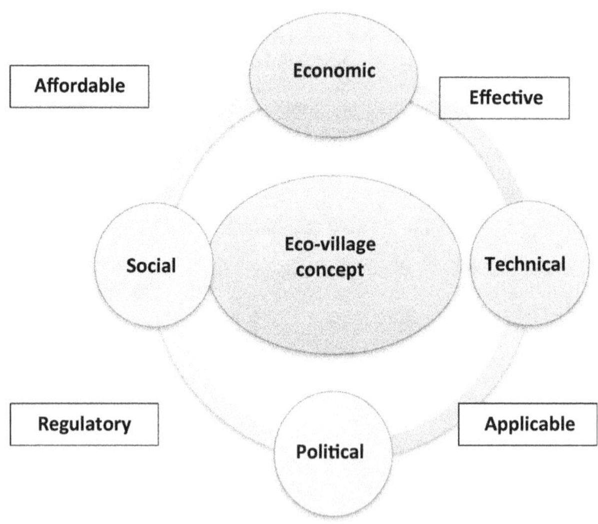


**FIGURE 1.10** A four-sphere diagram depicting the concepts of a sustainable eco-village.

management. Most of the MSW in India is dumped on land in an uncontrolled manner. Such inadequate disposal practices leads to problems that will impair human and animal health. Awareness should be created by inculcating the masses about the health hazards of untreated waste. If we recall our history, India is the country where the first sewage and drainage systems were built in the time of the Indus Valley civilization which has now at the back foot since last few centuries in waste management, water management and recycling. It is time to choose such technologies and concepts that convert waste into wealth. An integrated design approach, including the harmonized application of bionanotechnology with respect to the abovementioned five ideologies of sustainable development is critical to achieving the eco-village objectives. Green-synthesized and biogenic nanocatalysts and nanomaterials can cost-effectively and proficiently eliminate the inorganic, organic, pharmaceutical, and heavy metal pollutants from the aqueous streams. As low cost of production is imperative for their broader application in wastewater treatment, future studies should be dedicated to refining the economic viability of these nanomaterials and evaluation of their interactive mechanisms in water treatment systems.

## REFERENCES

Abdelbasir, S.M., McCourt, K.M., Lee, C.M., Vanegas, D.C. (2020). Waste-derived nanoparticles: synthesis approaches, environmental applications, and sustainability considerations. *Front. Chem.*, 8, 1–18 doi: 10.3389/fchem.2020.00782.

Abdin, A.R., Bakery, A.R., Mohamed, M.A. (2018). The role of nanotechnology in improving the efficiency of energy use with a special reference to glass treated with nanotechnology in office buildings. *Ain Shams Eng. J.* 9(4), 2671–2682.

Aitenneite, H., Abboud, Y., Tanane, O., Solhy, A., Sebti, S., Bouari, A.E. (2016). Rapid and green microwave-assisted synthesis of silver nanoparticles using aqueous *Phoenix dactylifera* L. (Date palm) leaf extract and their catalytic activity for 4-Nitrophenol reduction. *J. Mater. Environ. Sci.* 7, 2335–2339.

Arunachalam, R., Dhanasingh, S., Kalimuthu,B., Uthirappan, M., Rose, C., Mandal, A.B. (2012). Phytosynthesis of silver nanoparticles using *Coccinia grandis* leaf extract and its application in the photocatalytic degradation. *Colloids Surf. B Biointerfaces.* 94, 226–23.

Bassani D.G., Kumar R., Awasthi S., Morris S.K., Paul V.K., et al. (2010). Causes of neonatal and child mortality in India: A nationally representative mortality survey. *Lancet.* 376, 1853–60.

Beveridge, T., Murray, R. (1980). Sites of metal deposition in the cell wall of *Bacillus subtilis*. *J. Bacteriol.* 141, 876–887.

Bordbar, M., Mortazavimanesh, N. (2017). Green synthesis of Pd/walnut shell nanocomposite using *Equisetum arvense* L. Leaf extract and its application for the reduction of 4-nitrophenol and organic dyes in a very short time. *Environ. Sci. Pollut. Res. Int.* 24, 4093–4104.

Bordbar, M., Sharifi-Zarchi, Z., Khodadadi, B. (2017). Green synthesis of copper oxide nanoparticles/ clinoptilolite using *Rheum palmatum* L. Root extract: high catalytic activity for reduction of 4-nitro phenol, rhodamine B, and methylene blue. *J. Solgel Sci. Technol.* 81, 724–733.

Bremner, D.H., Molina, R., Martínez, F., Melero, J.A., Segura, Y. (2009). Degradation of phenolic aqueous solutions by high frequency sono-Fenton systems (US–Fe2O3/SBA-15–H2O2). *Appl. Catal. B.* 90, 380–388.

Chong, M.N., Jin, B., Chow, C.W., Saint, C. (2010). Recent developments in photocatalytic water treatment technology: a review. *Water Res.* 44(10), 2997–3027.

Chubar, N., Behrends, T., Behrends, P.V. (2008). Biosorption of metals ($Cu^{2+}$, $Zn^{2+}$) and anions ($F^-$, H2PO4$^-$) by viable and autoclaved cells of the Gram-negative bacterium *Shewanella putrefaciens. J. Colloid Interface Sci.* 65, 126–133.

Coker, V.S., Bennett, J.A., Telling, N.D., Henkel, T., Charnock, J.M., Gerrit, V.D.L., Pattrick, R.A., et al. (2010). Microbial engineering of nanoheterostructures: biological synthesis of a magnetically recoverable palladium nanocatalyst. *ACS Nano.* 25, 4(5), 2577–84.

Cui, H., Qian, Y., An, H., Sun, C., Zhai, J., Li, Q. (2012). Electrochemical removal of fluoride from water by PAO A modified carbon felt electrodes in a continuous flow reactor. *Water Res.* 46, 3943–3950.

Das, S.K., Liang, J., Schmidt, M., Laffir, F., Marsili, E. (2012). Biomineralization mechanism of gold by zygomycete fungi Rhizopusoryzae. *ACS Nano.* 24, 6(7), 6165–73.

Dauthal, P., Mukhopadhyay, M. (2015). Agro-industrial waste-mediated synthesis and characterization of gold and silver nanoparticles and their catalytic activity for 4-nitroaniline hydrogenation. *Korean J. Chem. Eng.* 32, 837–844.

Dhillon, G.S., Brar, S.K., Kaur, S., Verma, M. (2012). Green approach for nanoparticle biosynthesis by fungi: current trends and applications. *Crit. Rev. Biotechnol.* 32, 49–73.

Diallo, M., Fromer, N.A., Jhon, M.S. (2013). Nanotechnology for sustainable development: retrospective and outlook. *J. Nanopart. Res.* 15, 1–16.

Dinda, G., Halder, D., Mitra, A., Pal, N., Vázquez-Vázquez, C., López-Quintela, M.A. (2017). Study of the antibacterial and catalytic activity of silver colloids synthesized using the fruit of Sapindusmukorossi. *New J. Chem.* 41, 10703–10711.

Durán, M., Silveira, C.P., Durán, N. (2015). Catalytic role of traditional enzymes for biosynthesis of biogenic metallic nanoparticles: a mini-review. *IET Nanobiotechnol.* 9, 314–323.

Dutta, T., Kim, K.H., Deep, A., Szulejko, J.E., Vellingiri, K., Kumar, S. (2018). Recovery of nanomaterials from battery and electronic wastes: a new paradigm of environmental waste management. *Renew. Sustain. Energy Rev.* 82, 3694–3704.

Elsayed, D.M., Abdelbasir, S.M., Abdel-Ghafar, H.M., Salah, B.A., Sayed, S.A. (2020). Silver and copper nanostructured particles recovered from metalized plastic waste for antibacterial applications. *J. Environ. Chem. Eng.* 8(4), 103826.

Farhadi, S., Ajerloo, B., Mohammadi, A. (2017). Green biosynthesis of spherical silver nanoparticles by using date palm (phoenix dactylifera) fruit extract and study of their antibacterial and catalytic activities. *Acta Chim. Slov.* 64, 129–143.

Ferronato, N., Torretta, V. (2019). Waste mismanagement in developing countries: a review of global issues. *Int. J. Environ. Res. Public Health* 16, 1060.

Garai, C., Hasan, S.N., Barai, A.C., Ghorai, S., Panja, S.K., Bag, B.G. (2018). Green synthesis of *Terminalia arjuna*-conjugated palladium nanoparticles (TA-PdNPs) and its catalytic applications. *J. Nanostructure Chem.* 8, 465–472.

Gautam, P.K., Singh, A., Misra, K., Sahoo, A.K., Samanta, S.K. (2019). Synthesis and applications of biogenic nanomaterials in drinking and wastewater treatment. *J. Environ. Manage.* 231, 734–748.

Gopalakrishnan, R., Loganathan, B., Raghu, K. (2015). Green synthesis of Au–Ag bimetallic nanocomposites using *Silybum marianum* seed extract and their application as a catalyst. *RSC Adv.* 5, 31691–31699.

Gottardo, S., Mech, A., Drbohlavova, J. (2021). Towards safe and sustainable innovation in nanotechnology: State-of-play for smart nanomaterials. *Nano Impact.* 21, 100297. doi: 10.1016/j.impact.2021.100297.

Jaafara, A., Driouichb, A., Lakbaibib, Z., Ayouchiac, H.B.E., Azzaouid, K., Boussaouda, A., Jodehe, S. (2019). Central composite design for the optimization of basic red V degradation in aqueous solution using Fenton reaction. *Desalin. Water Treat.* 158, 364–371.

Jiang, Z., Liu, D., Jiang, D., Wei, W., Qian, K., Chen, M., Xie, J. (2014). Bamboo leaf assisted formation of carbon/nitrogen co-doped anatase $TiO_2$ modified with silver and graphitic carbon nitride: novel and green synthesis and cooperative photocatalytic activity. *J. Chem. Soc. Dalton Trans.* 43, 13792–13802.

Kabay, N., Arar, O., Samatya, S., Yuksel, U., Yuksel, M. (2008). Separation of fluoride from aqueous solution by electrodialysis: effect of process parameters and other ionic species. *J. Hazard. Mater.* 153, 107–113.

Karthiga, D.G., Senthil, K.P., Kumar, S.K. (2016). Green synthesis of novel silver nanocomposite hydrogel based on sodium alginate as an efficient biosorbent for the dye wastewater treatment: prediction of isotherm and kinetic parameters. *Desalin. Water Treat.* 57, 27686–27699.

Kathiravan, V. (2018). Green synthesis of silver nanoparticles using different volumes of *Trichodesma indicum* leaf extract and their antibacterial and photocatalytic activities. *Res. Chem. Intermed.* 44, 4999–5012.

Kaviya, S., Prasad, E. (2015). Biogenic synthesis of ZnO–Ag nano custard apples for efficient photocatalytic degradation of methylene blue by sunlight irradiation. *RSC Adv.* 5, 17179–17185.

Khoshnamvand, M., Huo, C., Liu, J. (2019). Silver nanoparticles synthesized using *Allium ampeloprasum* L. Leaf extract: characterization and performance in catalytic reduction of 4-nitrophenol and antioxidant activity. *J. Mol. Struct.* 1175, 90–96.

Kumar, B., Smita, K., Angulo, Y., Cumbal, L. (2016). Valorization of rambutan peels for the synthesis of silver-doped titanium dioxide (Ag/TiO2) nanoparticles. *Green Process. Synth.* 5, 371–377.

Kumar, B., Smita, K., Cumbal, L., Debut, A. (2015). Ultrasound agitated phytofabrication of palladium nanoparticles using Andean blackberry leaf and its photocatalytic activity. *J. Saudi Chem. Soc.* 19, 574–580.

Lopez-Miranda, J.L., González, M.V., Mares-Briones, F., Cervantes-Chávez, J., Esparza, R., Rosas, G., Pérez, R. (2018). Catalytic and antibacterial evaluation of silver nanoparticles synthesized by a green approach. *Res. Chem. Intermed.* 44, 7479–7490.

Meenakshi, Maheshwari, R.C. (2006). Fluoride in drinking water and its removal. *J. Haz. Mater.* 137, 456–463.

Miri, A., Mousavi, S.R., Sarani, M., Mahmoodi, Z. (2018). Using *biebersteinia multifida* aqueous extract, the photocatalytic activity of synthesized silver nanoparticles. *Orient. J. Chem.* 34, 1513–1517.

Mohapatra, M., Anand, S., Mishra, B.K., Giles, D.E., Singh, P. (2006). Review of fluoride removal from drinking water. *J. Environ. Mgt.* 91, 67–77.

Momeni, S.S., Nasrollahzadeh, M., Rustaiyan, A. (2016). Green synthesis of the Cu/ZnO nanoparticles mediated by Euphorbia prolifera leaf extract and investigation of their catalytic activity. *J. Colloid Interface Sci.* 472, 173–179.

Naraginti, S., Sivakumar, A. (2014). Eco-friendly synthesis of silver and gold nanoparticles with enhanced bactericidal activity and study of silver catalyzed reduction of 4-nitrophenol. *Spectrochim. Acta A. Mol. Biomol. Spectrosc* 128, 357–362.

Paixao, R.M., Reck, I.M., Bergamasco, R., Vieira, M.F., Vieira, A.M.S. (2018). Activated carbon of Babassu coconut impregnated with copper nanoparticles by green synthesis for the removal of nitrate in aqueous solution. *Environ. Technol.* 39, 1994–2003.

Parandhaman, T., Dey, M.D., Das, S.K. (2019). Biofabrication of supported metal nanoparticles: exploring the bioinspiration strategy to mitigate the environmental challenges. *Green Chem.* 21, 5469–5500.

Patil, S.S., Mali, M.G., Tamboli, M.S., Patil, D.R., Kulkarni, M.V., Yoon, H., Kim, H., Al-Deyab, S.S., Yoon, S.S., Kolekar, S.S. (2016). Green approach for hierarchical nanostructured Ag-ZnO and their photocatalytic performance under sunlight. *Catal. Today* 260, 126–134.

Peng, X., Bai, X., Cui, Z., Liu, X. (2019). Green synthesis of Pd truncated octahedrons using of firmiana simplex leaf extract and their catalytic study for electro-oxidation of methanol and reduction of p-nitrophenol. *Appl. Organomet. Chem.* 33(8), e5045.

Quintero, A.G., Palencia, M. (2021). A critical analysis of environmental sustainability metrics applied to green synthesis of nanomaterials and the assessment of environmental risks associated with the nanotechnology. *Sci. Total Environ.* 1(793), 148524.

Ramesh, A., Devi, D.R., Battu, G., Basavaiah, K. (2018). A Facile plant mediated synthesis of silver nanoparticles using an aqueous leaf extract of *Ficushispida* Linn. F. For catalytic, antioxidant and antibacterial applications. *South Afr. J. Chem. Eng.* 26, 25–34.

Robati, D., Mirza, B., Rajabi, M., Moradi, O., Tyagi, I., Agarwal, S., Gupta, V. (2016). Removal of hazardous dyes-BR 12 and methyl orange using graphene oxide as an adsorbent from aqueous phase. *Chem. Eng. J.* 284, 687–697.

Sengani, M., Grumezescu, A.M., Rajeswari, V.D. (2017). Recent trends and methodologies in gold nanoparticle synthesis: A prospective review on drug delivery aspect. *Open Nano.* 2, 37–46.

Sourirajan, S., Matsurra, T. (1972). Studies on reverse osmosis for water pollution control. *Water Res.* 6, 1073–1086.

Simons, R., (1993). Trace element removal from ash dam waters by nanofiltration and diffusion dialysis, *Desalination,* 89, 325–341.

Sundaram, C.S., Viswanathan N., Meenakshi S. (2008). Defluoridation chemistry of synthetic hydroxyapatite at nano scale: equilibrium and kinetic studies. *J. Hazard. Mater.* 155, 206–215.

Tamuly, C., Hazarika, M., Bordoloi, M., Bhattacharyya, P.K., Kar, R. (2014). Biosynthesis of Ag nanoparticles using pedicellamide and its photocatalytic activity: an ecofriendly approach. *Spectrochim. Acta A. Mol. Biomol. Spectrosc.* 132, 687–691.

Thapa, R., Bhagat, C., Shrestha, P., Awal, S., Dudhagara, P. (2017). Enzyme-mediated formulation of stable elliptical silver nanoparticles tested against clinical pathogens and MDR bacteria and development of antimicrobial surgical thread. *Ann. Clin. Microbiol. Antimicro.* 16, 39.

Tor, A., Danaoglu, N., Arslan, G., Cengeloglu, Y. (2009). Removal of fluoride from water by using granular red mud: batch and column studies. *J. Hazard. Mater.* 164, 271–278.

Tuan Zainazor, T.C., Fisal, A., Goh, E.G., CheSulaiman, N.F., Sarbon, N.M. 2020). Emerging of bio-nano composite gelatine-based film as bio-degradable food packaging: a review. *Food Res.* 4(4), 944–956.

UNEP (2020). Solid Waste Management, UNEP - UN Environment Programme. Available online at: https://www.unenvironment.org/explore-topics/resource-efficiency/what-we-do/cities/solid-waste-management (accessed May 10, 2020).

Vanaamudan, A., Soni, H., Sudhakar, P.P. (2016). Palm shell extract capped silver nanoparticles-as efficient catalysts for degradation of dyes and as SERS substrates. *J. Mol. Liq.* 215, 787–794.

Vennila, M., Prabha, N. (2015). Plant mediated green synthesis of silver nano particles from the plant extract of *Morindatinctoria* and its application in effluent water treatment. *Int. J. Chemtech Res.* 7, 2993–2999.

Wang, X., Zhang, D., Qian, H., Liang, Y., Pan, X., Gadd, G.M. (2018). Interactions between biogenic selenium nanoparticles and goethite colloids and consequence for remediation of elemental mercury contaminated groundwater. *Sci. Total Environ.* 613, 672–678.

Zaman, A.U., Lehmann, S. (2011). Challenges and opportunities in transforming a city into a "Zero Waste City". *Challenges.* 2, 73–93; doi:10.3390/challe2040073.

Zhang, P., Hou, D., O'Connor, D., Li, X., Pehkonen, S., Varma, R.S., Wang, X. (2018). Green and size-specific synthesis of stable Fe–Cu oxides as earth-abundant adsorbents for malachite green removal. *ACS Sustain. Chem. Eng.* 6, 9229–9236.

Zheng, Y., Wang, Z., Peng, F., Fu, L. (2017). Biosynthesis of silver nanoparticles by *Plectranthus amboinicus* leaf extract and their catalytic activity towards methylene blue degradation. *Rev. Mex. Ing. Quim.* 16, 41–45.

# 2 Scope and Application of Bionanotechnology for the Bioremediation of Emerging Contaminants Generated as Industrial Waste Products

*Md Shahid Alam, Surabhi Rode, Harry Kaur, Sapna Lonare, and Deena Nath Gupta**

**CONTENTS**



* All authors have contributed equally to this chapter.

DOI: 10.1201/9781003270959-2

## 2.1 INTRODUCTION

Since the 19th century, urbanization and industrialization have increased economic growth, technological advancements, and global improvement in people's living standards. The industrialization process is essential and continues to be critical for achieving self-sufficiency and improving the country's economy and development. At the same time, the amount of industrial and domestic by-products in the environment is mounting proportionally. The technological revolution hastened humanity's progress by permitting excessive resource extraction, product distribution, and waste disposal without caution in the environment. The combination of material generation and improper disposal has resulted in significant pollution issues that have posed challenges in the environmental, social, and economic spheres.

Industries discharge more than 10 million metric tons of contaminants into the environment each year (Avio et al., 2017; Thompson and Darwish, 2019). These emerging toxic contaminants may experience further complex reactions to form even more toxic compounds known as cross-contamination. The chemical and physical properties of emerging contaminants (ECs) vary greatly; these contaminants are highly cytotoxic and have various interactions with biotic (microorganisms, plants, animals) and abiotic environmental factors (water, minerals, and wind) (Hurtado et al., 2017). Emerging contaminants from diverse sectors pose severe ecological concerns and possible hazards to human health and aquatic life, even at low concentrations. There is a serious threat to the environment posed by over-discharges and long-term persistence of emerging contaminants, such as micropollutants, pesticides, medicines, hormones, toxins, and synthetic dyes containing hazardous pollutants. Traditional treatment methods cannot altogether remove or eliminate many emerging environmental contaminants from biological and environmental samples. The above environmental concerns have prompted great research efforts to develop more efficient remediation procedures and new methods to efficiently measure, detect, and treat emerging contaminants generated as industrial waste products. Hence, to bioremediate emerging pollutants in order to achieve environmental sustainability, other effective management techniques need to be implemented.

For the removal of hazardous waste from contaminated environments, bioremediation is an effective and popular cleaning technology as depicted in Figure 2.1. Bioremediation is the process of degrading, eradicating, immobilizing, or detoxifying various chemical wastes and harmful elements from the environment using the action of microorganisms (Sharma, 2020). Bioremediation includes biosorption, bioaccumulation, biotransformation, and biological stabilization (Ezziat et al., 2019). A variety of plants, bacteria, fungi,

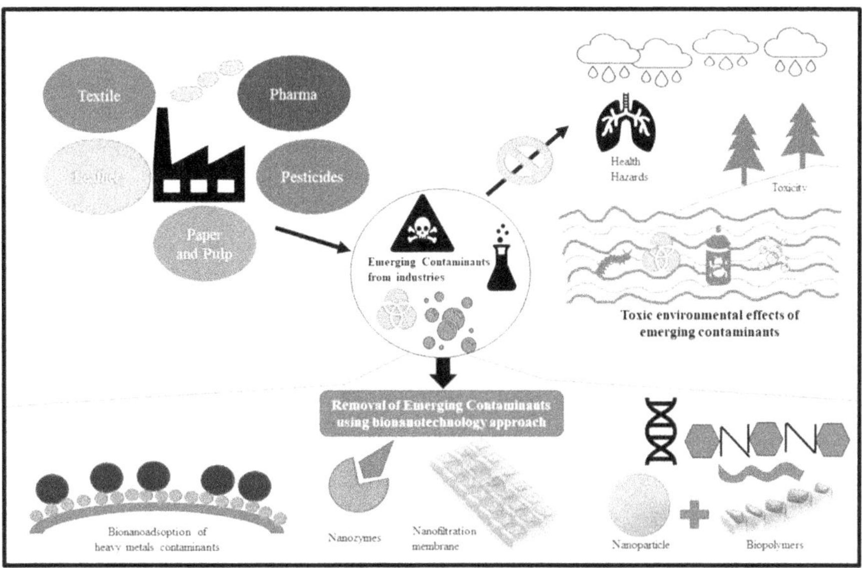


**FIGURE 2.1** Illustration of industrial emerging contaminants and remediation using the bionanotechnological approach.

and mixtures have been employed in these technologies. Bioremediation is a sustainable, clean, green, and nontoxic approach for the treatment of contaminated materials, as well as being considered an environmentally friendly process. An eco-friendly strategy based on biotechnology has already made environmental protection far more effective. Biological treatments have gained relevance among physical and chemical technologies for remediating ecological pollution because they are relatively cheap and have a wide range of applications (Kuppusamy et al., 2017). Despite their tremendous success, bio-based technologies are often time-consuming and sometimes show less effectiveness. Because traditional methods are ineffective at eliminating toxic pollutants, the focus of research has turned to newer techniques. The more recent development of biomimetic nanotechnology has allowed researchers to test this approach on a larger scale. However, these come at a heavy price in terms of installation and operation. As a result, there is a pressing need to create cost-effective and dependable technologies to deal with the growing problem of environmental pollutants.

Nanobiotechnology is a field that integrates science, technology, and living organisms (Fakruddin et al., 2012). In terms of improving global environmental conditions, nanobiotechnology has a lot of potentials. Consequently, nanobiotechnology and bioremediation can provide effective, efficient, and long-term solutions for a clean environment. Nanobiotechnology refers to applying nanotechnology to biological sciences. Moreover, recent advances in nanobiotechnology have made it possible to remediate emerging contaminants with environmentally friendly compounds. Nano-bioremediation is an excellent option for significantly cleaning up and reducing pollution in the environment. The term "nano-bioremediation" refers to nanoparticles combined with microbes or plants to sequester harmful contaminants from the environment. It may be feasible to eliminate emerging contaminants from the atmosphere using conventional bioremediation combined with nanobiotechnological methods. In recent years, biogenic nanoparticles made from microorganisms have gained much prominence. Because of their unique characteristics, such as high surface area and strong catalytic reactivity, they are a potential alternative for removing contaminants through biodegradation and biosorption processes. These biogenic nanoparticles are a growing alternative to chemically synthesized nanoparticles (NPs) due to their biological nature and nontoxicity.

Furthermore, enzyme immobilization on nanomaterials has garnered a lot of interest in the scientific community. With bionanotechnologies, remediation capabilities have improved greatly, avoiding process intermediates and accelerating degradation. Recent studies have combined nanomaterials with biological processes to remove hazardous chemicals from the environment more quickly and effectively.

## 2.2 BIONANOTECHNOLOGICAL APPROACH FOR THE MANAGEMENT OF ECs

Nano-bioremediation is used to break down or remove various industrial contaminants. The contaminants may be heavy metal ions and other toxins like herbicides, insecticides, dyes, and major emerging contaminants. These contaminants are generated from sectors like textile, agricultural, leather, pulp, pharmaceutical, and other industries. This chapter focuses on how bionanotechnology is applied to bioremediate the emerging contaminants in industrial waste generated by such industries to reduce their negative environmental impacts (Table 2.1).

### 2.2.1 Textile Industries

At the global level, textile industries contribute majorly to the development of the country. However, various health and environmental problems are associated with the unhealthy disposal of textile effluents in the ecosystem. The wastewater resulting

**TABLE 2.1**
**Representative Emerging Contaminants Generated as Industrial Waste Products from Major Industries**

| | Examples of emerging contaminants generated as industrial waste products | |
|---|---|---|
| S. No. | EC generator industries | Major emerging contaminants generated |
| 1 | **Textile** | Dyes waste, organic stabilizers, chemical solvents, peroxide, alkali, and heavy metals |
| 2 | **Agriculture and pesticides** | Copper sulphate, atrazine, organophosphates |
| 3 | **Leather** | Chromium, toluene, benzene, surfactants |
| 3 | **Paper and pulp** | Paint waste comprising heavy metals, chemical solvent, alkali |
| 4 | **Pharmaceuticals** | Pharmaceutical active compounds |

from textile effluents is the most significant source of water pollution due to its high organic content. They comprise a variety of synthetic dyes and compounds, including salts, heavy metals, surfactants, mineral oils, etc. Synthetic dyes include complex aromatic structures that are specifically developed for chemical stability, adaptability, and the capacity to withstand the effects of high temperature during wet processing procedures, making them highly recalcitrant. These dyes have become more prevalent in the textile and dyeing industries due to their cost-efficient nature, tolerance to temperature, light, microbial attack, etc., compared to natural dyes. Most of the synthetic dyes are released explicitly from the textile process, such as dyeing and printing. Thousands of different synthetic dyes are being created. The majority of them are "azo" dyes which are widely utilized in textile industries and other sectors for dyeing purposes since they are more cost-effective than natural dyes. The most significant group of synthetic dyes are the azo dyes and their structure is defined by the existence of azo (N=N) groups. They are toxic to the aquatic environment due to their bioaccumulation in water bodies. The metabolic products of azo dyes are toxic to humans causing carcinogenic and mutagenic effects (Saratale et al., 2011). One of the significant challenges is eliminating coloring components from textile wastewater effluents. Several methods, like chemical, physical, and biological methods have been used to overcome such problems. Physical methods include adsorption, coagulation, filtration, and flocculation; whereas chemical methods include chemical oxidation, electrolysis, etc. However, no such technology for treating textile wastewater was found to be effective and cost-efficient. Many biotechnological processes include bioremediation using fungi, bacteria, and plants that are eco friendly compared to chemical and physical methods. Enzymatic bioremediation is more effective against azo dyes; on the other hand, the isolation and characterization of the enzyme are very expensive and time-consuming (Lavanya et al., 2014). The advancement of nanotechnology has offered up new possibilities for water pollution cleanup. Researchers have concentrated on the development of nanoparticles to overcome the disadvantages of chemical and physical approaches. Nanoparticles (10 to 100 nm) contain unique features, such as smaller size, improved chemical properties, and the highest surface area to volume ratio, helpful in many biological applications. Nanotechnology is a cleaner and more environmentally friendly alternative to traditional physicochemical techniques (Khatoon and Sardar, 2017).

#### 2.2.1.1 Processing Steps in Textile Industries

Yarn formation, wet processing, and fabrication are the primary processes in the textile industry's production process (Madhu and Chakraborty, 2017). Wet processing of textile fabrics involves different steps as shown in Figure 2.2.

*Sizing and desizing* Sizing is the most crucial processing step in the textile industry. Yarn is coated throughout the sizing process to make weaving, knitting, and tufting easier. Before weaving, coating of warp yarn with sizing agents is done to reduce friction, minimize yarn breakage, and improve weaving production by enhancing insertion speeds. There are two types of sizing agents, natural and synthetic agents. The main natural sizing agents are starch (native and degraded), starch derivatives,

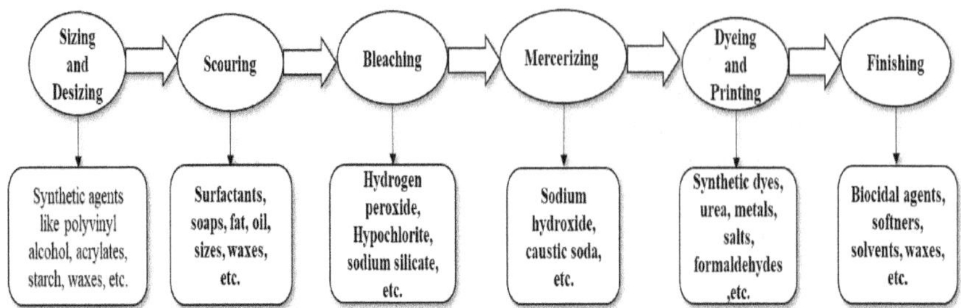


**FIGURE 2.2** The steps in textile wet processing showing the discharge of different pollutants.

modified starch, and cellulose derivatives; the main synthetic sizing agents are polyvinyl alcohols, acrylates, polyacrylates, and styrol or maleic acid copolymers. Yarns, mainly cotton, are often sized with starch-based sizing agents because they are cost-effective and provide satisfactory weaving performance. The effluent produced from a mill producing fabric of 60,000 meters is expected to contain roughly 750 kg of sizing material (Madhav et al., 2018).

Starch gets removed or changed into a water-soluble state during the desizing process by oxidation or hydrolysis (Madhu and Chakraborty, 2017). All the degraded sizes after that get separated through washing followed by thermal drying. Desizing agents include amylase, maltase, cellulose, and other enzymes derived from syrup extracts. Enzyme-based desizing agents are commonly employed in wet textile processing because they produce less waste and are also gentle on fabric texture. There are three steps involved in enzymatic desizing: enzyme application, starch digestion, and digestive products removal. The desizing process produces around half of the total volume of wastewater, which has a high biological oxygen demand (BOD) (Babu et al., 2007).

*Scouring* The scouring process is mainly used to remove impurities from textile materials without causing damage to the fabric. This process removes both natural (fat, oil, wax, etc.) as well as synthetic (added during fabrication process) impurities completely—mainly hydrophobic groups present in the fiber of fabric (Madhu and Chakraborty, 2017). After the scouring process, the material becomes suitable for the next bleaching process. This process discharges mainly surfactant, soap, fat, wax, oil, etc., pollutants in wastewater (Yaseen and Scholz, 2019).

*Bleaching* Bleaching helps decolorize the yarn to produce a white yarn that can be made into bright and light colors (Madhu and Chakraborty, 2017). Hydrogen peroxide, hypochlorite, sodium silicate, and oxalic acids are the most commonly used bleaching chemicals. The most widely used bleaching agent is hypochlorite. This process discharges mainly toxic pollutants in wastewater (Yaseen and Scholz, 2019).

*Mercerizing* After bleaching, cotton fiber fabrics are mercerized in the gray form to increase strength, add luster, and improve color absorption. The most popular mercerizing chemicals for cotton fiber are sodium hydroxide and caustic soda, which modify the cellulose and cotton fiber's crystalline character. This process discharges mainly sodium hydroxide and caustic soda in the wastewater (Ghosh et al., 2004).

*Dyeing and printing* Dyeing is an essential process of textile industries that imparts colors to the fiber using dyes (chemical pigments). Auxochrome groups (hydroxyl, carboxyl, amine) and chromophore groups (carbonyl, nitro, azo) in the dyes are responsible for imparting the colors. Synthetic dyes, mainly derived from petroleum sector intermediates and coal tar, are most commonly employed in the textile industries. A range of colored dye baths with high salts and organic compounds are discharged following the dying process. Dye baths, in general, are extremely toxic. Reactive dyes in dyeing wastewater are not fixed on the textile materials and get released in textile wastewater (20–30%). It is not possible to recycle organic substances or dyeing auxiliaries. As a result of their addition to effluents, the effluents have a high chemical oxygen demand (COD) and biochemical oxygen demand (BOD) (Babu et al., 2007).

The printing process is like the dyeing process except that, in printing, dyes are used in the form of a thick paste, while during dyeing, dyes are used in solution form. The reactive dye most commonly used in printing is urea, which is responsible for the highest wastewater discharge (Babu et al., 2007).

*Finishing* During the manufacturing of natural and synthetic textiles, various finishing procedures are required. The finishing processes include washing, drying, pressing, and conditioning. The fabric is given certain features such as softening, waterproofing, antimicrobial, and UV protection during this process. Finishing agents, especially biocidal agents, are used for providing antimicrobial characteristics to the final textile fabric. Such textile biocidal and dye carriers are associated with and contribute to the toxicity of the textile effluents (Holkar et al., 2016).

#### 2.2.1.2 Nanotechnology in Textile Wastewater Treatment

Nanomaterials are suitable for wastewater treatment because of their small size. They contain unique chemical, physical, and biological features that make them ideal for various applications. Using nanoparticles to remove harmful substances from wastewater is an environmentally friendly and cost-efficient method. Nanomaterials are microscopic particles having a diameter of a few nanometers that can be manufactured in various ways, including nanowires, nanotubes, films, etc. Nanomaterials can be divided into three categories based on their properties: nanoadsorbents, nanomembranes, and nanocatalysts (Anjum et al., 2019).

*Nanoadsorbents* Nanoparticles have also been used as adsorbents to remove toxic organic and inorganic substances from industrial effluents. Such nanoparticles have a higher affinity towards adsorbing agents. The process of sorption is the adsorption of sorbate material on another material called "sorbent" by utilizing physical and chemical methods. The most common types of nanoadsorbents are carbon nanoadsorbents, graphene, carbon nanotubes, activated carbon (AC), and metal oxide nanoadsorbents (Anjum et al., 2019).

*Carbon nanoadsorbents* Carbon-based nanomaterials have high adsorption capacity and are nontoxic, which helps in adsorption of toxic organic and inorganic substances from textile effluents. Most commonly synthesized nanoadsorbents are carbon nanotubes (CNTs), graphene, activated carbon, and fullerene. Fluoride ions have been removed from wastewater using activated carbon modified nanomagnets (Takmil et al., 2020). Nickle ions are removed by using a graphene-based nanocollector with the help of the ion floatation method (Hoseinian et al., 2020). CNTs have a well-defined and homogeneous atomic structure, which distinguishes them from ACs. To measure adsorption in ACs, some characteristics, such as adsorption energy and pore diameter, are required. In CNTs, fully specified adsorption sites existing on the adsorbed substances can be dealt with directly.

*Carbon nanotubes* Carbon nanotubes (CNTs) are employed as adsorbents for the toxic effluents discharged from textile, pharmaceutical, and various manufacturing industries. They are used to remove or degrade different dyes and their intermediates from textile industry wastewater. Single-walled nanotubes (SWNT) and multiwalled nanotubes (MWNT) are the two most used CNTs. They are beneficial in the adsorption of toxic dyes as the active site of CNTs has unique or enhanced physical, electrical, and thermal properties. The largest surface area to volume ratio makes them more effective as compared to other adsorbents. Different dyes are adsorbed differently depending upon the anionic and cationic nature of dyes. CNTs are good at degrading polycyclic aromatic compounds (PACs), making them particularly useful in the textile industry because dyeing emits the most aromatic compounds into the wastewater (Rajabi et al., 2017).

*Metal oxide nanoadsorbents* Nanoadsorbents that are metal oxide-based play a crucial role in the detachment of toxic compounds from industrial wastewater. Metal oxides are more widely used to detach heavy metals from textile effluents than available techniques like precipitation, membrane filtration, and ion exchange. They have the highest surface area to volume ratio, adsorption capacity, selectivity, and nanosized nature. The most common adsorbents used for effluent removal are iron oxide, zinc oxide, manganese oxide, aluminum oxide, titanium oxide, etc. Iron oxide (FeO) nanoparticles that are magnetic are used to get rid of heavy metals from industrial effluents (Kumar et al., 2017). Titanium oxide (TiO2) nanoparticles are used to eliminate many heavy metals like nickel, copper, zinc, etc., from wastewater. Zinc oxide (ZnO) nanoparticles are used as adsorbents to remove metals like $Ni^{2+}$, $Co^{2+}$, $Cu^{2+}$, etc. (Gupta et al., 2017). Manganese oxide (MnO) nanoparticles are accustomed to degrading heavy metals like arsenic (Anjum et al., 2019). Copper oxide (CuO) nanoparticles are used to get rid of heavy metals like $Fe^{3+}$ and $Cd^{2+}$ from effluents. Aluminum oxyhydroxide nanoparticles are used to remove methyl violet from wastewater (Kerebo et al., 2016).

*Nanomembranes* Nanomembranes play a vital role in getting rid of toxic dyes from wastewater using nanofiltration (NF), ultrafiltration (UF), microfiltration (MF), and reverse osmosis (RO). Different wastewater treatment requires other membranes having different pore sizes. The most significant advantage of this method is that it does not require any chemicals. MF and UF are used to remove colloids, debris, azo dyes, sulfur, etc. RO is used to take out many inorganic impurities from textile effluents (Amini et al., 2011). NF membranes (up to 2 nm pore size) are used to separate salt solutions from dye wastewater. The charges present on the membrane helps in permeation and rejection, resulting in the recovery of salts used in dyeing industries. Using nanofiltration membranes NF45 and DK1073, maximal dye rejection was shown to be as high as 95–99% (Lopes et al., 2005). Nanoparticles coating to nanomembranes helps increase permeability, resistance, and catalytic activity, ultimately enhancing bioremediation efficiency.

*Nanocatalysts* Nanoparticles as nanocatalysts have been used widely for wastewater treatment as they have small size, high catalytic activity, highest surface area to volume ratio, and enzyme storage activity. Nanocatalysts work on a photocatalytic principle, where a catalyst medium that is sensitive to light exposure has been used to degrade textile effluents. The nanoparticles that have high photoactivity and photostability and are chemically and biologically inert are ideal photocatalysts (Sharma et al., 2017).

Organic contaminants like alcohols, carboxylic acids, chlorinated aromatic compounds, and phenolic derivatives are the toxic compounds present in the wastewater and are converted into nontoxic substances by photocatalysis. The inorganic contaminants like ammonia, cyanides, nitrates, nitrites, and halides are also decomposed by this method. Heavy metals, mainly copper, mercury, arsenic, lead, and chromium, that are released from textile industries are also remediated using various nanocatalysts (Kumar et al., 2017).

#### 2.2.1.3 Microorganism-Based Nanotechnology

Microorganisms are also used to synthesize nanoparticles, which is an eco-friendly and cost-efficient method. There are some disadvantages to using chemical-based nanoparticles; but the nanoparticles which are synthesized from plants, fungi, and bacteria are green and eco-friendly solutions. Iron oxide nanoparticles biofabricated from *Aspergillus tubingensis* are used to remove 90% of various heavy metals, like nickel, copper, zinc, etc., from wastewater (Mahanty et al., 2020). Copper nanoparticles synthesized from *Escherichia* species SINT7 are found to degrade azo dyes and other textile effluents (Cheng et al., 2019). There are various mechanisms where microorganisms secrete many catalytically active enzymes, which, with nanoparticles, help to bioremediate textile effluents. The significant advantage of using microorganisms is that they also produce valuable products from industrial waste.

#### 2.2.1.4 Enzyme-Based Nanotechnology

Nanotechnology uses enzymes extensively, as they help make nanoparticles less toxic to the environment. Enzymes in nanoparticles make less cell interaction through steric hindrances, provide different catalytic activity, and are eco-friendly, making nanoparticles more efficient and effective for bioremediation. Enzymes are catalysts used in both free or immobilized forms. The crucial advantages of using immobilized enzymes are long-term stability, easy recovery, and reusability, which boost the wastewater treatment process and are very cost-efficient. The peroxidase enzyme is crucial in the breakdown of phenolic compounds like azo dyes (Darwesh et al., 2019). Laccases immobilization on magnetic nanoparticles biodegrades industrial effluents, including synthetic dyes, crystal violet, malachite green, brilliant green, etc. (Zhang et al., 2020b).

### 2.2.2 Introduction of ECs from the Agricultural Sector

The key performance indicators of the agricultural sector are high output at a cheap cost while posing minimal environmental and health risks. With the exponentially increasing demand for food, the agriculture sector has to accelerate productivity while maintaining economic viability. The Indian agriculture sector is dependent on mainly chemical insecticides. The reason behind this is the favorable climate suitable for insect breeding because of high temperature and humidity. With this, the use of pesticides becomes inevitable on the farmland, and its residue imposes a long-term threat to the surrounding ecosystem. Agricultural pollutants are mainly comprised of pesticides and their residues. Pesticides can be classified based on their usage; insecticide to control insect pests, nematicides to control nematodes, fungicides to control fungi, weedicides to control weed pests, etc. (Odukkathil and Vasudevan, 2013). The primary threat caused by the incorporation of pesticides in the environment depends on several factors: toxic characteristics, amount formulated, method of application, and, especially, its mobility and persistence in nature. The ecosystem is battling the harmful effects of the residues generated from the extensive use of chemical pesticides. The residues are in several forms, such as degradation products, metabolites, and congeners, all having toxicological effects. The direct or indirect exposure of pesticides reported abnormal physiological activities and dreadful symptoms in several biotic species. Chemical pesticides are a blessing, in terms of yield, to combat pests, and control insect-borne diseases (encephalitis, malaria, filariasis, dengue, etc.), but a curse to the ecosystem, causing accumulation of its residues in the environment and biomagnification in the food chain (Ali et al., 2019).

The use of hazardous chemicals has increased tremendously to protect economically essential crops and enhance crop productivity globally. Different pesticides, such as herbicides, insecticides, fungicides, etc., impose threats to different species. The pesticides can be categorized as depicted in Table 2.2, based on their function and the target pest they affect (Fishel and Ferrell, 2010).

Selecting a specific remediation mechanism that bestows both environmentally friendly and efficient outcome is the foremost priority for pesticide degradation. Researchers have reported several chemical and physical methods, but the nanobiotechnological strategy ensures safety and is highly efficient in eliminating unwanted chemicals. Nanotechnology-assisted, microbe-based, and enzymatic-based degradation are significantly advantageous approaches as they are amenable under ambient conditions. The biological entities are immobilized on the specific inert supports

**TABLE 2.2**
**Classification of Pesticides Based on the Target Pest**

| S. No. | Types of Pesticides | Target pest / Function |
|---|---|---|
| 1 | Acaricides | Kills mites and ticks or disrupts their growth or development |
| 2 | Algaecide | Kills or inhibits algae |
| 3 | Antifeedants | Prevents an insect or other pest from feeding |
| 4 | Avicides | Kills birds |
| 5 | Attractant | Attracts a wide range of pests |
| 6 | Bactericides | Kills or inhibits bacteria in plants or soil |
| 7 | Bird repellents | Repels birds |
| 8 | Fungicides | Prevents, cures, eradicates fungi |
| 9 | Fumigant | Wide range of organisms |
| 10 | Herbicides | Kills weeds and other unwanted plants |
| 11 | Insect attractants | Lures pests to a trap |
| 12 | Insect repellents | Deters an insect from landing on a human or an animal |
| 13 | Insecticides | Kills insects or disrupts their growth or development |
| 14 | Insect growth regulator | Insects |
| 15 | Lampricides | Targets the larvae of lampreys |
| 16 | Larvicides | Inhibits growth of larvae |
| 17 | Molluscicides | Kills slugs and snails |
| 18 | Nematicides | Controls nematodes |
| 19 | Ovicides | Inhibits the growth of insects' and mites' eggs |
| 20 | Piscicides | Kills fishes |
| 21 | Plant growth regulators | Alters the expected growth, flowering or reproduction rate of plants |
| 22 | Predacide | Kills mammal predators |
| 23 | Rodenticides | Kills rats and related animals |
| 24 | Synergists | Enhances the toxicity of a pesticide to a pest |
| 25 | Silvicides | Acts against woody vegetation |
| 26 | Termiticides | Kills termites |
| 27 | Virucides | Destroys or inactivates viruses |

to enhance the efficiency of the enzyme. The immobilization further accelerates the stability, efficacy, applicability, and durability of the enzymes and microbes. Furthermore, the functionalization of nanoparticles intensifies the operational and thermal stability for chemical degradation.

*Emerging strategies for pesticide degradation* With the advent of bionanotechnology, several strategies have evolved that decontaminate various types of agricultural toxic waste. Combining the biological system and nano-based method ensures a sustainable approach for dealing with released pollutants. Using an optimized medium, this interdisciplinary approach at the nanoscale level provides the best solution for monitoring the generated waste and degrading it. Figure 2.3 discusses some of the emerging nano-based strategies.

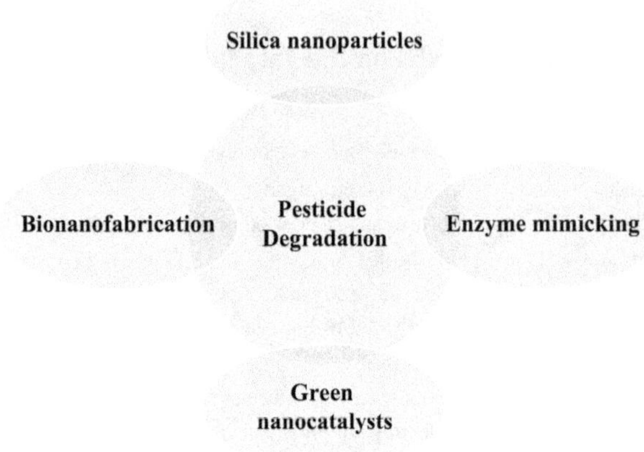


**FIGURE 2.3** Several emerging strategies for the detoxification of agricultural pollutants.

#### 2.2.2.1 Silica Nanoparticles as Immobilization Matrix

Materials with greater porosity have significant applications for the development of nanoparticles. Among the several materials, silica-based nanoparticles have excellent potential as they degrade several classes of pesticides. The silica-based nanoparticles are employed to immobilize catalytic species such as enzymes. This adsorbent has a greater range of potential due to its flexibility, porosity, and stability. Silica nanoparticles (SiNPs) possess a larger surface area and are functionalized by conjugating with different biochemical entities. SiNPs are employed either as nanosheets or in the form of spheres. Owing to larger surfaces, the tailoring of enzyme loading densities dramatically aids in the remediation of environmental contaminants such as pesticides. Researchers have designed mesoporous silica with desirable pore diameter, particle morphology, and size for several catalytic reactions. Different entities can be associated with SiNPs, such as enzymes or microorganisms, to degrade the environmental contaminants.

*Enzyme-mediated pesticide degradation* Carboxylesterase (CbE) hydrolyses carboxyl esters and degrades various insecticides such as organophosphates, pyrethroids, and carbamates. The combination of nanotechnology and the enzymatic approach is well established due to its recyclable and robust characteristics. Diao et al. (2013) immobilized insect (*Spodoptera litura*) CbE on mesoporous silica nanoparticles (SBA-15 and MCM-41) (Diao et al., 2013). Taking advantage of mesoporous material, the enzyme was immobilized to enhance its insecticide degrading property. The SBA-15 immobilized enzyme degraded 60% of Malathion (organophosphate) and 50% of Diethofencarb (carbamate) in the initial two hours of exposure. The enzyme exhibited its activity even in organic solvents and a denaturing agent such as urea. Based on the stability, catalytic efficiency, and lower loading time, SBA-15 is a preferable platform over MCM-41. The larger pores of SBA-14 rendered more significant immobilization potential and thus ensured a higher rate of insecticide degradation.

El-Boubbou et al., 2012 employed acid-prepared mesoporous Silica (APMS) for pesticide hydrolysis (El-Boubbou et al., 2012). Organophosphate hydrolase (OPH) associated with the detoxification of organophosphate pesticides, such as paraoxon, was immobilized within the particles. The pore diameter of APMS is similar to the enzyme and substrate size, with a range of 2–20 nm, and hence provides a suitable confined environment for effective interaction within the pores. Their study revealed the interactions responsible for detoxification of paraoxon and its correlation with pore diameter and various functionalization entities. APMS was covalently amended with ammonium, amine, or carboxyl groups organosilanes along with varied pore diameter, and then the hydrolysis efficacy was monitored. The ammonium functionalized NPs having an optimum pore diameter of 6nm proved to be the most effective system for paraoxon degradation. The toxin was 100% decontaminated in the initial two minutes of the reaction. Thus, immobilization within nanoparticles with specific chemical modification enhances the degradation efficacy of the enzyme. Different enzymes can be availed with the aid of nanotechnology to degrade different classes of pesticides.

Bisphenol A (4,40′-isopropylidenediphenol, BPA) is a phenol-containing contaminant that acts as a pesticide. It is released in large amounts from polycarbonate and epoxy resin production sources and accumulates in water sources, contaminating it. This toxic compound is reported to cause hormonal imbalance leading to an endocrine disorder. Laccase, an oxidase enzyme, is well-known for the degradation of phenolic compounds. Galliker et al. (2010) immobilized the enzyme to escalate the enzymatic activity and reinforce its stability (Galliker et al., 2010). They designed a new nanoparticulate system contriving the linkage of laccase enzyme isolated from a fungal source (*Coriolopispolyzona*) on SiNPs. The enzyme was covalently anchored on the silica particles using glutaraldehyde as a cross-linker. The enzymatic degradation was accessed using several analytical techniques, such as HPLC, LC-MS, and GC-MS. The immobilized enzyme exhibited increased stability and, hence, rendered higher catalytic efficiency in the degradation of the BPA.

*Microorganism mediated pesticide degradation* To control broadleaf weeds, during the early days of the plantation, atrazine (2-chloro-4-ethylamine-6-isopr opylamino-s-triazine), an herbicide, is applied, which by eventual runoff contaminates water bodies. To detoxify the harmful effect of this compound, the whole-cell potential microorganism was utilized in the past. Reategui et al. (2012) used SiNPs to immobilize *Escherichia coli* which recombinantly expressed the Atrazine-dechlorinating enzyme, AtzA (Reátegui et al., 2012). The bacterial cells were encapsulated in the silica-based matrix, comprising silicon oxide precursors (silica nanoparticles, alkoxides) and polyethylene glycol as the cross-linker. The porous gel allowed the entrance of atrazine and water into it and released hydroxatrazine out of the gel. The immobilized cells expressed stability at 23 and 45°C and were proved to be more efficient than the free cells. This nano-based hybrid biomaterial degraded the pesticide for four months without any regeneration requirement.

Fungicides such as dimethomorph are used on farmland to control several pathogenic attacks, like late blights, downy mildew, etc., in a number of crops. The remnants of this fungicide pose a threat to the ecosystem due to its toxicity. To combat the toxic effect, a microbial-assisted degradation strategy is preferred due to environmental friendliness. For the degradation of dimethomorph *in situ*, a nano-based immobilization system is reported (Zhang et al., 2020). The system consists of a bamboo charcoal (BC) and sodium alginate (SA) matrix. *Bacillus cereus* WL08, was isolated from the contaminated fungicide site and thereby utilized for decontamination of dimethomorph in both the water and soil systems. They demonstrated the degradation efficiency of the bacteria in the water system by illustrating in the continuous reactor and by immobilization for the decontamination of outdoor soil conditions. In the reactor system, 85.61% of the dimethomorph got degraded with an influent concentration of 50–100 mg/L by the immobilized bacteria in 30 days. In the outdoor study for seven days, the free WL08 and the immobilized WL08 degraded 68.51% and 95.34% of dimethomorph, respectively. In 72 hours, the free and the immobilized WL08 decontaminated 66.95% and 96.88% of dimethomorph (50 mg/L), respectively. Thus, the immobilized WL08 efficiently decontaminated the fungicide both in the water and in the soil system. The tolerance towards environmental changes increases due to immobilization which, in turn, enhances the stability and degradation efficiency.

#### 2.2.2.2 Enzyme Mimicking Nanocatalysts

Organophosphate (OP) is indispensable as it is associated with several biological activities such as signaling, regulation, synthesis, and transduction of energy. On the contrary, they are used in the agriculture field as pesticides and chemical warfare due to their stability. The model substrates of OP, such as paraoxon, are studied to check the toxicity and the mechanism behind their activity. Owing to the catalytic and degradation efficiency, several detoxifying strategies are conducted by mimicking the enzymes. Silva et al. (2017) developed imidazole-derived nanocatalysts due to their resemblance with the enzymes' active site and their ability to act both as acid-base or nucleophilic catalysts (Silva et al., 2017). The NPs of Ag, Au, and bimetallic AgAu, in the form of AgSiO2, AuSiO2, and AgAuSiO2, were designed to degrade organophosphate (OP) pesticides. The NPs were functionalized with methimazole (MTZ), consisting of thiol and imidazole entities. The MTZ reacted via Sulfur moiety with AuNPs and through the nitrogen group of imidazole with AgNPs. Due to the free imidazole group, these nanocatalysts were highly reactive and efficient in the degradation process. Paraoxon, the model substrate of OPs, and the simulant diethyl-2,4-dinitrophenyl phosphate (DEDNPP) were considered for the detoxification study. Generally, paraoxon takes 600 years for its 50% degradation, but with the advent of imidazole-coated nanocatalysts, very rapid degradation was reported. Within seven days, 50% of paraoxon was degraded.

#### 2.2.2.3 Green Nanocatalysts

Another recyclable nanocatalyst designed from plant fiber has been reported recently. The use of turmeric root powder as the green nanocatalysts performing photocatalysis has emerged as the noble strategy for pesticide degradation. Kumar and Luxmi (2020) developed green nanocatalysts, which are nanosheets undergoing photocatalytic phenomena without any chemicals (Kumar and Luxmi, 2020). The turmeric root powder was characterized using several analytical techniques such as XPS, UV-vis spectroscopy, SEM, TEM, BET, and PL. The morphology of turmeric root powder nanosheets revealed pore size (~2 nm) and greater surface area and bandgap in the visible range (~2.52 eV). Moreover, the optical properties were analyzed in great detail. The presence of turmeric catalyst proved to be quite effective for pesticide degradation in an aqueous medium. This green photocatalyst degraded carbofuran and malathion pesticides up to 77% and 95%, respectively, in 140 min. Owing to the advantages such as high efficiency in detoxification, recyclable and green approach, this proves to be the sustainable technology for decontamination of water pollutants. The photocatalytic reaction carried by these noble nanocatalysts is the best alternative to photocatalytic chemical reactions as it is an eco-friendly and sustainable technology.

#### 2.2.2.4 Bio-nanofabrication

The NPs synthesized using biological entities such as microbes, enzymes, and other biodegradable polymers are highly advantageous. It is a rapid and eco-friendly strategy to remediate several environmental contaminants. Several NPs have been synthesized using biological systems, such as bacteria, yeast, fungi, viruses, etc. Researchers have fabricated living sources in different nanomaterials that are safer and more efficient than other chemical methods. The method of biofabrication includes the synthesis of nanoparticles with the aid of biological systems and functioning as a bionanofactory. This is associated with the quicker remediation of contaminants into less toxic forms through the enzymatic process. The decontamination process can be either intracellular or extracellular. Koul et al., 2021 have reviewed the synthesis of biofabricated NPs using biological entities and further discussed its widespread applications (Koul et al., 2021).

In the case of intracellular fabrication, the internal mechanism of the microbial system is employed for NP generation. The microbial population is allowed to interact with several metals in the aqueous media. After an incubation period, the color of the NPs starts developing. NPs of different metals exhibit different morphology and appearance. To cite some, yellow to whitish-yellow specify ZnNPs and MnNPs, AuNPs exhibits pinkish to pale yellowish color, and brownish color specify AgNPs. The metal ions, which are positive in charge, get entrapped inside the negatively charged cell wall of the microorganism. The metals get reduced by the enzymatic reaction in the cell wall forming nanoclusters. These NPs get released to the solution, which is thereby used for decontamination purposes.

In extracellular fabrication, the microbes are grown and then centrifuged, the pellet is discarded, and the supernatant is processed for NP synthesis. In another setup, the supernatant-containing enzymes are allowed to react with the metal ions. Thus, the NPs get synthesized due to the onset of the enzymatic reaction in the incubated vessel. The NPs thus generated are analyzed by several analytical techniques to characterize the particles morphology, size, shape, color, uniformity, and stability.

The biophysical characterization of the generated NPs is carried out using SEM, FTIR, TEM, XRD, XPS, etc. The biofabricated NPs have several applications; they are used to decontaminate pesticides, used in pharmaceuticals, medicine, and diagnostics, etc. Owing to the unique characteristics of this reusable catalyst, they are used as sensors to detect and thereafter to decontaminate the pollutants.

The hazardous compounds are released from agricultural lands in the form of pesticides accumulated in the soil, underground water, or water bodies. Living systems are hugely affected due to the prolonged use and accumulation of hazardous pesticides and are primarily detected in soil, food chains, and water bodies (Anju et al., 2010). These chemicals take years for their natural breakdown, and some residues do not get easily degraded. Due to the rising need for pesticides, the threat to the ecosystem has increased tremendously.

Bionanotechnology is the most innovative way to detect and degrade environmental contamination. It ensures sustainable and innovative techniques to deal with the emerging chemicals released by runoff from agricultural lands. The uniqueness of nano-based technology is its durability, stability, and availability. Moreover, it is the best alternative for several chemical and physical methods for combating the waste management of both the soil and water system.

### 2.2.3 Leather Industry

The leather industry is one of the oldest and demanding sectors worldwide. The turnover of the leather industry across the world is around US$80 billion per year, which makes it very important commercially and economically. Products from the leather industry, such as leather footwear, leather bags, and leather clothing are used in day-to-day life. Making leather from the raw skin in the industry is broadly categorized into three sections, i.e., preparation of the raw material, tanning of the leather, and "crusting" in which dyes and other chemical processing are involved.

#### 2.2.3.1 Stages of Leather Processing and the ECs Released During Every Step

*Preparatory step* The raw material used for leather products is obtained mainly from the meat industry, like the skin of sheep, goats, pigs, and other animals. In the preparatory step of hide processing, the raw skin of animals is cleaned by soaking, liming, and deliming. Removal of hair is done during the liming process by using sodium sulfide, one of the major chemical contaminants released from the leather industry. Ammonium sulfate, used for deliming to maintain the basic pH during the process, but when released untreated, it increases the alkalinity of water. It is very toxic to the aquatic ecosystem. Higher ammonium contamination in water enhances the growth of aquatic ferns and microorganisms, resulting in foul odor, and the color of which makes the water unusable. The final step in the preparation of material is degreasing which removes oils and fats by using surfactants. The release of surfactants in water resources affects the aquatic ecosystem due to their toxicity to plants and invertebrates. Surfactants also inhibit the growth of microorganisms in bioremediation plants due to toxicity to the cells (Ivankovic and Hrenovic, 2010). Effluents released during these steps contain acids, chemicals, salts, surfactants, etc. that are highly responsive to enhance the BOD and COD of water bodies.

*Tanning and crusting steps* Many toxic heavy metals are released in the effluent during the process of tanning. Tanning can be done in two different ways, i.e., by the use of vegetable tannins extracted from various plant sources, which is an eco-friendlier way, or by using chromium sulfate which, while it is a much more efficient and time-saving method, poses life-threatening side effects. Chromium contamination in water bodies is observed due to the mixing of effluents in groundwater and water resources. Phenolic compounds from vegetable tanning increase the COD of water bodies. Crusting includes the use of various organic compounds and dyes in a stepwise manner. The organic compounds and water-soluble dyes released in effluents produced during crusting are also a matter of great concern.

#### 2.2.3.2 Adverse Effects of the ECs from the Leather Industry on the Environment and Human Health

Various types of hazardous chemicals and waste products are released during the significant stages of hide processing up to the finished leather product. 70–80% of the residual parts of the hide after removal of raw skin is discarded as waste that cannot be reused (Cabeza et al., 1998). It contains dead body parts of animals, hair, tissue, protein, and fats. If unprocessed, this type of solid waste can create a very unhealthy environment due to the growth of various pathogenic strains. It increases the biological oxygen demand of water bodies if the effluents containing dead tissue, protein, and fat are discharged without proper processing. Approximately 600 kg of solid waste is generated during the pre-tanning and tanning process in the leather industry (Ozgunay et al., 2007). A large amount of water is used in the tanneries for different treatments, and about 30–35 $m^3$ of wastewater is discharged in the environment containing various chemical contaminants. Chemicals like chromium sulfide are one of the most toxic and carcinogenic agents used during tanning of the leather. Chromium is a heavy metal used in chrome tanning, and about 40% of the chromium is discharged into the sludge as waste after the tanning process (Nur-E-Alam et al., 2020). Exposure to chromium can induce skin cancer, respiratory, kidney, liver, and other types of cancer in leather industry workers. It also affects the aquatic ecosystem when released into rivers or other water bodies. Various types of volatile compounds and toxic gases are also released from the leather industry like ammonia, hydrogen sulfide, fumes of acids, chlorides, etc. These are very toxic to the ecosystem and responsible for lung diseases and fetal neuronal toxicity. Decolorization of dyes and degradation of phenolic compounds in the effluents of the leather industry are also challenging tasks due to the shortage of cost-effective and efficient technologies.

#### 2.2.3.3 Strategies for Detoxification of Pollutants and Potential Applications of Bionanotechnology

Chromium exists in different oxidation states, namely Cr (III), Cr (IV), and Cr (VI), out of which Cr (VI) is the most toxic and hazardous form with life-threatening side effects on human health and the ecosystem. There are some methods used to remove chromium from the effluent of the leather industry, like adsorption, accumulation, precipitation, electro-dialysis, etc. Chemical precipitation of Cr from the wastewater includes various chemical and precipitating agents, like sodium hydroxide, calcium hydroxide, or magnesium oxides, etc. These are additional chemical components to the wastewater. Some advanced techniques, like electrodialysis, are very expensive to apply on a larger scale. Hence, we need some cost-effective and efficient methods for the detoxification or removal of Cr contamination from effluents of the leather processing industry. Nowadays, various attempts are being made with the help of green technology for the bioremediation of chromium contamination as an eco-friendly and cost-effective alternative. In order to minimize the contamination of toxic components in the environment, the first step for the minimization is to reduce the use of toxic components, which eventually will reduce its disposal in the environment. Hence use of bionanotechnology to increase the efficiency of the alternative, less effective yet eco-friendly methods of tanning can show a remarkable difference in the release of chromium in the ecosystem. Application of plant phenolic compounds along with the silver nanoparticles in tanning the leather can augment the efficacy of the process. Recently some reports have been made of designing silver nanoparticles to enhance the efficacy of the vegetable tanning method. A huge amount of solid waste, including dead animal body parts, tissue, waste hide residuals, etc., is generated in the leather industry. Its removal can be facilitated by biodegradation using various enzymes associated with nanoparticles (Koyani et al., 2017). A combination of a different set of enzymes, such as lipase, collagenase, and proteinase nanoparticles can be used to degrade the proteins and fat-rich effluents. Dye is also a serious type of contaminant found in the effluents released from the leather industry. Degradation of Congo red and malachite green dyes using copper nanoparticles synthesized through microorganisms has been observed to be a very efficient method with 97.07% and 90.55% degradation, respectively, at a lower concentration (Noman et al., 2020). Application of nanofiltration technique has also been reported for the removal of heavy metal contaminants from industrial effluents (Liu et al., 2019). Li et al. (2019) developed a superparamagnetic micro-bio-nano-adsorbent for the removal of Chromium Cr(VI) contamination using polydopamine, biochar, and $Fe_3O_4$ for enhancing the efficiency of adsorption of chromium (Li et al., 2019). This new method has been observed to remove the Cr(VI) contamination very efficiently from 20 mg/L to 0.2 mg/L in the water. Besides all these technological advancements and attempts, there is a vast scope for study in the field of bionanotechnology for the removal of emerging contaminants to clean up the effluents and maintain the balance of the ecosystems.

### 2.2.4 Paper and Pulp Industry

The paper industry is one of the oldest and most widely known industries in the world since ancient times. The raw material used in the paper and pulp industry is mainly obtained from forest and agricultural resources. As per reports in 2018, around 422 million metric tons of paper are utilized globally. The United States, the world's largest paper manufacturer, followed by China, has reported generating a revenue of more than US$20 billion annually. A huge amount of solid waste material is disposed of by this industry, which is a serious concern for government agencies and environmentalists. The raw material used in the paper and pulp industry is wood from which lignin and cellulose fibers are separated. There are two major steps involved in manufacture of paper, i.e., making pulp out of the raw material and bleaching. Various types of chemical and biological waste are generated during these two steps. Water consumption during manufacturing is very high, up to 200 $m^3$/ton of paper production (Cecen et al., 1992). It ultimately leads to the discharge of very heavily contaminated effluent containing chlorine, sodium carbonate, acids, halides, other organic solvents, dyes, and toxic pollutants. Other than these chemical contaminants, the effluents released from the paper and pulp industry contain waste from the residual raw material, as a much smaller amount of raw material is actually converted into paper—most of the raw material is discarded as waste. The effluents contain wood particles, some cellulose fibers, tannins, resins, lignin (30–45%), saccharinic acid (25–35 %), formic acid, and acetic acid (10%), etc. (Singh et al., 2017). These contaminants, if untreated, disturb the balance of the aquatic ecosystem leading to increases of the BOD and COD of water. Toxic compounds, like furans and dioxins, cause carcinogenic and mutagenic effects on the fauna and flora in the water bodies and lead to bioaccumulation and biomagnification of the toxic compounds. Hence, removing or detoxifying these emerging contaminants released from the paper and pulp industry is essential.

#### 2.2.4.1 Methods for Removing Contaminants Using Bionanotechnology

Disposal of the solid waste from the paper and pulp industry is generally done by anaerobic treatment, enzymatic digestion, activated sludge treatment, and land disposal. Despite multiple available methods for the removal of the solid waste, efficient disposal and detoxification methods need to be developed to remove and reduce the release of emerging contaminants in the environment.

*Enzyme-linked nanoparticles* Bionanotechnology has a great potential to meet the challenges in reaching the goal of zero waste to make the environment clean and healthy. Degradation of lignin in the wastewater and solid waste is a challenging task. It is difficult to degrade and is generated in a high amount giving the dark brown color and smell to the effluent. Immobilizing ligninolytic enzymes on the magnetic nanoparticles is a novel strategy to increase the availability and efficiency of enzymatic degradation (Johnson et al., 2008). Magnetic nanoparticles conjugated with enzymes have advantages in that they have very high specificity of substrates to which enzyme molecules will bind, nanoparticles prolong the reaction time. Their recovery can be made from the effluent after the treatment due to the magnetic property. The enzyme is covalently linked to the nanoparticle through carbodiimide linkage then the efficiency of the enzyme is evaluated to degrade the pollutants in the effluent.

*Carbon nanotubes* Carbon nanotubes are used to enhance the efficiency of adsorption of dyes, toxic pollutants, or heavy metals on the surface to enhance the degradation or detoxification process. Silva AR. et al. has used the technology of nanotubes along with microorganisms for anaerobic reduction of the dyes from effluent (Silva et al., 2020). The study shows that the carbon nanotube significantly enhanced the biodegradation process.

The application of nanotechnology in bioremediation is an eco-friendly and effective green alternative to the traditional methods; hence further studies and development of novel techniques are critical.

### 2.2.5 Pharmaceuticals

Pharmaceuticals are used for human health and personal care as well as veterinary purposes. Pharmaceuticals are chemicals that are toxic, biologically active, and can disrupt hormones. The presence of unused or non-assimilated pharmaceuticals in drinking water, wastewater, and sludge have been reported publicly by researchers, environmentalists, and government experts over the last two decades. After exposure, these chemical compounds continue to increase and affect human health and other organisms in the local ecosystem. Globally, pharmaceutical compounds are released by single or miscellaneous sources from treated wastewater and sewage, industrial effluent, sewer overspill, animal feed lots, etc. The harmful effect and speedy passage of pharmaceuticals depend on water treatment quality, chemical characteristic, concentration, life span of compounds, etc. The US Geological Survey identified drugs in the aquatic life of US water systems through its 1999–2000 report on organic wastewater contaminants (Kolpin et al., 2002). In the last two decades, advancements in research and technology, including analytical improvements, suitable sample preparation methods, and most detective sensors, have made it possible to detect the presence of ECs chemicals contaminants in aquatic systems at parts per trillion to parts per billion levels.

#### 2.2.5.1 Pharmaceutical Compounds

In recent years, hospitalization has increased due to urbanization, industrialization, human population expansion, and the COVID-19 pandemic scenario, leading to increased pharmaceutical use and raising the number of ECs globally. The ECs produced by pharmaceuticals can lead to resistance in living organisms and microbes present in the environment. The presently used techniques are insufficient to remove ECs from wastewater produced by pharmaceutical compounds; therefore, there is a need to focus on cost-effective, sustainable advanced technology to manage ECs in the environment. It includes human and veterinary prescribed drugs, such as blood pressure regulators, antidepressants, analgesics, antibiotics, antifungals, antimicrobials, anti-inflammatory, antiallergy, bactericides, etc. The most commonly used drugs are discussed in detail.

*Analgesic drugs* Diclofenac (DCF) is the most recommended nonsteroidal and anti-inflammatory (NSAID) drug that reduces body pain and inflammation. The reported concentration of DCF found in drinking water is below 10 ng/L (Rigobello et al., 2013). DCF is used to cure mild to moderate pain, such as rheumatoid arthritis or osteoarthritis, and its powder form is also used for migraine headache outbreaks. The consecutive rise of concentration of DCF in water bodies could be described by its excessive use in human and veterinary medicine. DCF has harmful effects on living species of the aquatic ecosystem. The European Union have agreed to include DCF in the first watch list of the Water Framework Directive (EU2015/495, European Commission) to know more about the occurrence and its effect on the environment. In Europe, the EU Water Framework Directive emphasizes major water bodies, including the coastline and transitional waters. In the near catchment area, many wastewater treatment plants are established that also affect the ecosystem. The degradation of DCF depends on the technology used in wastewater treatment, but the low biodegradability commonly results in low removal rates through the treatment. In the marine ecosystem, the DCF concentration from a few ng/L to 15 ug/L is found. Marine organisms are used as an indicator of contamination in water as various toxic compounds accumulate and result in biomagnification. The marine organism can metabolize DCF and improve its excretion, which means it does not have any adverse effects on such organisms. In contrast, several studies have introduced DCF effects in the marine organism under research laboratory situations.

*Antibiotic drugs* The continuous exposure of wastewater to antibiotics has potentially harmful effects on aquatic organisms as well as the environment. Due to the increase in pharmaceutical compounds in the environment, it has become the focus of emerging concern in recent years. Antibiotics are chemical compounds that can kill or inhibit bacterial growth, and they are extensively used to inhibit and cure human and cattle disease. The most commonly recommended antibiotics are tetracyclines, penicillins, and sulfonamides in the treatment and control of communicable diseases. More than 30 antibiotics have been found in samples from sewage, drinking, and groundwaters. The different kinds of antibiotics serve as a growth promoter in animal and aquaculture systems. Antibiotic utilization increases every year. In 1998, according to the European Animal Health Association, about ten million kilograms of antibiotics were used, and 50% of them were used for veterinary purposes (Kummerer, 2008).

*Antipyretic drugs* Antipyretic drugs are given to control body temperature, which induces the hypothalamus to enhance prostaglandin secretion. The use of antipyretic drugs is increasing exponentially because of the Covid-19 pandemic situation and other medical conditions. The most usable antipyretics are paracetamol, ibuprofen, aspirin, etc. Antipyretic drugs are not entirely metabolized by humans or animals and are excreted into wastewater or sludge. The increased concentration of antipyretic medications in wastewater induces harmful effects and risks on aquatic life forms. Their elevated level in recent years is an emerging concern, and an appropriate strategy needs to be developed.

#### 2.2.5.2 Approaches towards the Removal of Waste Pharmaceutical Compounds

Wastewater treatment plants are considered the leading and conventional modules to eliminate or reduce pollutants or ECs. The biological treatment process for removing pharmaceuticals from wastewater or sludge have been in common practice for many years. The demand for emerging technologies such as bionanotechnology, microbial degradation, enzyme technology, etc., has increased to boost the process, accuracy, and efficiency of pharmaceutical treatment.

*Biological degradation* Biodegradation plays a potential role in the removal of pharmaceutical residues in wastewater. Microbes have different metabolic abilities to degrade chemical components, personal care products, and pharmaceuticals. Micro-organisms have potential enzymes that participate in the biodegradation of DCF, transforming toxic to a nontoxic structurally altered form. The use of selected strains *Bacillus subtilis* and *Brevibacillus laterosporus* have produced particular degradation enzymes essential for the biodegradation and metabolization of DCF. After 17 hours of treatment, more than 95% of DCF was successfully degraded (Grandclément et al., 2020). The parent compound, DCF, is degraded into the 4′-hydroxy-diclofenac metabolite by hydroxylation, which is screened by liquid chromatography with tandem mass spectrometry (LC-MS/MS) analysis. Fungal strains, like *Trametes versicolor* and *Ganoderma lucidum* and algae, such as *Chlorella sorokiniana*, are recognized to bio-transform or biodegrade different categories of drugs.

*Enzyme immobilization of nanomaterials* Nanomaterial immobilized with enzymes is very stable due to unfolding confrontation, use of multiple cycles, and improved activity. Due to the enzyme immobilization of a nanomaterial, the improved surface area increases efficiency and enhances enzyme stacking. Immobilized oxidoreductase can play a vital role in the successful removal of pharmaceuticals from wastewater. The pharmaceutical compounds can be removed with the aid of enzyme oxidoreductase immobilization. These drugs, such as anti-inflammatory, antibiotics, and antidepressants, were removed by immobilizing laccase enzyme from *Trametes versicolor* and covalent binding on polyacrylonitrile biochar membrane. The immobilized enzyme showed more than 90% activity after one month of storage, although un-immobilized enzymes have up to 32% activity (Zdarta et al., 2021). Another study suggests that the laccase graphene oxide as a biocatalyst can be used to effectively degrade diclofenac and ibuprofen. It also reduced the carbamazepine and showed more than 80% elimination of pharmaceutical compounds from wastewater. In other studies, tetracycline was eliminated by laccase, immobilized with the bentonite-derived mesoporous component. The physical adsorption analysis shows the improved enzyme stability and has more potential with a broad range of temperatures and pH than the un-immobilized enzyme. The tetracycline removal capability of the biocatalyst is around 60% after the three-hour degradation process in the presence of 1-hydroxy-benzotriazole. 50% or more of the biocatalyst retained its initial activity after the repeated four cycles.

*Biogenic nanoparticles* Biogenic materials are synthesized by living organisms through metabolism and generate particular composites that have toxic effects on other organisms. Metallic nanoparticles are developed through chemical, physical, and biological methods. In biological methods, the utilization of biological mass is either microbial or plant source as a reductive agent via intracellularly or extracellularly. Several research groups have developed cost-effective and sustainable biogenic nanoparticles globally to manage pharmaceuticals. Green technology is used to synthesize zirconia nanoparticles from microbial sources that participate in the bioremediation of tetracycline from pharmaceutical waste (Debnath et al., 2020). The different types of nanoparticles are developed in the presence of biologically active molecules. Some of biological origin, like enzymes, peptides, phenolics, alkaloids, vitamins, etc., are attached to the metal salt solution under different conditions. It depends on the precursor materials and future use of nanoparticles (Aguilar-Pérez et al., 2020). The details of some pharmaceutical compound treatments that use biogenic nanoparticles are discussed in Table 2.3.

*Nanofiltration membranes* The membrane-based filtration in which excellent quality of filtrate is allowed to pass with pressure through small pores in nanometer ranges (1–10 nm) falls under the nanofiltration process. This membrane type is used to filter drinking water, groundwater, surface area water, and also wastewater. It is most accepted because of its biocompatibility, low energy intake, and cost-effective nanofiltration. Several important pharmaceutical compounds are found in industrial waste and surface water as ECs, which directly or indirectly affect the environment. Nanofiltration membrane can recollect the chemicals released by pharmaceutical industries through biological persistence and the membrane's hydrophilic property. Usually, a nanofiltration membrane comprises a thin-film multilayer composite that supports layers and permits the stream and the active layer to perform separation with high selectivity (Ouyang et al., 2019). In nanofiltration, the electric charge and pore size are directly related to the efficient removal of a pharmaceutical contaminant. Several studies suggest that the percentage of pharmaceutical compound removal increases if nanoparticles and nanofilter membrane are employed together. It has been reported that the highly water-soluble pharmaceutical compounds, such as Triclosan and sulfamethazine, get separated more than 90%, while ordinarily they can achieve around 40% (Bandehali et al., 2020; Lin et al., 2014). Recent studies show that the carbon structure incorporated with nanofiltration membrane-like graphene or composed of completely graphene oxide is used to make a thin graphene nanofiltration membrane (Li et al., 2018). This kind of development supports mechanical strength, simple accessibility, chemical stability, and high surface area. Due to graphene nanofiltration membrane development, the surface becomes lipophilic, attracts hydrophobically, and participates in more than 60–95% removal of contaminants released from pharmaceutical industries (Li et al., 2018; Zhang and Chung, 2017). The treatment of pharmaceuticals by nanofiltration membrane are summarized in Table 2.4.

*Nanozyme treatments* Nanozymes are nanomaterials with unique physiochemical characteristics to form a particular structural assembly. They mimic enzymes and hence perform essential biochemical reactions. The natural enzyme has some limitations (such as low thermal stability, narrow working pH range and reaction environment, storage issues, stability, etc.) for treating pharmaceutical or chemical compounds present in wastewater (Wong et al., 2021). The nanozymes are like

**TABLE 2.3**
**The Treatment of Pharmaceutical Compounds from Wastewater by Biogenic Nanoparticles**

| S. No. | Contaminant | Synthesis process | Nanomaterial | Removal amount | Reference |
|---|---|---|---|---|---|
| 1 | 17b-estradiol, Ibuprofen, sulfamethoxazole and ciprofloxacin | Synthesis from microbial extracellular region | The Desulfovibrio vulgaris used with palladium and platinum nanoparticles | 85, 94, and 70% | (Martins et al., 2017) |
| 2 | Diclofenac | Synthesis from microbial intracellular region | The Streptomyces griseobrunneus used with selenium nanoparticles | 97.43% | (Ameri et al., 2020) |
| 3 | Tetracycline | Synthesis from microbial extracellular region | The bacteria *Pseudomonas aeruginosa* used with Zirconium oxide nanoparticles | 98% | (Debnath et al., 2020) |
| 4 | Imipenem and imatinib | Chemical reduction | The Carumcarvi L. seeds used with nanocomposite of $Fe_3O_4$/Au form | 96 and 92% | (Mirsadeghi et al., 2020) |
| 5 | Naproxen | Hydrothermal path | The Corchorusolitorius used with nano composite of $SnO_2$/activated carbon form | 94% | (Begum and Ahmaruzzaman, 2018) |
| 6 | Paracetamol | Synthesis from microbial extracellular region | *Aspergillus flavus* used with fluorescent Zinc sulphide nanoparticles | 51% | (Uddandarao et al., 2019) |

**TABLE 2.4**
**The Treatment of Pharmaceutical Compounds from Wastewater by Nanofiltration Membrane**

| S. No. | Contaminant | Synthesis process | Nanomaterial | Pore size (nm) | Removal amount | Reference |
|---|---|---|---|---|---|---|
| 1 | Amoxicillin | Phase inversion via immersion precipitation process | Fe-based nanoparticles with polyacrylonitrile | 2.13 | 92% | (Balarak et al., 2017) |
| 2 | Ibuprofen, Ketoprofen, Fenoprofen, Diclofenac, Naproxen, Carbamazepine, and Indomethacin | Not applicable | Membrane from cellulose acetate | Not applicable | 80–95% | (Zhang and Chung, 2017) |
| 3 | Ibuprofen, carbamazepine, sulfadiazine, triclosan, sulfamethoxazole, and sulfamethazine | Not applicable | Alginate, humic acid, and silica with fully aromatic polyamide thin film composite nanofiltration | .034 | >90% | (Jafari and Aghamiri,2011) |
| 4 | Carbamazepine, atenolol, and ibuprofen | In this technique used layer by layer | Polyethersulfone, quaternate chitosan, and polydopamine with dually charged multilayer membrane | Not applicable | 92.5%, 81.67%, and 89.85% | (Bandehali et al., 2020) |
| 5 | Betamethasone and Fluconazole | Piperazine | Piperazine | 0.76 | 70 and 60% | (Li et al., 2018) |

natural enzymes, but they have properties such as catalytic efficiency, specificity, wide temperature and pH range for activity, and stability to decompose pollutants such as coloring dyes, agrochemical, pesticide, industrial waste, etc. They play a vital role in the degradation of drug molecules and waste released from pharmaceutical industries in water bodies as ECs.

## 2.3 CONCLUSIONS

The prevalence of emerging contaminants is a major issue in many nations. In addition, the fate and behavior of emerging contaminants in the environment are generally unknown and usually underappreciated. It is critical to closely monitor emerging contaminants to stop them from creating a problem. The most novel method for bioremediating emerging contaminants is to use bionanotechnology. It is the best option for various chemical and physical waste treatment procedures and for emerging contaminants generated as industrial waste products. Hybrid technologies, such as nano-bioremediation, are necessary to convert pollutants that damage the environment into harmless forms. *In situ* or *ex-situ* methods are suitable for large-scale cleanup efforts, with low-cost implications and few hazards. Compared to traditional technologies, nano-bioremediation stands out for its effectiveness, durability, steadiness, speed, and affordability. Therefore, this technology can and should be applied to decontaminate emerging contaminants from various industries.

## 2.4 FUTURE PERSPECTIVE

Evaluation of emerging contamination threats and analytical monitoring can safeguard ecosystems from potential risks. Based on recent research in nanobiotechnology, it may be predicted that, in the future, nanobiotechnology will become indispensable in the bioremediation of emerging contaminants generated as industrial waste products. However, considering the rapidly evolving field of nanotechnology, bioremediation technology is still very naive and needs to be upgraded. Future research on both basic and applied aspects of nano-bioremediation is encouraged. Furthermore, emerging contaminant management should be supplemented with additional methods, such as improved chemical use and its management, waste minimization, suitable waste disposal, and chemical discharge reduction in the environment. Also, in evaluating new methods, risk-reduction and cost-effectiveness are suggested.

## REFERENCES

Aguilar-Pérez, K. M., Avilés-Castrillo, J. I., & Ruiz-Pulido, G. (2020). Nano-sorbent materials for pharmaceutical-based wastewater effluents: An overview. *Case Stud. Chem. Environ. Eng.*, 2, 100028.

Ali, N., Khan, S., Li, Y., Zheng, N., & Yao, H. (2019). Influence of biochars on the accessibility of organochlorine pesticides and microbial community in contaminated soils. *Sci. Total Environ.*, 647, 551–560.

Ameri, A., Shakibaie, M., Pournamdari, M., Ameri, A., Foroutanfar, A., Doostmohammadi, M., & Forootanfar, H. (2020). Degradation of diclofenac sodium using UV/biogenic selenium nanoparticles/H2O2: Optimization of process parameters. *J. Photochem. Photobiol., A*, 392, 112382.

Amini, M., Arami, M., Mahmoodi, N. M., & Akbari, A. (2011). Dye removal from colored textile wastewater using acrylic grafted nanomembrane. *Desalination*, 267(1), 107–113.

Anju, A., Ravi S, P., & Bechan, S. (2010). Water pollution with special reference to pesticide contamination in India. *J. Water Resour. and Prot.*, 2, 432–448.

Anjum, M., Miandad, R., Waqas, M., Gehany, F., & Barakat, M. A. (2019). Remediation of wastewater using various nano-materials. *Arabian J. Chem.*, 12(8), 4897–4919.

Avio, C. G., Gorbi, S., & Regoli, F. (2017). Plastics and microplastics in the oceans: From emerging pollutants to emerged threat. *Mar. Environ. Res.*, 128, 2–11.

Babu, B. R., & Parande, A. K., Raghu, S., & Kumar, T. P. (2007). Cotton textile processing: Waste generation and effluent treatment. *J. Cotton Sci.*, 11, 110–122.

Balarak, D., Mostafapour, F. K., & Joghataei, A. (2017). Application of single-walled carbon nanotubes for removal of aniline from industrial waste water. *Biosci. Biotechnol. Res. Commun.*, 10(2), 311–318.

Bandehali, S., Parvizian, F., Moghadassi, A., & Hosseini, S. M. (2020). High water permeable PEI nanofiltration membrane modified by L-cysteine functionalized POSS nanoparticles with promoted antifouling/separation performance. *Sep. Purif. Technol.*, 237, 116361.

Begum, S., & Ahmaruzzaman, M. (2018). Biogenic synthesis of SnO2/activated carbon nanocomposite and its application as photocatalyst in the degradation of naproxen. *Appl. Surf. Sci.*, 449, 780–789.

Cabeza, L. F., Taylor, M. M., DiMaio, G. L., Brown, E. M., Marmer, W. N., Carrio, R., & Cot, J. (1998). Processing of leather waste: Pilot scale studies on chrome shavings. Isolation of potentially valuable protein products and chromium. *Waste Manag.*, 18(3), 211–218.

Cecen, F., Urban, W., & Haberl, R. (1992). Biological and advanced treatment of sulfate pulp bleaching effluents. *Water Sci. Technol.*, 26(1–2), 435–444.

Cheng, S., Li, N., Jiang, L., Li, Y., Xu, B., & Zhou, W. (2019). Biodegradation of metal complex Naphthol Green B and formation of iron–sulfur nanoparticles by marine bacterium Pseudoalteromonas sp CF10-13. *Bioresource Technol.*, 273, 49–55.

Darwesh, O. M., Matter, I. A., & Eida, M. F. (2019). Development of peroxidase enzyme immobilized magnetic nanoparticles for Bioremediation of textile wastewater dye. *J. Environ. Chem. Eng.*, 7(1), 102805.

Debnath, B., Majumdar, M., Bhowmik, M., Bhowmik, K. L., Debnath, A., & Roy, D. N. (2020). The effective adsorption of tetracycline onto zirconia nanoparticles synthesized by novel microbial green technology. *J. Environ. Manage.*, 261, 110235.

Diao, J., Zhao, G., Li, Y., Huang, J., & Sun, Y. (2013). Carboxylesterase from Spodoptera Litura: Immobilization and use for the degradation of pesticides. *Procedia Environ. Sci.*, 18, 610–619.

El-Boubbou, K., Schofield, D. A., & Landry, C. C. (2012). Enhanced enzymatic activity of OPH in ammonium-functionalized mesoporous silica: Surface modification and pore effects. *J. Phys. Chem. C.*, 116(33), 17501–17506.

Ezziat, L., Elabed, A., Ibnsouda, S., & El Abed, S. (2019). Challenges of microbial fuel cell architecture on heavy metal recovery and removal from wastewater. *Front. Energy Res.*, 7, 1.

Fakruddin, M., Hossain, Z., & Afroz, H. (2012). Prospects and applications of nanobiotechnology: A medical perspective. *J. Nanobiotechnol.*, 10(1), 1–8.

Fishel, F. M., & Ferrell, J. A. (2010). Managing pesticide drift. *EDIS*, 7.

Galliker, P., Hommes, G., Schlosser, D., Corvini, P. F. X., & Shahgaldian, P. (2010). Laccase-modified silica nanoparticles efficiently catalyze the transformation of phenolic compounds. *J. Colloid Interface Sci.*, 349(1), 98–105.

Ghosh, P., Samanta, A. K., & Basu, G. (2004). Effect of selective chemical treatments of jute fibre on textile-related properties and processibility. *Indian J. Fibre Text. Res.*, 29(1), 85–99.

Grandclément, C., Piram, A., Petit, M. E., Seyssiecq, I., Laffont-Schwob, I., Vanot, G., & Doumenq, P. (2020). Biological removal and fate assessment of diclofenac using Bacillus subtilis and brevibacillus laterosporus strains and ecotoxicological effects of diclofenac and 4′-Hydroxy-diclofenac. *J. Chem.*, 2020, 1–12

Gupta, V. K., Tyagi, I., Sadegh, H., Ghoshekandi, R. S., & Makhlouf, A. H. (2017). Nanoparticles as adsorbent; a positive approach for removal of noxious metal ions: A review. *Sci. Technol. Soc.*, 34(3), 195–214.

Holkar, C. R., Jadhav, A. J., Pinjari, D. V., Mahamuni, N. M., & Pandit, A. B. (2016). A critical review on textile wastewater treatments: Possible approaches. *J. Environ. Manage.*, 182, 351–366.

Hoseinian, F. S., Rezai, B., Kowsari, E., Chinnappan, A., & Ramakrishna, S. (2020). Synthesis and characterization of a novel nanocollector for the removal of nickel ions from synthetic wastewater using ion flotation. *Sep. Purif. Technol.*, 240, 116639.

Hurtado, C., Montano-Chávez, Y. N., Domínguez, C., & Bayona, J. M. (2017). Degradation of emerging organic contaminants in an agricultural soil: Decoupling biotic and abiotic processes. *Water Air Soil Pollut.*, 228(7), 1–8.

Ivanković, T., & Hrenović, J. (2010). Surfactants in the environment. *Arh. za Hig. Rada Toksikol.*, 61(1), 95–110.

Jafari, M., & Aghamiri, S. F. (2011). Evaluation of carbon nanotubes as solid-phase extraction sorbent for the removal of cephalexin from aqueous solution. *Desalination Water Treat.*, 28(1–3), 55–58.

Johnson, A. K., Zawadzka, A. M., Deobald, L. A., Crawford, R. L., & Paszczynski, A. J. (2008). Novel method for immobilization of enzymes to magnetic nanoparticles. *J. Nanopart. Res.*, 10(6), 1009–1025.

Kerebo, A., Desta, A., & Duraisamy, R. (2016). Removal of methyl violet from synthetic wastewater using nano aluminium oxyhydroxide. *Int. J. Eng. Res. Dev.*, 12(8), 22–28.

Khatoon, N., & Sardar, M. (2017). Efficient removal of toxic textile dyes using silver nanocomposites. *Curr. Nanosci.*, 2, 1–5.

Kolpin, D. W., Furlong, E. T., Meyer, M. T., Thurman, E. M., Zaugg, S. D., Barber, L. B., & Buxton, H. T. (2002). Pharmaceuticals, hormones, and other organic wastewater contaminants in US streams, 1999–2000: A national reconnaissance. *Environ. Sci. Technol.*, 36(6), 1202–1211.

Koul, B., Poonia, A. K., Yadav, D., & Jin, J. O. (2021). Microbe-mediated biosynthesis of nanoparticles: Applications and future prospects. *Biomolecules.*, 11(6), 886.

Koyani, R., Pérez-Robles, J., Cadena-Nava, R. D., & Vazquez-Duhalt, R. (2017). Biomaterial-based nanoreactors, an alternative for enzyme delivery. *Nanotechnol. Rev.*, 6(5), 405–419.

Kumar, A., & Luxmi, V. (2020). Novel green photo-catalyst 'turmeric roots' for pesticides degradation: Preparation and characterizations. *Mater. Lett.*, 262, 127030.

Kumar, P. S., Narayan, A. S., & Dutta, A. (2017). Nanochemicals and effluent treatment in textile industries. *Cloth. Text. Res. J.*, 57–96.

Kuppusamy, S., Thavamani, P., Venkateswarlu, K., Lee, Y. B., Naidu, R., & Megharaj, M. (2017). Remediation approaches for polycyclic aromatic hydrocarbons (PAHs) contaminated soils: Technological constraints, emerging trends and future directions. *Chemosphere.*, 168, 944–968.

Lavanya, C., Rajesh, D., Sunil, C., & Sarita, S. (2014). Degradation of toxic dyes: A review. *Int. J. Curr. Microbiol. Appl.*, 3(6), 189–199.

Li, B., Cui, Y., Japip, S., Thong, Z., & Chung, T. S. (2018). Graphene oxide (GO) laminar membranes for concentrating pharmaceuticals and food additives in organic solvents. *Carbon.*, 130, 503–514.

Li, L., Zhong, D., Xu, Y., & Zhong, N. (2019). A novel superparamagnetic micro-nano-bio-adsorbent PDA/Fe 3 O 4/BC for removal of hexavalent chromium ions from simulated and electroplating wastewater. *Environ. Sci. Pollut. Res.*, 26(23), 23981–23993.

Lin, Y. L., Chiou, J. H., & Lee, C. H. (2014). Effect of silica fouling on the removal of pharmaceuticals and personal care products by nanofiltration and reverse osmosis membranes. *J. Hazard. Mater.*, 277, 102–109.

Liu, W., Wang, D., Soomro, R. A., Fu, F., Qiao, N., Yu, Y., & Xu, B. (2019). Ceramic supported attapulgite-graphene oxide composite membrane for efficient removal of heavy metal contamination. *J. Membr. Sci.*, 591, 117323.

Lopes, C. N., Petrus, J. C. C., & Riella, H. G. (2005). Color and COD retention by nanofiltration membranes. *Desalination.*, 172(1), 77–83.

Madhav, S., Ahamad, A., Singh, P., & Mishra, P. K. (2018). A review of textile industry: Wet processing, environmental impacts, and effluent treatment methods. *Environ. Qual. Manag.*, 27(3), 31–41.

Madhu, A., & Chakraborty, J. N. (2017). Developments in application of enzymes for textile processing. *J. Clean. Prod.*, 145, 114–133.

Mahanty, S., Chatterjee, S., Ghosh, S., Tudu, P., Gaine, T., Bakshi, M., & Chaudhuri, P. (2020). Synergistic approach towards the sustainable management of heavy metals in wastewater using mycosynthesized iron oxide nanoparticles: Biofabrication, adsorptive dynamics and chemometric modeling study. *J. Water Process. Eng.*, 37, 101426.

Martins, M., Mourato, C., Sanches, S., Noronha, J. P., Crespo, M. B., & Pereira, I. A. (2017). Biogenic platinum and palladium nanoparticles as new catalysts for the removal of pharmaceutical compounds. *Water Res.*, 108, 160–168.

Mirsadeghi, S., Zandavar, H., Yousefi, M., Rajabi, H. R., & Pourmortazavi, S. M. (2020). Green-photodegradation of model pharmaceutical contaminations over biogenic Fe3O4/Au nanocomposite and antimicrobial activity. *J. Environ. Manage.*, 270, 110831.

Noman, M., Shahid, M., Ahmed, T., Niazi, M. B. K., Hussain, S., Song, F., & Manzoor, I. (2020). Use of biogenic copper nanoparticles synthesized from a native Escherichia sp. as photocatalysts for azo dye degradation and treatment of textile effluents. *Environ. Pollut.*, 257, 113514.

Nur-E-Alam, M., Mia, M. A. S., Ahmad, F., & Rahman, M. M. (2020). An overview of chromium removal techniques from tannery effluent. *Appl. Water Sci.*, 10(9), 1–22.

Odukkathil, G., & Vasudevan, N. (2013). Toxicity and bioremediation of pesticides in agricultural soil. *Rev. Environ. Sci. Biotechnol.*, 12(4), 421–444.

Ouyang, Z., Huang, Z., Tang, X., Xiong, C., Tang, M., & Lu, Y. (2019). A dually charged nanofiltration membrane by pH-responsive polydopamine for pharmaceuticals and personal care products removal. *Sep. Purif. Technol.*, 211, 90–97.

Ozgunay, H., Çolak, S. E. L. İ. M. E., Mutlu, M. M., & Akyuz, F. (2007). Characterization of leather industry wastes. *Pol. J. Environ. Stud.*, 16(6).

Rajabi, M., Mahanpoor, K., & Moradi, O. (2017). Removal of dye molecules from aqueous solution by carbon nanotubes and carbon nanotube functional groups: Critical review. *RSC Adv.*, 7(74), 47083–47090.

Reátegui, E., Reynolds, E., Kasinkas, L., Aggarwal, A., Sadowsky, M. J., Aksan, A., & Wackett, L. P. (2012). Silica gel-encapsulated AtzA biocatalyst for atrazine biodegradation. *Appl. Microbiol. Biotechnol.*, 96(1), 231–240.

Rigobello, E. S., Dantas, A. D. B., Di Bernardo, L., & Vieira, E. M. (2013). Removal of diclofenac by conventional drinking water treatment processes and granular activated carbon filtration. *Chemosphere.*, 92(2), 184–191.

Saratale, R. G., Saratale, G. D., Chang, J. S., & Govindwar, S. P. (2011). Bacterial decolorization and degradation of azo dyes: A review. *J. Taiwan Inst. Chem. Eng.*, 42(1), 138–157.

Sharma, I. (2020). *Bioremediation Techniques for Polluted Environment: Concept, Advantages, Limitations, and Prospects. Trace Metals in the Environment-New Approaches and Recent Advances.* IntechOpen.

Sharma, N., Bhatnagar, P., Chatterjee, S., John, P. J., & Soni, I. P. (2017). Bio nanotechnological intervention: A sustainable alternative to treat dye bearing waste waters. *J. Pharm. Biol. Res.*, 5(01), 17–24.

Silva, A. R., Soares, O., Pereira, M., Alves, M. M., & Pereira, L. (2020). Tailoring carbon nanotubes to enhance their efficiency as electron shuttle on the biological removal of acid orange 10 under anaerobic conditions. *Nanomaterials*, 10(12), 2496.

Silva, V. B., Rodrigues, T. S., Camargo, P. H., & Orth, E. S. (2017). Detoxification of organophosphates using imidazole-coated Ag, Au and AgAu nanoparticles. *RSC Adv.*, 7(65), 40711–40719.

Singh, P., Srivastava, N., & Singh, P. (2017). Analysis of various iron nanoparticles and compounds in pulp & paper mill waste water treatment. *Integr. Res. Adv.*, 4(2), 24–28.

Takmil, F., Esmaeili, H., Mousavi, S. M., & Hashemi, S. A. (2020). Nano-magnetically modified activated carbon prepared by oak shell for treatment of wastewater containing fluoride ion. *Adv. Powder Technol.*, 31(8), 3236–3245.

Thompson, L. A., & Darwish, W. S. (2019). Environmental chemical contaminants in food: Review of a global problem. *J. Toxicol.*

Uddandarao, P., Hingnekar, T. A., Balakrishnan, R. M., & Rene, E. R. (2019). Solar assisted photocatalytic degradation of organic pollutants in the presence of biogenic fluorescent ZnS nanocolloids. *Chemosphere.*, 234, 287–296.

Wong, E. L., Vuong, K. Q., & Chow, E. (2021). Nanozymes for environmental pollutant monitoring and remediation. *Sensors.*, 21(2), 408.

Yaseen, D. A., & Scholz, M. (2019). Textile dye wastewater characteristics and constituents of synthetic effluents: A critical review. *Int. J. Environ. Sci. Technol.*, 16(2), 1193–1226.

Zdarta, J., Jankowska, K., Bachosz, K., Degorska, O., Kaźmierczak, K., Nguyen, L. N., & Jesionowski, T. (2021). Enhanced wastewater treatment by immobilized enzymes. *Curr. Pollut. Rep.*, 1–13.

Zhang, C., Li, J., Wu, X., Long, Y., An, H., Pan, X. & Zheng, Y. (2020). Rapid degradation of dimethomorph in polluted water and soil by Bacillus cereus WL08 immobilized on bamboo charcoal–sodium alginate. *J. Hazard. Mater.*, 398, 122806.

Zhang, K., Yang, W., Liu, Y., Zhang, K., Chen, Y., & Yin, X. (2020). Laccase immobilized on chitosan-coated Fe3O4 nanoparticles as reusable biocatalyst for degradation of chlorophenol. *J. Mol. Struct.*, 1220, 128769.

Zhang, Y., & Chung, T. S. (2017). Graphene oxide membranes for nanofiltration. *Curr. Opin. Chem. Eng.*, 16, 9–15.

# 3 Bio(nano)cleaning: Bio(nano)purification and Bio(nano)remediation

*Biljana S. Maluckov*

## CONTENTS



DOI: 10.1201/9781003270959-3

## 3.1 INTRODUCTION

Diversiform toxic waste and emitted harmful gases and particles (Tasic et al., 2014; Tasic et al., 2017) from different industries pollute water, land, and air. The removal of pollutants from different media i.e. cleaning of media can be before and after release of pollutants into the environment. When pollutants are removed from media before their release into the environment is called purification, and when they are removed after having been released into the environment, it is called remediation (Figure 3.1).

There are different methods for removing various contaminants. Conventional methods for heavy metal removal from contaminated effluents are: chemical precipitation, coagulation/flocculation, ion exchange, membrane technologies, and electrochemical methods (Zamora-Ledezma et al., 2021; Abdi and Kazemi, 2015). For remediation of soils contaminated with heavy metals, the following conventional methods are used: physical separation, soil replacement, thermal treatment, vitrification, immobilization, stabilization/solidification, washing, and electrokinetic methods (Raffa et al., 2021). However, conventional methods can be expensive and inefficient, and they can induce the appearance of other new pollutants, such as for example carcinogenic and toxic polycyclic aromatic hydrocarbon compounds (Alagic et al., 2015) and particles during incineration, which is often used for the treatment of solid waste. Nanoparticles have great potential for improving already existing conventional methods for purification and remediation of different media and development of new methods as an integral part of the circular economy.

In addition to being able to remove contaminants, nanoparticles can also have antimicrobial effects, allowing that they can be used to prevent additional microbiological contamination, as for the removal of already present pathogenic microorganisms. Their antimicrobial effect is also applicable in pharmaceutical, textile, agricultural, cosmetics industries, etc. They can also have an anticancer effect (Akintelu et al., 2020, Das et al., 2022). They were effective in the occlusion of dentinal tubules and remineralizing of intertubular dentin (Abou Neel and Bakhsh, 2021) and they can be used as a UV filter (Dreno et al., 2019). They can also be used as fertilizers (Singh et al., 2018), etc.

Nanoparticles can be synthesized physically, chemically, and biogenically. Biogenic nanoparticles (Figure 3.2) can be completely bionatural or combined biologically with chemical and physical (UV-irradiation, microwave, ultrasound, electric current) synthesized.

In the metabolic reactions of microorganisms and plants, the biohost itself synthesizes a nanoproduct that serves to remove contaminants, or the bioproduct forms a nanoproduct with the pollutant. Bionanopurification and bionanoremediation are methods that use biosynthesized nanoproducts for the removal of contaminants. Biosynthesized nanoproducts which can be used for bioremediation and biopurification are: biopolymers, metals, metal oxides, and metal sulfides. They can effect as

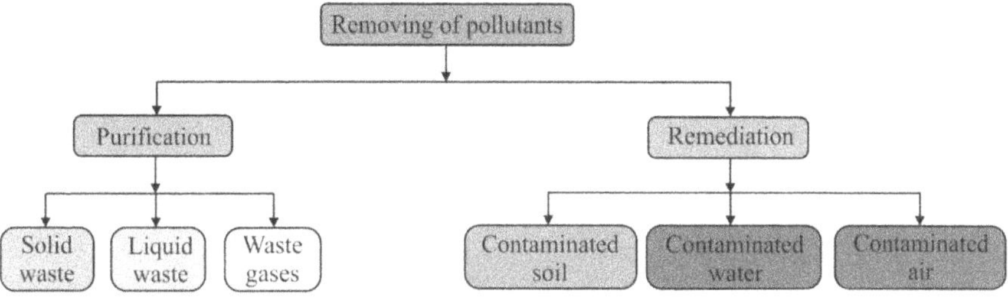


**FIGURE 3.1** Procedures for removing pollutants.

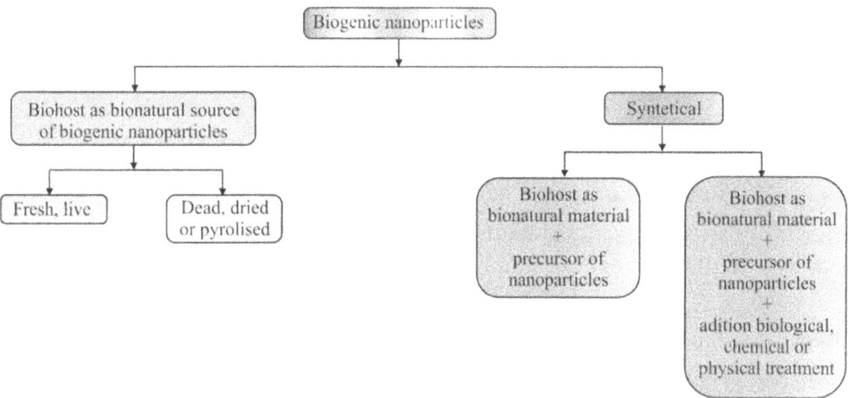


**FIGURE 3.2** Kinds of biogenic nanoparticles.

catalysts and adsorbents in pollutant removal procedures. A bionanoadsorbent which is rich in carbon so-called a biochar can be obtained by ball-milling of pyrolyzed herbal biomass.

Depending on which products of the biohost are used, there are several procedures (Figure 3.3) for removing contaminants:

1. Already synthesized nanoproduct is isolated from biohosts and used for biopurification or bioremediation.
2. Part or all of the biohost is used for the synthesis of bionanoproducts in laboratory conditions.
3. Pollutant is absorbed (microorganisms), i.e., uptake (plants) by the biohost and a bionanoproduct is formed which is sequestered in the cell wall, organelles, or tissue of the biohost.
4. A certain metabolic reaction is stimulated in which the release of the metabolic product from the biohost that is then placed in contact with the pollutant and so the bionanoproduct is rewarded.

In the first two methods, the removal of contaminants is performed by nanoparticles obtained biologically or by combined biosynthesis. In the third process, the pollutant

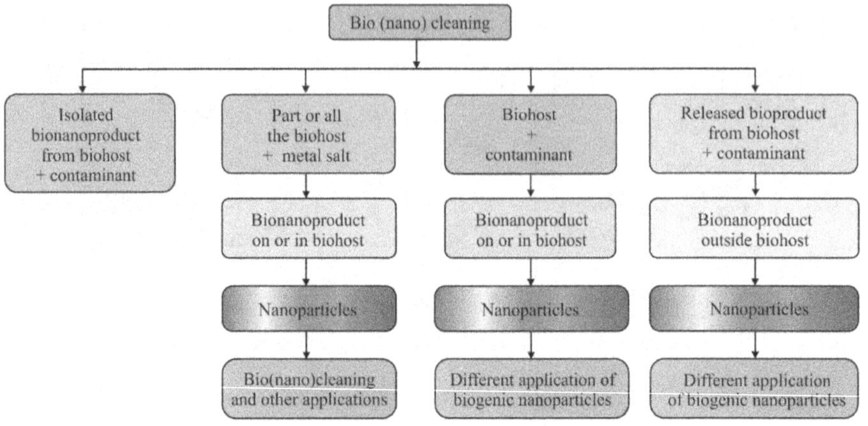


**FIGURE 3.3** Ways of bio(nano)cleaning.

is directly transformed into bionanoparticles that can be used for bionanocleaning, and in the fourth, by removing pollutants, bionanoparticles are obtained indirectly biologically and chemically from the pollutants that can also be used for bionanocleaning. All the different ways for removing contaminants will be discussed in this chapter.

## 3.2 BIONANOCLEANING WITH ISOLATED NATURAL SELF-SYNTHESIZED NANOPRODUCTS FROM BIOHOSTS

Isolated plant or microorganism nanoproducts that they normally produce themselves can be used for cleaning. Shredded to nano size, biomass can be used for cleaning also. Biomass can be treated thermally and then shredded to nano size. Straw is obtained in the thermal process of drying, and biochar by the thermal process of pyrolysis.

Cellulose and lignin are polymers that plants synthesize themselves, and biochar is a product obtained from biomass. They can be used as bioadsorbents. Metals-based nanoparticles can be isolated from hyperaccumulator plants, but also from microorganisms in which they are formed in the biochemical process of biomineralization. Metals-based nanoparticles can be bioadsorbents and photocatalysts.

Most of the functionalized nanocellulose-based adsorbents showed better adsorption performance for heavy metals, chemical dyes, and organic oil contaminants than other commercially available adsorbents (Norrrahim et al., 2021).

The uniform lignin nanoparticles can be produced from a wide range of technical lignins, despite the varied lignocellulosic biomass and the pretreatment methods/conditions applied (Tian et al., 2017). Combining lignin with other polymeric materials as polypropylene, polyvinyl alcohol, starch, cellulose, chitosan, can be prepare functional composite membranes (Li et al., 2021). The surface-functionalized porous lignin has a surface area 12 times larger than that of lignin and a high density of dithiocarbamate groups, which is why its saturated adsorption capacity for lead ions is 13 times that of the original lignin and 7 times that of activated carbon (Li et al., 2015).

Biochar particles as small as 10 microns can be obtained by the ball milling process (Amusat et al., 2021). Ball-milling alters the structures and size distributions of the pores and increases specific surface area. Biochar can be physicochemically modified by other methods to improve its adsorption capacity. Others modification methods commonly used are acid–base modification, metal oxide and metal salt modification, and clay mineral modification (Cheng et al., 2021).

Biochars made from woody residues, such as waste timber, can be effective sorbents to reduce the leaching of per- and polyfluoroalkyl substances (PFAS) from contaminated soils. The effect is improved by using activated biochar, with either a steam or $CO_2$ activation agent. The nonactivated biochar was reduced leaching 23–78%, while activated biochar was reduced leaching 63–95%. Sorption of short chain PFAS (CF3–CF4) was lower than for longer chain PFAS (>CF5), and sorption of PFSA (sulfonates) was stronger than that of PFCA (carboxylic acids). Waste timber-activated biochars as sorbents for soil PFAS are good alternatives for sorbents from fossil feedstock (Sørmo et al., 2021).

In combining the biochar and nano-metal oxides, nano-metal oxide-biochar composites formed in which are improve the dispersity and stability of nano-metal oxides and enhance the chemical adsorption capacity of modified biochar by giving it more different functional groups to produce more chemical bonds (Amusat et al., 2021, Cheng et al., 2021, Zhao et al., 2021).

The nanoparticles of activated banana peel carbon (ABPC) are potential cost-effective absorbents for the removal of Rhodamine B dye from aqueous media in acidic and basic conditions. It used the nanoparticles of size 75 μm. A slightly higher adsorption was at pH 2. They remove 81.9–85.6% of dye by the batch method. Initially the adsorption rate is rapid, then it slows down. Quantitative adsorption of Rhodamine B was at a contact time of 60 min and 12 mg adsorbent dose (Singh et al., 2020).

## 3.3 BIONANOCLEANING WITH BIOHOST-MEDIATED BIOSYNTHESIZED NANOPARTICLES IN LABORATORY CONDITIONS

By using different parts of plants and different microorganisms in the presence of a suitable metal chemical precursor, metals-based nanoparticles (Dhanker et al. 2021; Dikshit et al., 2021, Hussain et al., 2016, Jadoun et al., 2020, Tanwar et al., 2021), as nanocomposites (Obayomi et al., 2021) can be obtained. Biological capping and stabilization agents in the biohosts act on growth and inhibit agglomeration processes of nanoparticles (Hussain et al., 2016).

### 3.3.1 Bionanocleaning with Phytosynthesed Nanoparticles in Laboratory Conditions

Due to the rapidity and ease of obtaining nanoparticles using plant parts in the presence of a suitable metal salt precursor, there are a large number of experimental syntheses of nanoparticles in laboratory conditions with different plants and later testing of their catalytic potential for degradation and removal by adsorption different contaminants as azo dyes, antibiotics, phenols, heavy metals, metalloids etc. Dyes are used in various industries. They and antibiotics are persistent pollutants. Metal oxides are mainly used for their photocatalytic decomposition, but can also be used as adsorbents. In Table 3.1 are shown experimental results of efficiency of phytosynthesized nanocatalyst in laboratorial conditions for removing of different contaminants.

The nanoparticles are promising as disinfectants, so some researchers, in addition to the photocatalytic and adsorptive potential of biogenic synthesized nanoparticles, also examined their possible for use as a disinfectant to destroy or prevent the growth of microorganisms. Their antimicrobial effect is potentially applicable in medicine, cosmetics, textile industry, etc. Table 3.2 showed experimental results of the antimicrobial efficiency of phytosynthesized nanoparticles in laboratorial conditions. For comparison, the results of antimicrobial action of some conventional antimicrobial drugs are shown in Table 3.3. Some authors also examined whether biogenically synthesized nanoparticles have anticancer potential and in addition to the antimicrobial effect they presented and those results in their paper, but this is not considered in this chapter.

#### 3.3.1.1 Phytosynthesized Ti-Based Nanoparticles in Laboratory Conditions

Due to its biocompatibility and corrosion resistance, titanium has found application in implant fabrication (Maluckov, 2014). Titanium-based nanoparticles can be used for catalytic degradation (Krakowiak et al., 2021) and synthesis of organic compounds (Joshi et al., 2017; Carlucci et al., 2019). They are used in cosmetics as a UV filter (Dreno et al., 2019), as antimicrobials in toothpastes, and have the potential for tooth regeneration (Abou Neel et Bakhsh, 2021). Titanium dioxide was used as a coloring agent E171, but after reevaluation it was concluded that E171 can no longer be considered safe as a food additive (Younes et al., 2021).

Pure anatase phase titanium dioxide nanoparticles synthesized using solution of titanium butoxide and leaf extract of Citrus Limetta. The citric acid from the Citrus Limetta extract acted as a reducing and capping agent for the nanoparticles. Pure $TiO_2$ nanoparticles were approximately 80–100 nm in size, spherical in shape, with almost uniform distribution all over the sample. More than 90% Rhodamine B dye degraded within 80 min of UV irradiation by pure anatase phase titanium dioxide (Nabi et al., 2021).

The property of chemically and green synthesized $TiO_2$ nanoparticles was similar in both processes. Randomly arranged, the spherical-shaped rutile phase of $TiO_2$ nanoparticles had an average crystalline size of 31–42 nm. Phyto-mediated synthesized $TiO_2$ has a higher degradation efficiency of Methylene blue under UV-visible irradiation compared to chemically synthesized $TiO_2$ nanoparticles. The degradation efficiency of Methylene blue by chemically synthesized $TiO_2$ nanoparticles by hydrothermal method with titanium tetra isopropoxid and ethanol was 82%. Alkaloids, coumarins, and flavonoids from jasmine flower extract acted as reducing and stabilizing agents for nanoparticles. The presence of the hydroxyl group in jasmine flower extract increased the degradation efficiency. The degradation efficiency of Methylene

**TABLE 3.1**
**Experimental Results of Efficiency of Phytosynthesized Nanocatalyst in Laboratory Conditions for Removal of Different Contaminants**

| Kind of plant material | Kind of nanocatalyst | Kind of contaminant | Efficiency of degradation (%) | References |
|---|---|---|---|---|
| Jasmine flower extract | $TiO_2$ | Methylene blue | 89 % | Aravind et al., 2021 |
| Citrus limetta leaf extract | $TiO_2$ | Rhodamine B | More than 90% | Nabi et al., 2021 |
| Prickly pear leaf extract | CuO | 4-nitrophenol | 99% | Badri et al., 2021 |
| *Cinnamon bark* extract | Cu | Methylene blue<br>Methyl orange | More than 80%<br>More than 80% | Sarwar et al., 2021 |
| *Withania coagulans* extract | ZnO | Methylene blue | 90% | Saif et al., 2021 |
| *Ficus carica* leaf extract | ZnO | Methylene blue | 95.35 % | Arumugam et al., 2021 |
| *Acacia caesia* Bark Extract | ZnO | Methylene blue | 92.2% | Ashwini et al., 2021 |
| *Gynostemma pentaphyllum* extracts | ZnO | Malachite Green | 89% | Park et al., 2021 |
| *Rosmarinus offcinalis* leaves | Zn0.95Ag0.05O | Methylene blue | 98.5% | Hojjati-Najafabadi et al., 2021 |
| *Tithonia diversifolia* | Nanocomposite ZnO / activated carbon | Methylene blue | 95% | Obayomi et al., 2021 |
| *Ricinus communis* leaves | Nano-zero valent iron | Tetracycline | 98% | Abdelfatah et al., 2021 |
| *Trigonella foenum-graecum* seed extract | Nano-zero valent iron<br>+<br>UV light | Methyl orange | 95% | Radini et al., 2018 |
| *Canthium coromandelicum* leaf | Iron oxide | Janus green B | 97.23% | Sudhakar et al., 2021 |
| *Calotropis gigantea* flower extract | Nano-zero valent iron/biomaterial/ chitosan/1/1/2 | Methylene blue<br>Anilin | 85.5%<br>74.8% | Sravanthi et al., 2018 |
| *Trigonella foenum-graecum* seeds extract | Ag | Methyl orange<br>Methylene blue<br>Eosin Y | 100%<br>100%<br>100% | Vidhu and Philip, 2014 |

*(Continued)*

**TABLE 3.1 (CONTINUED)**
**Experimental Results of Efficiency of Phytosynthesized Nanocatalyst in Laboratory Conditions for Removal of Different Contaminants**

| Kind of plant material | Kind of nanocatalyst | Kind of contaminant | Efficiency of degradation (%) | References |
|---|---|---|---|---|
| *Dracocephalum kotschy* leaf extract | Ag | Methylene blue | 67% | Chahardoli et al., 2020 |
| *Nervalia zeylanica* leaf extract | Ag | Methyl orange | 100% | Vijayan et al., 2019 |
| | | Rhodamine B | 100% | |
| *Mussaenda glabrata* leaf extract | Ag | Rhodamine B | 100% | Francis et al., 2017 |
| | | Methyl orange | 100% | |
| *Lythrum salicaria L.* aerial parts | Ag | Congo red | 100% | Srećković et al., 2021 |
| *Citrus reticulata Blanco* peel extract | Ag | Malachite Green | 100% | Jaast and Grewal, 2021 |
| *Mussaenda glabrata* leaf extract | Au | Rhodamine B | 100% | Francis et al., 2017 |
| | | Methyl orange | 100% | |
| *Atriplex halimus* leaves | Pt | Methylene blue | 100% | Eltaweil et al., 2022 |

**TABLE 3.2**
**Experimental Investigations of the Antimicrobial Effect of Phytosynthesized Nanocatalysts**

| Kind of plant material | Kind of nanocatalyst | Kind of microorganism/Zone of inhibition/(diameter in mm) | References |
|---|---|---|---|
| Jasmine flower extract | $TiO_2$ | *Escherichia coli*/12<br>*Klebsiella pneumonia*/11<br>*Staphylococcus aureus*/8 | Aravind et al., 2021 |
| *Cinnamon bark* extract | Cu | *Staphylococcus aureus*/21<br>*Escherichia coli*/19 | Sarwar et al., 2021 |
| *Ficus carica* leaf extract | ZnO | *Escherichia coli*/17<br>*Klebsiella pneumonia*/17<br>*Bacillus cereus*/15<br>*Staphylococcus aureus*/13 | Arumugam et al., 2021 |
| *Acacia caesia* bark extract | ZnO | *Escherichia coli*/18<br>*Candida albicans*/15<br>*Aspergillus niger*/14<br>*Staphylococcus aureus*/13 | Ashwini et al., 2021 |
| *Trigonella foenum-graecum* seed extract | Nano-zero valent iron 50 µl<br>Nano-zero valent iron 100 µl | *Escherichia coli*/16<br>*Staphylococcus aureus*/12<br>*Escherichia coli*/22<br>*Staphylococcus aureus*/19 | Radini et al., 2018 |
| *Canthium coromandelicum* leaf | Iron oxide | *Staphylococcus aureus*/18<br>*Salmonella typhi*/16.5 | Sudhakar et al., 2021 |

*(Continued)*

**TABLE 3.2 (CONTINUED)**
**Experimental Investigations of the Antimicrobial Effect of Phytosynthesized Nanocatalysts**

| Kind of plant material | Kind of nanocatalyst | Kind of microorganism/Zone of inhibition/(diameter in mm) | References |
|---|---|---|---|
| *Clematis orientalis* leaf extract | CuO/0,25M | *Escherichia coli*/10 | Nadeem et al., 2021 |
| | | *Klebsiella pneumonia*/23 | |
| | | *Staphylococcus aureus*/19 | |
| | | *Pseudomonas aeruginosa*/13 | |
| | | *Bacillus subtilis*/11 | |
| | CuO/0,50M | *Escherichia coli*/12 | |
| | | *Klebsiella pneumonia*/15 | |
| | | *Staphylococcus aureus*/18 | |
| | | *Pseudomonas aeruginosa*/16 | |
| | | *Bacillus subtilis*/13 | |
| *Clematis orientalis* leaf extract | MgO/0,25M | *Escherichia coli*/13 | Nadeem et al., 2021 |
| | | *Klebsiella pneumonia*/11 | |
| | | *Staphylococcus aureus*/11 | |
| | | *Pseudomonas aeruginosa*/11 | |
| | | *Bacillus subtilis*/10 | |
| | MgO/0,50M | *Escherichia coli*/17 | |
| | | *Klebsiella pneumonia*/14 | |
| | | *Staphylococcus aureus*/13 | |
| | | *Pseudomonas aeruginosa*/18 | |
| | | *Bacillus subtilis*/10 | |

(*Continued*)

**TABLE 3.2 (CONTINUED)**
**Experimental Investigations of the Antimicrobial Effect of Phytosynthesized Nanocatalysts**

| Kind of plant material | Kind of nanocatalyst | Kind of microorganism/Zone of inhibition/(diameter in mm) | References |
|---|---|---|---|
| *Clematis orientalis* leaf extract | FeO/0,25M | *Escherichia coli*/12<br>*Klebsiella pneumonia*/11<br>*Staphylococcus aureus*/10<br>*Pseudomonas aeruginosa*/20<br>*Bacillus subtilis*/11 | Nadeem et al., 2021 |
| | FeO/0,50M | *Escherichia coli*/11<br>*Klebsiella pneumonia*/10<br>*Staphylococcus aureus*/10<br>*Pseudomonas aeruginosa*/10<br>*Bacillus subtilis*/11 | |
| *Dracocephalum kotschyi* leaf extract | Ag | *Escherichia coli*/20<br>*Staphylococcus aureus*/18<br>*Pseudomonas aeruginosa*/15<br>*Bacillus subtilis*/13<br>*Staphylococcus epidermidis*/13<br>*Serratia marcescens*/13 | Chahardoli et al., 2020 |
| *Moringa oleifera lam* aerial part | Ag | *Escherichia coli*/6<br>*Salmonella typhy*/5<br>*Staphylococcus aureus*/9 *Streptococcus pneumonia*/8 | Neupane et al., 2022 |
| *Humulus lupulus* wasted, crushed, and unusable extract | Ag | *Escherichia coli*/15<br>*Staphylococcus aureus*/12 | Das et al., 2022 |
| *Conocarpus Lancifolius* fruit extract | Ag | *Staphylococcus aureus*/24<br>*Streptococcus pneumonia*/18 | Oves et al., 2022 |

(*Continued*)

**TABLE 3.2 (CONTINUED)**
**Experimental Investigations of the Antimicrobial Effect of Phytosynthesized Nanocatalysts**

| Kind of plant material | Kind of nanocatalyst | Kind of microorganism/Zone of inhibition/(diameter in mm) | References |
|---|---|---|---|
| *Cleistocalyx operculatus* leaf extract | Ag nanoparticle/grapheme nanoplatelet<br>50/100/200 µl | *Escherichia coli*/8<br>*Escherichia coli*/12<br>*Escherichia coli*/15 | Thi et al., 2022 |
| *Plantago major* leaf extract | Ag<br>10 µl | *Escherichia coli*/9.12<br>*Staphylococcus aureus*/5.76<br>*Pseudomonas aeruginosa*/7.42 | Sukweenadhi et al., 2021 |
| | Ag<br>20 µl | *Escherichia coli*/ 9.91<br>*Staphylococcus aureus*/7.31<br>*Pseudomonas aeruginosa*/8.59 | |
| *Piper nigrum* seed extract | Ag-based chitosan nanocomposite<br>50 µl | *Escherichia coli*/21.8<br>*Bacillus subtilis*/17.3 | Kanniah et al., 2021 |
| | Ag-based chitosan nanocomposite<br>100 µl | *Escherichia coli*/22.1<br>*Bacillus subtilis*/20.3 | |

*(Continued)*

**TABLE 3.2 (CONTINUED)**
**Experimental Investigations of the Antimicrobial Effect of Phytosynthesized Nanocatalysts**

| Kind of plant material | Kind of nanocatalyst | Kind of microorganism/Zone of inhibition/(diameter in mm) | References |
|---|---|---|---|
| *Hagenia Abyssinica* plant leaf | Ag 50 µl | *Klebsiella pneumonia*/10<br>*Salmonella typhimurium*/14<br>*Streptococcus pneumonia*/6.4 | Melkamu and Bitew, 2021 |
| | Ag 100 µ | *Klebsiella pneumonia*/10.6<br>*Salmonella typhimurium*/16.3<br>*Streptococcus pneumonia*/7 | |
| | Ag 150 µl | *Klebsiella pneumonia*/12<br>*Salmonella typhimurium*/17.3<br>*Streptococcus pneumonia*/7.3 | |
| | Ag 200 µl | *Klebsiella pneumonia*/14<br>*Salmonella typhimurium*/18.3<br>*Streptococcus pneumonia*/8.6 | |
| *Aloe vera* leaf extract | Ag | *Staphylococcus aureus*/21<br>*Escherichia coli*/20<br>*Pseudomonas aeruginosa*/14<br>*Enterobacter sps*/32 | Anju et al., 2021 |
| *Ocimum canum* leaf extract | Ag 10/20/30 ppm | *Escherichia coli*/17<br>*Escherichia coli*/17.5<br>*Escherichia coli*/24 | Tailor et al., 2020 |
| *Atriplex halimus* leaves | Pt | *Escherichia coli*/at all<br>*Klebsiella pneumonia*/17<br>*Bacillus subtilis*/resistant<br>*Staphyllococcus aureus*/resistant | Eltaweil et al., 2022 |

**TABLE 3.3**
**Antimicrobial Activity (Zone of Inhibition; mm) of Standard Antibiotics**

| Antibiotic concentration (μg) | Zone of inhibition/(diameter in mm) | References |
|---|---|---|
| Streptomycin 100 | *Escherichia coli*/25<br>*Staphylococcus aureus*/25 | Ashwini et al., 2021 |
| Streptomycin 200 | *Escherichia coli*/30<br>*Staphylococcus aureus*/34<br>*Staphylococcus epidermidis*/25<br>*Slmonella enterica*/27<br>*Porteus mirabilis*/33 | Ibrahim et al., 2021 |
| Streptomycin<br>100 | *Escherichia coli*/11.33<br>*Staphylococcus aureus*/10.33 | Das et al., 2022 |
| Gentamycin<br>30 | *Bacillus cereus*/23<br>*Staphylococcus aureus*/26<br>*Escherichia coli*/26<br>*Klebsiella pneumoniae*/24 | Arumugam et al., 2021 |
| Amoxicillin | *Escherichia coli*/7<br>*Salmonella typhi*/6<br>*Staphylococcus aureus*/ 8<br>*Streptococcus pneumonia*/7 | Neupane et al., 2022 |
| Roxithromycin | *Bacillus subtilis*/18<br>*Staphylococcus aureus*/24<br>*Escherichia coli*/25<br>*Klebsiella pneumoniae*/19<br>*Pseudomonas aeruginosa*/21 | Nadeem et al., 2021 |
| Ciprofloxacin | *Klebsiella pneumonia*/21<br>*Salmonella typhimurium*/20<br>*Streptococcus pneumonia*/19 | Melkamu and Bitew, 2021 |
| Clotrimazole 100 | *Aspergillus niger*/29<br>*Candida albicans*/27 | Ashwini et al., 2021 |

blue by biosynthesized $TiO_2$ nanoparticles using jasmine flower extract was 89%. Both nanoparticles showed a good antimicrobial effect. The zone of inhibition for gram-negative bacteria such *Escherichia coli* and *Klebsiella pneumonia* and gram-positive *Staphylococcus aureus* are 12, 11, and 8 mm with chemically $TiO_2$; and 14, 12, and 7 mm with phyto-mediated $TiO_2$, respectively (Aravind et al., 2021).

#### 3.3.1.2 Phytosynthesized Cu-Based Nanoparticles in Laboratory Conditions

The principal application of copper oxide nanoparticles is in biomedical and waste treatment. Biomolecules of plants can enhance the toxicity effect of copper oxide nanoparticles against microbes and degradation effectiveness of copper oxide nanoparticles synthesized by the biological method from plant sources (Akintelu et al., 2020).

Copper oxide nanoparticles that synthesized using peel of prickly pear fruit aqueous extract with copper sulfate pentahydrate as a precursor crystallized in the monoclinic C2/c space group. They were spherical morphology with uniform symmetry and in a 20–40 nm size range. Their reduction effciency of 4-nitrophenol in the presence of sodium borohydride was 96%, 97%, and 99% for temperatures of 298 K, 308 K, and 318 K, respectively. Copper oxide nanoparticles can be reused and the conversion rate of 4-nitrophenol was 93.2% after five successful cycles at room temperature (Badri et al., 2021).

In the synthesis of copper nanoparticles using copper nitrate as a precursor and cinnamon extract from cinnamon bark as a reducing agent, the reducing action of the cinnamon was synergistically stimulated by the addition of citric acid. More than 80% of Methylene blue and Methyl orange degradated with synthesized copper nanoparticles in conjunction with sodium hypophosphite. Synthesized copper nanoparticles were effective against both gram-positive and gram-negative bacteria with remarkable inhibition of *Staphylococcus aureus* and *Escherichia coli.* The zone inhibition layer for *Staphylococcus aureus* and *Escherichia coli* without citric acid was 21 and 19 mm, and with citric acid 28 and 23 mm, respectively (Sarwar et al., 2021).

Green synthetized copper oxide nanoparticles by using the extracts of mint leaf and orange peels as green reducing agents were investigated for lead, nickel, and cadmium removal from contaminated water. The maximum uptake capacity of both kinds of copper oxide nanoparticles followed the order of Pb(II) > Ni(II) > Cd(II). The optimum uptake capacity of 0.33 g/L copper oxide nanoparticles at pH 6 were 88.80, 54.90, and 15.60 mg/g for Pb(II), Ni(II), and Cd(II), respectively (Mahmoud et al., 2021).

#### 3.3.1.3 Phytosynthesized Zn-Based Nanoparticles in Laboratory Conditions

Plant-mediated biosynthesized zinc oxide nanoparticles can be effective catalysts in the photocatalysis process for removing dyes from wastewater (Fagier, 2021; Tanwar et al., 2022). Zinc oxide nanoparticles also have potential for use as biofertilizers and pesticides (Singh et al., 2018).

In green synthesis of zinc oxide nanoparticles by fresh and dry alhagi plant with zinc nitrate formed zinc oxide nanoparticles in the range of 25–100 nm and 35–100 nm, respectively. The ZnO nanoparticles synthetased by fresh plant extract were almost spherical in shape with weak agglomeration, and the particles ZnO nanoparticles that formed with dry plant were spherical shape without any agglomeration (Falih et al., 2022).

Degradation of Methyl blue was 78% with zinc oxide nano-flowers (NFs) with an average size of 30 nm that chemically synthesized via reduction–precipitation method. When flower-like ZnO nanostructure with average size of 25 nm were phytosynthesized with fruit *Withania coagulans* extract and zinc acetate, then degradation of 50mg/l Methyl blue was 90% at 120 min. Antibacterial efficiency by phytosynthesized ZnO nano-flowers for gram-positive *Staphylococcus* aureus and gram-negative *Pseudomonas aeruginosa* was 85% and 94%; and with chemical it was 78% and 88%. In the case of anti-fungal efficiency of phytosynthesized ZnO NFs from *Aspergillus niger* and *Candida albicans* was 91% and 100%; and 78% and 80% with chemically synthesized ZnO NFs (Saif et al., 2021).

Small spherical zinc oxide nanoparticles, the wurtzite-type hexagonal crystals with high crystallinity and an average diameter of 35 nm that were synthesized with leaf extract of *Ficus carica* tree and zinc nitrate hexahydrate, showed 95.34% photocatalytic efficiency of degradation of Methylene blue under direct sunlight irradiation time of 70 min. The photocatalytic effciencies for the second, third, fourth, and fifth reuse cycles was 94.28, 92.68, 90.87, and 89.12% respectively. Zinc oxide nanoparticles showed the highest bactericidal performance against both *Escherichia coli* and *Klebsiella pneumoniae* bacteria with a zone of inhibition of 17 mm for the same ZnO NPs dosage concentration of 100 μL. Less bactericidal activity against *Bacillus cereus* bacteria and *Staphylococcus aureus* bacteria was detected with zones of inhibition of 15 and 13 mm for ZnO NPs dosage concentration of 100 μL. High sensitivity of *Escherichia coli* bacterial strain can be attributed to its thin cell wall. Because of that the penetration in the cell of $Zn^{2+}$ ions is facilitated (Arumugam et al., 2021).

Zinc oxide nanoparticles of the crystalline core size about 35.41 nm synthetased from *Gynostemma pentaphyllum* plant extract with zinc nitrate using the co-precipitation method exhibited 89% degradation efficiency of Malachite green after 180 min under UV illumination. The hexagonal wurtzite structure creates a more active site to interact with molecules of dye. Gynostemma zinc oxide nanoparticles could use more than five times with a slight reduction of activity (Park et al., 2021).

Methyl blue degradation was 92.2% within 40 min by hexagonal ZnO nanoparticles with the average crystalline size of 32.32 nm that synthetased using *Acacia caesia* bark extract and zinc nitrate hexahydrate. Gram-negative strain *Escherichia coli* was more susceptible, with a zone of inhibition of 18 mm, than gram-positive strain *Staphylococcus aureus* with zone of inhibition of 13 mm at a concentration of 1000 μg/mL of these ZnO nanoparticles. Investigation of their antifungal activity showed that *Candida albicans*, with a zone of inhibition of 15 mm, was more susceptible than *Aspergillus niger*, with a zone of inhibition of 14 mm (Ashwini et al., 2021).

Biosynthesized spherical ZnO nanoparticles using *Cinnamomum camphora* leaf extracts and zinc acetate inhibited the spore germination and mycelial growth of *Alternaria alternata*. They induced lipid peroxidation of *Alternaria alternata*, resulting in cell membrane damage and cellular leakage. The average particle of synthesized ZnO nanoparticles synthesized at pH 7, pH 8, and pH 9 were about 13.92, 15.19, and 21.13 nm, respectively. The synthesized ZnO nanoparticles at pH 7 showed the best antifungal effect with minimum inhibitory concentration value recorded as 20 mg/L (Zhu et al., 2021).

Nanoparticles of Zn0.95Ag0.05O (ZnAgO) with quasi-spherical configuration and the mean particle size in the range of 22–40 nm was synthesized with *Rosmarinus offcinalis* (rosemary) leaf, zinc acetate, and silver nitrate. The photocatalytic degradation of Methylene blue in the presence of Ag-doped ZnO nanoparticles, and the ultraviolet lamp as the light source, is nearly 98.5% after exposure for 100 min. The synthesized Zn0.95Ag0.05O nanoparticles were effcient in inhibition of growth of *Staphylococcus aureus* and *Escherichia coli* (Hojjati-Najafabadi et al., 2021).

Nanocomposites with biosynthesized ZnO and activated carbon can be effective in adsorption for dyes. Nanocomposite which is made from biosynthesized zinc oxide nanoparticles using *Tithonia diversifolia* leaf extract and zinc nitrate was loaded on the surface of flamboyant pods (*Delonix regia*)-activated carbon removed 99.07% Methylene blue in solution pH 10. The maximum electrostatic attraction between positively charged dye and negatively charged adsorbent surfaces because of more negative ions on the adsorbent surface at higher pH, promoted maximum dye removal. This ZnO-activated carbon nanocomposite was used again for the adsorption of Methylene blue repeatedly for five cycles (Obayomi et al., 2021).

#### 3.3.1.4 Phytosynthesized Fe-Based Nanoparticles in Laboratory Conditions

Iron-based nanoparticles, such as different polymorphs of iron oxides, oxyhydroxides, iron hydroxide, and zero-valent iron nanoparticles are promising, effective nanomaterials for the treatment of wastewater. The mechanisms of removal of contaminants from the wastewater by iron-based nanoparticles are based on adsorptive mechanisms and photocatalytic degradation. Adsorptive mechanisms included electrostatic interaction, ligand combination, and binding on the surface. Surface modifications of iron-based nanoparticles can reduce aggregation and increase resistance to corrosion (Aragaw et al., 2021).

The biogenic synthesized iron oxide nanoparticles using *Canthium coromandelicum* leaf extract and iron (II) chloride tetrahydrate showed degradation efficiency of 97.23% for Janus green B at 180 min under sunlight irradiation. Polyphenols present in the *Canthium coromandelicum* leaf extract was involved in the synthesis process. These iron oxide nanoparticles showed the zone of inhibition at $18 \pm 0.32$ mm and $16.5 \pm 0.34$ mm for gram-negative *Salmonella typhi* and gram-positive *Staphylococcus aureus*, respectively, at 100 μg/mL concentrations (Sudhakar et al., 2021).

Nano-zero valent iron (nZVI) synthesized with *Ricinus communis* leaf extracts and ferric chloride removed 98% of tetracycline from aqueous solution by adsorption. Tannins in plant extract play an quintessential role in the reduction and stabilization of nZVI. The synthesized nZVI retained its adsorption ability. Their regeneration efficiency decreased to 84% after five cycles. This decrease in tetracycline removal percentage could be attributed to the loss of nZVI and/or the irreversible occupation of partial adsorption sites (Abdelfatah et al., 2021).

Zero-valent iron nanoparticles formed in an aqueous solution of $FeCl_3$ with *Trigonella foenum-graecum* seed extract as a reducing agent as well as an effective nanoparticle surface stabilization agent. The degradation of Methyl orange by these nanoparticles under UV irradiation was 95% in 90 min, and the degradation efficiencies for three consecutive cycles were 87%, 79%, and 76%, respectively. Zone of inhibition was 16 mm for *Escherichia coli*, and for *Staphylococcus aureus*, it was 12 mm at concentration of 50 μl nanoparticles (Radini et al., 2018).

The polyphenols present in the flower extract of *Calotropis gigantea* are responsible for the formation of zero-valent iron nanoparticles nZVI average size 50–90 nm with ferric nitrate nonahydrate. Synthesized nZVI are immobilized on biomaterial with the help of chitosan, because nZVI tend to agglomerate, resulting in a significant loss of reactivity. Immobilized nZVI efficiently adsorbed Methylene blue and aniline. Removal rates of Methylene blue of blank, synthesized nZVI, and sorbent materials (with different ratios of biomaterial, chitosan, and nZVI) were 29.4%, 63.1%, 80.5% (1:1:0.5), 83.9%(1:1:1), and 85.5% (1:1:2), respectively, within the first 30 min. Removal rates of aniline of blank, synthesized nZVI and sorbent materials were17.1%, 49.4%, 56.3% (1:1:0.5), 59.1% (1:1:1), and 74.8% (1:1:2), respectively, after 12 h (Sravanthi et al., 2018).

The magnetite and maghemite ($Fe_3O_4$ and $\gamma$-$Fe_2O_3$) synthesized with a *Cymbopogon citratus* extract and iron(III) chloride hexahydrate and sodium carbonate as the reducing and pH stabilizer agents, respectively. The *Cymbopogon citratus* extracts allows the reduction of iron(III) to iron(II) using $H^+$ radicals produced from hydroxyl and carbonyl groups, whereas the $Na_2CO_3$ acted as a pH stabilizer agent. The presence of phytochemicals on the FeO nanoparticles enhanced their biocompatibility and acted as a secondary food source in *Caenorhabditis elegans* nematodes (Patino-Ruiz et al., 2020).

Metal oxide nanoparticles of copper (CuO), magnesium (MgO) and iron oxide (FeO) nanoparticles synthesized using *Clematis orientalis* leaf extract and aqueous solution of corresponding metal salts copper acetate, magnesium sulfate, and ferric chloride separately. Nanoparticles of CuO and FeO were crystalline in nature, and MgO were crystalline cum amorphous in nature. The concentration of salt used for the synthesis of nanoparticles effect on their bio-physicochemical properties. They showed good antibacterial effect on *Klebsiella pneumoniae*, *Pseudomonas aeruginosa*, *Staphylococcus aureus*, *Escherichia coli*, and *Bacillus subtilis.* The largest zones of inhibition were 23 mm for *Klebsiella pneumoniae* by CuO nanoparticles at 0.5 M, 20 mm for *Pseudomonas aeruginosa* by FeO nanoparticles et 0.25 M, 19 mm for *Staphylococcus aureus* by CuO nanoparticles at 0.25 M, 17 mm for *Escherichia coli* by MgO nanoparticles et 0.5 M, 13 mm for *Bacillus subtilis* by CuO nanoparticles at 0.5 M (Nadeem et al., 2021).

#### 3.3.1.5 Phytosynthesized Ag-Based Nanoparticles in Laboratory Conditions

Silver nanoparticles synthesized using *Trigonella foenum-graecum* seeds and silver nitrate are effective in the degradation of Methyl orange, Methylene blue and Eosin Y in present $NaBH_4$. Samples of silver nanoparticles synthesized by adding 2 mL, 5 mL, 8 mL, 10 mL, and 15 mL extract of *Trigonella foenum-graecum.* The size of the synthesized nanoparticles increased with increasing amount of added extract, and the reaction time of degradation increased with increase of particle size for all tested harmful compounds. The complete reduction of Methyl orange was 8 min, for Methylene blue 6 min and for Eosin Y 10 min with nanoparticles that synthesized with 2ml of extract (Vidhu and Philip, 2014).

Spherical silver nanoparticles were synthesized using the leaf extract of *Dracocephalum kotschyi* and silver nitrate under sunlight irradiation. Flavonoids, a phenolic compound (rosmarinic acid) and proteins from *Dracinfusion* are responsible for the stabilization of silver nanoparticles. In synthesis without sunlight, the color change started from colorless to pale yellow after 4 h and gradually turned to yellowish and dark brown after 15 and 22–24 h. But under sunlight irradiation, a color change to dark brown is confirmation of the bioreduction of silver ions into silver nanoparticles occurred quickly in a few minutes and reached an optimum and stable level within 2 h. These Ag nanoparticles were degradated 67% of Methylene blue after 48 h. They had the greatest antibacterial effect against *Escherichia coli* with a zone of inhibition of 20 mm, and then against *Staphylococcus aureus* with 18 mm, *Bacillus subtilis* with 15 mm, and for others, *Staphylococcus epidermidis*, *Serratia marcescens*, and *Pseudomonas aeruginosa*, the zone of inhibition was 13 mm (Chahardoli et al., 2020).

Green synthesized silver nanoparticles of spherical shape with average particle size of 34.2 nm using *Nervalia zeylanica* leaf extract and silver nitrate by microwave-assisted synthesis completely degradated Methyl orange and Rhodamine B by $NaBH_4$ within 10 min. The nanoparticles with intermediate reduction potential of dyes and borohydride make the electron transfer very easy between them. These green synthesized silver nanoparticles showed antimicrobial activity against two types of gram-positive bacteria (*Staphylococcus aureus* and *Lactobacillus brevis*), gram-negative bacteria (*Pseudomonas putida* and *Pseudomonas* sp.), and fungi (*Penicillium chrysogenum* and *Penicillium citrinum*). The leaf extract of *Nervalia zeylanica* also showed low antimicrobial properties (Vijayan et al., 2019).

Gold and silver nanoparticles were synthesized by the microwave-assisted synthesis with *Mussaenda glabrata* and chloroauric acid and silver nitrate separately. Gold nanoparticles (5 µg/mL) and silver nanoparticles (0.02 mg/ mL) activated the complete removal of Rhodamine B by 5 and 9 min. The glabrata-reduced gold and silver nanocolloids degraded of Methyl orange within a period of 4 and 7 min, respectively, in a heterogeneous catalytic pathway. The synthesized gold and silver nanoparticles revealed their potency to inhibit pathogenic microorganisms *Bacillus pumilus*, *Staphylococcusaureus*, *Pseudomonas aeruginosa*, *Escherichia coli*, *Aspergillus niger*, and *Penicillium chrysogenum* (Francis et al., 2017).

Silver nanoparticles AgNPs synthesized using *Lythrum salicaria* aerial parts (LSA-AgNPs) and root extract (LSR-AgNPs) and silver nitrate can be used for catalytic degradation of Congo red and 4-nitrophenol in the presence of $NaBH_4$. The effectiveness of LSA-AgNPs and LSR-AgNPs was higher for the degradation of Congo red than the reduction of 4-nitrophenol to 4-aminophenol. Investigations of microbiocidal activity on 11 species of bacteria and 9 species of fungi showed that obtained nanoparticles have lower antifungal activity than antibacterial activity, especially LSA-AgNPs (Sreckovic et al., 2021).

The Malachite green dye was completely degradated after 120 h of incubation in the presence of sunlight with synthesized spherical silver nanoparticles by *Citrus reticulata* Blanco (Kinnow mandarin hybrid) peel extract and silver nitrate (Jaast and Grewal, 2021).

Biosynthesized silver nanoparticles with aerial part of *Moringa oleifera lam* and silver nitrate were spherical with less than 50 nm size. These silver nanoparticles inhibit the growth of the gram-positive and gram-negative bacteria with more efficiency against gram-positive bacteria (*Staphylococcus aureus* and *Streptococcus pneumonia*) than gram-negative bacteria (*Escherichia Coli* and *Salmonella Typhi*). Zones of inhibition were 9 mm for *Staphylococcus aureus*, 8 mm for *Streptococcus pneumonia*, 6 mm for *Escherichia Coli*, and 5 mm for *Salmonella Typhi* at 250 µg/ml silver nanoparticles. The zone of inhibition was similar to the standard drug amoxicillin (Neupane et al., 2022).

The metallic silver nanoparticles biosynthesized using *Sambucus ebulus* extract and silver nitrate salt were of cubic structure, spherical morphology with an average size of 35–50 nm. Volumes of 10 µl synthesized silver nanoparticles degraded 95.89% of Methyl orange after 11 min, and volumes of 15 µl degraded 95.47% of dye after 7 min under sunlight irradiation in the presence of $NaBH_4$. The synthesized silver nanoparticles showed greater activity against *Staphylococcus aureus*, *A. baumannii*, *Escherichia Coli*, and *Klebsiella pneumonia* than to *Proteus mirabilis*, *Enterococcus faecalis*, and *Pseudomonas aerigunosa* (Hashemi et al., 2022).

Using wasted and unusable *Humulus lupulus* (hops) extract and silver nitrate spherical silver nanoparticles are synthesized with an average size of 17.40 nm. Solution with concentration of 500 µg/mL silver nanoparticles have exhibited a zone of inhibition $12 \pm 0.81$ mm against *Staphylococcus aureus*, and for *Escherichia coli* a zone of inhibition of $15.33 \pm 0.94$ mm (Das et al., 2022).

*Conocarpus Lancifolius* fruit aqueous extract and silver nitrate formed silver nanoparticles of spherical shape with an average size of 26.28 nm. They act on the bacterial surface and inhibit its growth. The zones of inhibition were 18 mm against the *Streptococcus pneumonia* and 24 mm against the *Staphylococcus aureus* when treated with silver nanoparticles at a concentration of 50 µg/ml. Biogenic silver nanoparticles showed also strongly activity against fungal pathogen *Rhizopusus stolonifera* and *Aspergillus flavus*. The colony number and size steadily decreased with a growing concentration of silver nanoparticles. The diameter decreased to 4 mm at 100 µg/mL silver nanoparticles (Oves et al., 2022).

Silver nanoparticles, decorating the surfaces of graphene nanoplatelets, were synthesized with the *Cleistocalyx operculatus* leaf extract as a green reductant, with silver nitrate and graphene nanoplatelets. These Ag nanoparticle/graphene nanoplatelet nanocomposites showed inhibitory activity against *Escherichia coli* compared with the samples containing 50, 100, and 200 µl of 0.1% nanoparticle/graphene nanoplatelet nanocomposite solution creating inhibited growth zones of 8, 12, and 18 mm. The Ag nanoparticle/graphene nanoplatelet nanocomposite also shows good antimicrobial activity against *Staphylococcus aureus*, *Bacillus subtilis*, and *Pseudomonas aerigunosa*. The Ag nanoparticle/graphene nanoplatelet nanocomposite was stabilized by a protective coating of polysaccharides from the *Cleistocalyx operculatus* extract. The enhanced antibacterial activity of the Ag nanoparticle/graphene nanoplatelet nanocomposite may be attributed to the uniformly distributed Ag nanoparticles with a diameter range of 20–40 nm on the surface of the graphene nanoplatelets, enabling more of the former to come into contact with the bacteria, thereby improving the antimicrobial activity (Thi et al., 2022).

The spherical silver nanoparticles with a size range of 10–20 nm obtained with *Plantago major* leaf extract and silver nitrate showed antibacterial activity equal compared to the chloramphenicol and gentamicin. At their concentration of 10 µg/mL, zones of inhibition was 5.76, 7.42 and 9.12mm for *Staphylococcus aureus*, *Pseudomonas aeruginosa*, and *Escherichia coli*, respectively, and at concentration of

20 μg/mL zones of inhibition was 7.31, 8.59 and 9.91 also respectively (Sukweenadhi et al., 2021).

The growth of bacterial cultures *Bacillus subtilis* and *Escherichia coli* were effectively suppressed by green synthesized silver nanoparticles and silver-based chitosan nanocomposite with seed extract of *Piper nigrum* and silver nitrate. The synthesized silver nanoparticles were polydispersed with a size range of 15–38 nm, mostly spherical, and few undefined large sized. The inhibition zones were 17.3 ± 1.19 mm and 21.8 ± 1.30 mm of *Bacillus subtilis* and *Escherichia coli*, respectively, obtained at the concentration of 50 μg/ml. For the concentration of 100 μg/mL, the inhibition zone was 20.3 ± 1.13 mm for *Bacillus subtilis* and 22.1 ± 1.73 mm for *Escherichia coli*. When the cotton fabric material was coated with nanocomposite particles, the zone of inhibition growth of *Bacillus subtilis* was 41.7 ± 1.04 mm, and for *Escherichia coli*, it was 36.4 ± 1.17 mm (Kanniah et al., 2021).

The synthesized silver nanoparticles with *Hagenia Abyssinica* plant leaf and silver nitrate showed significant antibacterial activity against both gram-negative (*Klebsiella pneumonia* and *Salmonella typhimurium*) and gram-positive bacteria (*Streptococcus pneumonia*). The antibacterial activity was at its maximum at the highest concentration of 200 μg/mL. For *Salmonella typhimurium*, the zone of inhibition was 18.3 mm. The zone of inhibition for *Klebsiella pneumoniae* was 14 mm, and for *Streptococcus pneumoniae*, it was 8.6 mm (Melkamu and Bitew, 2021).

Silver nanoparticles were synthesized using different ratios of aqueous extract of *Acacia cyanophylla* and silver nitrate and for different temperatures and time of exposition. When the extract concentration increased, the particle size of the sample increased. The smallest size of 86.98 nm was for the ratio of 9:1 (extract solution: silver nitrate). For the ratio of 9:2, the size was 132.1 nm; and when using the ratio of 5:5, the size was 347.2 nm. The sizes were 121.5 nm after 24 hours, while the sizes increased significantly after 48 hours to 198.6 nm with the ethanolic extract. With the aqueous extract, size of 62.41nm was observed after 24 hours and 86.98 nm after 48 hours. The size increased with the increase in the reaction time. The highest size of 788.3 nm was at 72 hours for the 5:5 ratio. For the ratio 9:1 in aqueous extract, the silver nanoparticles did not form at low temperatures of 20°C, after 24 hours at 35°C their the size was relatively small of 104.2 nm, and they formed immediately upon conducting the reaction at high temperatures but the size was relatively large 182.2 nm.

The functional groups in acemannan, the major bioactive polysaccharide of *Aloe vera* gel, acted as a reducing and stabilizing agent in the formation of silver nanoparticles using the inner gel extract of *Aloe vera* leaf, silver nitrate, and sodium borohydride. *Aloe vera* extract alone and green synthesized silver nanoparticles showed good antibacterial activity against both gram-negative and gram-positive bacterial strains. A higher inhibition zone was for *Escherichia coli*, *Staphylococcus aureus*, and *Enterobacter* sps. with green synthesized silver nanoparticles than *Aloe vera* extract alone, whereas the inhibition zone for *Pseudomonas aeruginosa* between silver nanoparticles and extract alone has no significant variation. The inhibition zone diameter were 20, 21, and 32 mm for *Escherichia coli*, *Staphylococcus aureus*, and

*Enterobacter*, respectively, and *Pseudomonas aeruginosa* was most resistant with a minimum inhibition zone of 14 mm with silver nanoparticles (Anju et al., 2021).

Synthesized uniform and spherical particles with an average diameter of 8.67 of silver nanoparticles using safflower (*Carthamus tinctorius* L.) aqueous extract from waste and silver nitrate inhibited the growth of *Staphylococcus aureus* (gram-positive) and *Pseudomonas fluorescens* (gram-negative). Proteins, polysaccharides, and flavonoids present in the aqueous extract of safflower waste were capping and stabilizing nanoparticles by inclusion on the surface of the nanoparticles mainly through amino, hydroxyl, and aromatic ring groups. Nanoparticles at the lowest concentration evaluated of 0.9 μg/mL inhibited the growth of both types of bacteria (Rodríguez-Felix et al., 2021).

Prepared silver nanoparticles of the spherical and rod-like morphological shapes with an average size of 15.72 nm using leaf extract of *Ocimum canum* Sims and silver nitrate showed the antibacterial activity against pathogenic bacterium *Escherichia coli*. The zone of inhibition of the bacterium was 17 mm at 10 ppm concentration, at 20 ppm concentration was 17, 5 mm, and it was 24.5 mm at 30 ppm concentration of silver nanoparticles. The carbohydrates, flavonoids, and polyphenol compounds from leaf extract acted as the surface active stabilizing molecules of silver nanoparticles (Tailor et al., 2020).

#### 3.3.1.6 Phytosynthesized Au-Based Nanoparticles in Laboratory Conditions

The flavonoids, phenolic compounds, and amino acids from the extract were most likely involved in the reduction of Au ions and/or are complexed (or adsorbed) on the surface of gold nanoparticles in the synthesis of spherical gold nanoparticles with an average size of $16.88 \pm 5.47$~$29.93 \pm 9.80$ nm using a water extract of *Artemisia capillary* and hydrochloroauric acid trihydrate. The average size of the gold nanoparticles decreased as the extract concentration during the synthesis was increased. Small nanoparticles possess a large surface-area-to-volume ratio, increasing their catalytic activities. The removal of extracts enhanced their catalytic activity of dissolution 4-nitrophenol reduction reaction in the presence of sodium borohydride by up to 50.4% and shortened the time required for completion (Lim et al., 2016).

The microwave-assisted synthesized gold nanoparticles, with an average particle size of 18.72 nm, that used leaves of *Myristica fragran* and auric trichloride completely degraded of Methyl orange dye and reducted 4-nitrophenol in the presence of $NaBH_4$. The microwave helped to complete the phyto-mediated synthesis within a few minutes. The complete degradation of Methyl orange dye occurred within 11 min, and the complete reduction of 4-nitrophenol was within 9 min. Gold nanocatalyst acted as the electron mediator between electron donor borohydride ions and electron acceptor Methyl orange. The phytocompounds surrounding the nanosurface bring all the reactants much closer to them by electrostatic forces and facilitate the electronic transfer to the acceptor 4-nitrophenolate ions (Punnoose et al., 2021).

Synthesized silver, gold, and palladium nanoparticles with *Solanum nigrum* leaf extract and silver nitrate, chloroauric acid, and palladium chloride separately were spherical in shape with an average size of 3.46 nm, 9.39 nm, and 21.55 nm, respectively. The luteolin, kaempferol, and gentisic acid from *Solanum nigrum* leaf extract acted as good reducing agents, and the polyphenols acted as capping agents and prevented agglomeration of the particles. Silver nanoparticles are capped more effectively than gold and palladium nanoparticles, which is why the antimicrobial and catalytic effects of silver nanoparticles were better than gold and palladium nanoparticles. The reduction time for the complete reduction of 4-Nitrophenol with the presence of $NaBH_4$ was 3, 4, and 6 min for Ag, Au, and Pd nanoparticles, respectively. The zone of inhibition of the bacterium *Escherichia coli* was 23 and 20 mm, 19.2 and 20 mm, and 18 and 19 mm for Ag, Au, and Pd nanoparticles, respectively (Vijilvani et al., 2020).

#### 3.3.1.7 Phytosynthesized Pt-Based Nanoparticles in Laboratory Conditions

Green platinum nanoparticles are mainly spherical with ultra-small particle size (1–3 nm) and are synthesized using an aqueous extract of *Atriplex halimus* leaf as a reductant and hexachloroplatinic acid. They were stable for up to three months. It took approximately five min for complete degradation of 100 ppm Methylene blue by these nanoparticles and sodium borohydride. The removal efficiency of 60 ppm Methylene blue remained 100% after three cycles of regeneration, and it started to diminish slightly to 93% in the fourth cycle. Eventually, it reached 91.1% after five cycles of reuse. These nanoparticles were able to stop the growth of *Escherichia coli* completely for *Klebsiella pneumonia* at an inhibition zone of 17 mm, and gram-positive bacteria *Bacillus subtilis* and *Staphyllococcus aureus* were resistant to them (Eltaweil et al., 2022).

### 3.3.2 Bionanocleaning with Microbiosynthesized Nanoparticles in Laboratory Conditions

Biogeochemical transformation processes caused by microorganisms can be used for metal production, purification, and remediation (Maluckov, 2015; Maluckov, 2017; Maluckov, 2021).

A poorly crystalline iron oxyhydroxysulfate schwertmannite biosynthesized using *Acidithiobacillus ferrooxidans* and $FeSO_4$ solution had the degradation efficiency of phenanthrene of 99.0% within 3–5 h reaction in schwertmannite/$H_2O_2$ system. Phenanthrene was readily adsorbed on the surface of schwertmannite and then oxidized by a hydroxyl radical produced from $H_2O_2$ decomposition. The degradation efficiency of phenanthrene decreased to 80.0% after schwertmannite was reused for 12 cycles, and its mineral structure did not change during repeated use (Meng et al., 2017).

Metal-reducing bacterium *Geobacter sulfurreducens* can transform soluble Cu(II) to $Cu_2S$ nanoparticles that were associated with the cell surface (Kimber et al., 2020).

Spherical, well-dispersed MgO nanoparticles with a size range of 8.0–38.0 nm magnesium oxide nanoparticles synthesized using the fungal strain *Aspergillus terreus* and magnesium nitrate hexahydrate. They adsorbed 97.5% chromium ions from the tanning effluent and exhibited antimicrobial activity. The zones of inhibiton of growth for *Candida albicans*, *Escherichia coli*, *Pseudomonas aeruginosa*, *Staphylococcus aureus*, and *Bacillus subtilis* were $12.8 \pm 0.3$ mm, $11.3 \pm 0.6$ mm, $14.7 \pm 1.9$ mm, $11.3 \pm 0.6$ mm, and $13.3 \pm 1.9$ mm, respectively (Saied et al., 2021).

The extracellularly synthesized silver nanoparticle from the filamentous fungi *Penciillium Citreonigum Dierck* and *Scopulaniopsos brumptii Salvanet-Duval* and isolated from Lake Burullus showed excellent antibacterial property on gram-positive and gram-negative bacterial strains. The maximum zone of inhibition was 25 mm for *Staphylococcus aureus* followed by *Escherichia coli* with 17 mm and *Pseudomonas aeruginosa* 17 mm zone of inhibition at 500 μg/ml concentration of nanoparticles synthesized by *Penciillium Citreonigum Dierck*. The maximum zone of inhibition was 15 mm for *Escherichia coli* at 500 μg/mL concentration, followed by *Staphylococcus aureus* showing 17 mm and *Pseudomonas aeruginosa* showing 17 mm zone of inhibition nanoparticles synthesized by *Scopulaiopsos brumptii Salvanet-Duval*. Polyurethane foam was used as a silver carrier, and nano-silver solution was used for the removal of pathogenic bacteria from wastewater. The highest antibacterial effect with a removal efficiency of 100% was observed with the concentration 676.9 mg/l of silver nanoparticle at contact time of 1 h (Moustafa, 2017).

Silver nanoparticles were synthesized using the natural polysaccharide produced by *Bacillus subtilis sp. suppress* and silver nitrate with exposure to UV radiation for 24 h. Linen fabrics were treated with silver nanoparticles before the dyeing process and after the dyeing process. The results showed their antibacterial activities against *Escherichia coli* and *Staphylococcus aureus* for linen fabrics, both undyed and dyed. Zone of inhibition for *Escherichia coli* and *Staphylococcus aureus* after one laundering cycle was 3.98mm and 3.76 mm for undyed linen. The zone of inhibition before dyeing with Reactive red 195 was for 4.35mm *Escherichia coli* and 3.73mm for *Staphylococcus aureus*; after dyeing, it was 3.95 mm and 3.775 mm respectively. For before dyeing with Reactive black 5 was for 3.975 mm *Escherichia coli* and 3.75 mm for *Staphylococcus aureus*, for after dyeing was 4.05 mm and 3.72 mm respectively. After 5th laundering cycles, the antibacterial activity of after dyeing treated fabrics lost about 8–10%, after the 10th laundering cycles, they lost about 11–13%, and after the 20th laundering cycles, they lost about 12–15%. While the antibacterial activity of before dyeing treated fabrics, after 5th and 10th laundering cycles, they lost about 2–3%, and after the 20th laundering cycles, they lost about 4–7%. Silver nanoparticles increasing dyeability, also (Abdelghaffar et al., 2021).

The biogenically formed silver nanoparticles by using *Bacillus cereus* and silver nitrate were small sized, ranging from 5 to 7.06 nm. The zone of inhibition for *Staphylococcus epidermidis*, *Staphylococcus aureus*, *Escherichia coli*, *Salmonella enterica*, *Porteus mirabilis* was 32.12±0.55, 38.25±0.5, 33.05±1.33, 30.73±0.25, 35.44±1.08, respectively (Ibrahim et al., 2021).

With silver nitrate and aqueous extracts of cyanobacteria *Synechococcus elongatus* and *Microcystis aeruginosa* and microalgae, the chlorophytes *Coelastrum astroideum* and *Desmodesmus armatus*, and the charophytes *Cosmarium punctulatum* and *Klebsormidium flaccidum* were synthesized silver nanoparticles. In all syntheses, nanoparticles were of well dispersed, crystalline, spherical shape with a size range of 1.8 to 5.4 nm and with no agglomerate formed. The presence of hydroxyl groups of peptidoglycan nature, acting as stabilizing agents in the surface of the nanoparticles, determines the formation of nanoparticles with a rather narrow size dispersity and prevents agglomeration (Moraes et al., 2021).

Silver nanoparticles and gold nanoparticles synthesized using the marine algae extract *Sargassum serratifolium* at 80° C showed excellent catalytic activities for reducing Methylene blue, Rhodamine B, and Methyl orange by $NaBH_4$ at room temperature (Kim et al., 2021).

For the synthesis of gold nanoparticles with microorganisms, $HAuCl_4$ is usually used (Srinath et al., 2017; Saitoh et al., 2018). Microorganisms degrade aurocomplex in the gold-rich solutions, releasing gold cations, which seek an electron from closer surroundings to be stabilized in the form of metal because of its galvanic properties (Aitimbetov et al., 2005). However, instead of the usual $HAuCl_4$, the gold-thiosulfate complexes can use for bionanosynthesis, which can be formed by the use of non-ammoniacal thiosulfate solutions or thiosulfate synthesized by thiosulfate producing microorganisms (Maluckov, 2021).

## 3.4 BIOCLEANING WITH SIMULTANEOUS SYNTHESIS OF NANOPARTICLES WITH CONTAMINANTS ON OR IN A BIOHOST

There are plants and microorganisms that have a great ability to absorb, uptake and accumulate large amounts of pollutants called hyperaccumulators. In biological remediation processes of soil (Maluckov, 2015), water (Hussain et al., 2018) and air (Alagić et al., 2015) using hyperaccumulator plants and microorganisms that are adapted, highly resistant to large amounts of pollutants in their environment, pollutants are absorbed and uptaked, which can be stored inside them as nanoparticles. Thus, in these bioremediation procedures there is a simultaneous synthesis of nanoparticles on or in the biohost. This can further be used as an integral part of the circular economy to isolate the synthesized nanoparticles and use them for remediation.

### 3.4.1 Phytocleaning with Simultaneous Synthesis of Nanoparticles with Contaminants

Phytocleaning with simultaneously synthesis of nanoparticles in plants is a procedure (Figure 3.4) in which simultaneous decontamination of the medium from metals and metalloids is performed and nanoparticles of the same are formed in plants, that when its be isolated from the plant they can be used to remove contaminants. The use of plant hyperaccumulators to recover gold from gold deposits has developed a special area of phytomining, where after harvesting, in the incineration process, gold is obtained from plants. After harvesting the plants, the incineration process can be replaced by pyrolysis in order to obtain biochar or by the process of isolating nanoparticles from plants that can be used to remove contaminants.

During phytoremediation and phytomining, microorganisms can be added to improve the plant's uptake of metals and metalloids. With the alternating use of

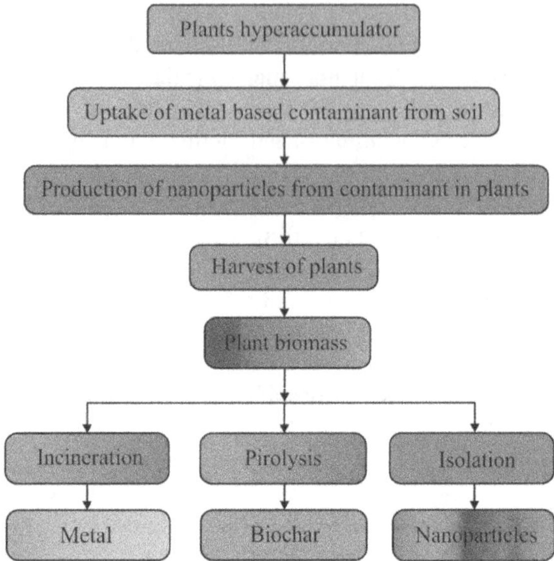


**FIGURE 3.4** Simultaneous phytocleaning/phytomining and synthesis of nanoparticles.

hyperaccumulating microorganisms and plants, it is possible to perform both remediation soils from metals and metalloids and biomining of precious metal from deposit (Maluckov, 2015).

Dead plant biomass can also be used as a successful bioadsorbent. Cellulose, hemicellulose, lignin, protein, etc. are present in straws, enable efficient absorption of harmful metals and nanoparticle formation.

Wheat straw and corn straw are ground in a ball mill and sieved to particle sizes of 100 mesh for investigation of their adsorption properties for Cr(VI) and Cr(III) in solution. Investigations show that the saturated adsorption capacity of wheat straw for Cr(VI) and Cr(III) can reach 125.6 and 68.9 mg/g and that of corn straw for Cr(VI) and Cr(III) can reach 87.4 and 62.3 mg/g, respectively, and that the straws possess excellent reusability. The adsorbed straw could be completely desorbed (Chen et al., 2020).

### 3.4.2 Microbialcleaning with Simultaneous Synthesis of Nanoparticles with Contaminants

The use of appropriate microorganisms and the application of appropriate potential can favor the mineralization and formation of certain nanocompounds (Mauckov, 2020) which can later be used to remove contaminants. In the mines, so-called acid mine drainage (AMD) waters are formed with a high content of metals and metalloids, the remediation of which can be done with the help of sulfate-reducing bacteria.

Zinc in concentration of 260 mg/L is completely removed by sulfate-reducing bacteria as a precipitate of sphalerite and wurtzite, two polymorphs of ZnS, forming <2.5 μm-diameter spherical aggregates (Castillo et al., 2012). When the acid mine drainage from Sarcheshme copper complex with a copper concentration of about 60 mg/l was added to biomss of *Fusarium oxysporum* and incubated for four days, it formed nanoparticles composition like covelite with an average size of 10–40 nm. The properties of the nanoparticles from AMD were the same as the nanoparticles synthesized from the pure copper sulfate solution (Schaffie and Hosseini, 2014).

Biomass of sulfate-reducing bacteria can remove heavy metals from simulated wastewater. Metals recovered from bioprecipitates in the form of nanopowder and the metal recovery efficiency was in the order Cu > Pb > Cd > Zn > Ni > Fe > Mn. Nanopowder of CuS from bioprecipitates of 20 mg/L Direct Red 80 dye was degraded within 10 min with a removal efficiency of 96% (Kumar and Pakshirajan, 2020).

Arsenic and zinc were totally removed from natural AMD water from the Carnoulès mine by incubations with an indigenous consortium of sulfate-reducing bacteria. Arsenic first precipitated as amorphous orpiment (am-$As^{III}{}_2S_3$) (33–73%), and realgar ($As^{II}S$) (0–34%). A minor fraction of As was also found as thiol-bound $As^{III}$ (14–23%). Orpiment ($As_2S_3$) further transformed into AsS nanowires. Zinc precipitated as ZnS nanoparticles (Le Pape et al., 2017).

When using microorganisms for soil remediation, remediation processes can be improved by acting with electricity (Yan and Reible, 2015). Dead microbial biomass can also be used successfully in bioadsorbtion and nanoparticle formation.

The investigation showed that dead biomass of *Hypocrea lixii* can be used for the uptake of copper from wastewater and production of copper nanoparticles. Dead biomass removed a higher percentage of copper than live and dried biomass. Adsorbed metal ions can be desorbed and dead biomass can be reutilized. A maximum copper biosorption by dead biomass of 19.0 mg/g after 60 min of contact, at pH 5.0, temperature of 40°C, and agitation speed of 150 rpm was achieved. Mainly spherical nanoparticles, with an average size of 24.5 nm, were synthesized extracellularly in the cell wall. Proteins were released from the cell of the fungus during the autoclaving process, which bound on the surface cell interacting with the particles as stabilizing and capping agents surrounding the copper nanoparticles (Salvadori et al., 2013).

## 3.5 BIOCLEANING BY STIMULATION OF SYNTHESIS A SPECIFIC METABOLIC EXOPRODUCT OF BIOHOST THAT IN CONTACT WITH THE CONTAMINANTS FORMS NANOPARTICLES

Natural process of bioleaching in mines is undesirable due to the formation of acid mine drainage waters (AMD), while industrial bioleaching is very desirable and it used for the biohydrometalurgical production of metals (Maluckov, 2017). As it has been noticed that amino acids can be both corrosion inhibitors and activators, research is being done on their effects on minerals (Maluckov et al., 2017; Maluckov and Mitrić, 2018). During remediation, the acidic mine waters can be a significant source of metals, and their remediation, in addition to its ecological significance, is an integral part of the circular economy.

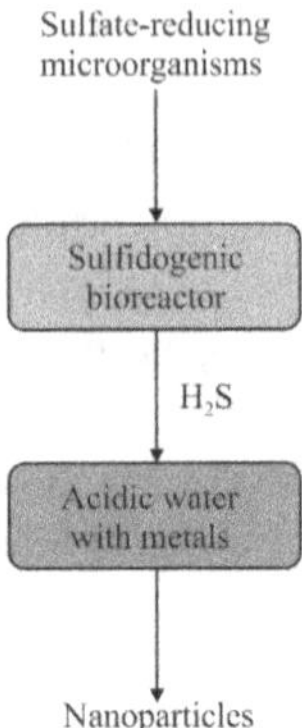


**FIGURE 3.5** Simultaneous remediation of acid mine drainage and synthesis of nanoparticles.

Stimulated biogen-produced hydrogen sulfide can be used for the bioprecipitation of metals and metalloids from AMD. In these procedures (Figure 3.5) in the bioreactor with sulfate-reducing bacteria, their metabolism is stimulated, in which $H_2S$ is formed, and the formed $H_2S$ in contact with polluted acidic mine waters forms sulfide nanoparticles. Unlike the procedure previously described in section 3.4.2, where sulfate-reducing bacteria are mixed with acidic water, here it is not necessary to separate microbial biomass from newly formed nanoparticles.

Covellite (CuS) nanoparticles were synthetized from an AMD sample after exposure to hydrogen sulfide ($H_2S$), produced in a bioreactor containing acidophilic sulfate-reducing bacteria (SRB) (Colipai et al., 2018; Silva et al. 2019). Biogenic hydrogen sulfide, produced from the dissimilatory reduction of sulfate, is used for the removal of copper and to synthesize copper sulfide nanoparticles from synthetic acid mine water. After vigorous shaking, large agglomerates with mean hydrodynamic diameters of 192, 596, and 667 nm occurred at 15, 30, and 45 min exposure times, respectively. Suspended material of a mean diameter of 28 nm was obtained at 60 min without shaking; but when the same sample was subjected to shaking, the mean diameter was 684 nm. Nanoparticles did not appear in the samples without shaking at 5, 15, 30, and 45 min (Colipai et al., 2018). The higher exposure time of the AMD sample (collected from a Brazilian copper mine) to biogenic $H_2S$, produced in a bioreactor containing acidophilic sulfate-reducing bacteria, resulted in smaller covellite particles. After 48 h of exposure the sample had a size of 43.4 μm or less, whereas for 96 h exposure, 90% of the particles had a size of 22 μm or less (Silva et al., 2019).

Nanocrystals of covellite, prepared by precipitating copper with biologically produced sulfide, were of spheroid morphology with a mean size of about 3.5 nm. The synthesis, in the presence of $TiO_2$ and $SiO_2$ submicron particles as substrates, resulted in the respective composite materials (da Costa et al., 2013).

The bioremediation process using real acid mine drainage, was coupled with the synthesis of ZnS nanoparticles and/or ZnS/$TiO_2$ nanocomposites. The bioremediation process was performed for 339 days in a start-up Up-flow Anaerobic Packed

Bed reactor system with AMD from the S. Domingos mine, with ethanol as a carbon source for the SRB. Investigations showed that the dimensions of the particles produced can be controlled by manipulating the system's operating conditions and/or reaction parameters (Vitor et al., 2015).

Therefore, in the two-stage cleaning process, with a separate stage of interaction of metabolic products with the contaminant, apart from the fact that it is not necessary to separate nanoparticles from biomass, it is easier to control the conditions for the synthesis of nanoparticles and, thus, their physicochemical characteristics.

## 3.6 CONCLUSION

In cleaning processes of different media from contaminants, nanoparticles can be used as a catalyst and adsorbent for degradation, an adsorptive removal of pollutants, and as a disinfectant.

Biogenic nanoparticles have great potential in the circular economy. According to their origin, biogenic nanoparticles can be completely bionaturally or synthetically formed in combined biological and chemical syntheses, which can be assisted with physical treatment. Both plants and microorganisms can be used as a natural source for the synthesis of nanoparticles. Natural biogenic nanoparticles can be obtained from fresh, live, dead, or thermally treated biomaterial. In laboratory conditions, they can be synthesized with the biohost and the corresponding nanoparticle precursor. The physicochemical characteristics of the synthesized biogenic nanoparticles, as well as their catalytic, absorptive, antimicrobial, and other properties, depend on temperature, pH, precursor concentration in which the extract is dissolved, exposure time, whether it additionally acts on some physical treatment with UV irradiation, ultrasound, microwaves, etc. The main roles of the various biocompounds present in the biohost, in the synthesis processes of nanoparticles, are reducing and stabilizing. The synthesized biogenic nanoparticles can be used in several cycles in pollutant removal procedures. They can also be synthesized with the pollutant itself during a bioremediation or biopurification process that can later be used for the same purpose to remove a contaminant or can be used in some other branches such as medicine, agro-industry, cosmetics, textiles, etc. Thus, synthetic biogenic nanoparticles for cleaning can be formed by the desired intentional synthesis or as a by-product in some other cleaning process. Biohosts can uptake, i.e., absorb, dissolved forms of harmful metals and then self-synthesize nanoparticles on their own. Microbiological treatment of cleaning can be with mixing microbial biomass and the contaminant and that their metabolite is first isolated and then placed in contact with the contaminant. In the two-stage cleaning process, with a separate stage of interaction of metabolic products with the contaminant, it is not necessary to separate nanoparticles from biomass, and it is easier to control the conditions for the synthesis of nanoparticles, and thus their properties, than when microbial biomass and pollutant are mixed.

A large number of studies have shown high efficiency of biogenic nanoparticles for the degradation of azo dyes, but they are also effective in the degradation of antibiotics, phenols, etc. Studies also show their high antimicrobial effect, which is comparable to antimicrobial drugs that are already used. Most studies show that

*Escherichia coli*, the main fecal pathogenic bacterium, is very sensitive to them, so they can be used to effectively disinfect hospitals and fecal wastewater. They can also be used to protect textiles and prevent the development of microorganisms in wastewater from the textile and other industries in which pathogenic bacteria can develop.

## REFERENCES

Abdelfatah, A.M., Fawzy, M., El-Khouly, M.E. and Eltaweil, A.S. (2021). Efficient adsorptive removal of tetracycline from aqueous solution using phytosynthesized nano-zero valent iron. *J. Saudi Chem. Soc.* 25, 101365.

Abdelghaffar, F., Mahmoud, M.G., Asker, M.S. and Mohamed, S.S. (2021). Facile green silver nanoparticles synthesis to promote the antibacterial activity of cellulosic fabric. *J. Ind. Eng. Chem.* 99, 224–234.

Abdi, O., and Kazemi, M. (2015). A review study of biosorption of heavy metals and comparison between different biosorbents. *J. Mater. Environ. Sci.* 6(5), 1386–1399.

Abou Neel, E.A. and Bakhsh, T.A. (2021). An eggshell-based toothpaste as a cost-effective treatment of dentin hypersensitivity. *Eur. J. Dent.* 15, 733–740.

Aitimbetov, T., White, D.M. and Seth, I.(2005). Biological gold recovery from gold–cyanide solutions. *Int. J. Miner. Process.* 76, 33–42.

Akintelu, S.A., Folorunso, A.S., Folorunso, F.A. and Oyebamiji, A.K. (2020). Green synthesis of copper oxide nanoparticles for biomedical application and environmental remediation. *Heliyon.* 6, e04508.

Alagić, S.Č., Maluckov, B.S. and Radojičić, V.B. (2015). How can plants manage polycyclic aromatic hydrocarbons? May these effects represent a useful tool for an effective soil remediation? A review. *Clean Techn. Environ. Pol.* 17(3), 597–614.

Amusat, S.O., Kebede, T.G., Dube, S. and Nindi, M.M. (2021). Ball-milling synthesis of biochar and biochar-based nanocomposites and prospects for removal of emerging contaminants: A review. *J. Water Process Eng.* 41, 101993.

Anju, T.R., Parvathy, S., Veettil, M.V., Rosemary, J., Ansalna, T.H., Shahzabanu, M.M. and Devika, S. (2021). Green synthesis of silver nanoparticles from Aloe vera leaf extract and its antimicrobial activity. *Mater. Today: Proc.* 43, 3956–3960.

Aragaw, T.A., Bogale, F.M. and Aragaw, B.A. (2021). Iron-based nanoparticles in wastewater treatment: A review on synthesis methods, applications, and removal mechanisms. *J. Saudi Chem. Soc.*25, 101280.

Aravind, M., Amalanathan, M. and Sony Michael Mary, M. (2021). Synthesis of TiO2 nanoparticles by chemical and green synthesis methods and their multifaceted properties. *SN Appl. Sci.* 3, 409.

Arumugam, J., Thambidurai, S., Suresh, S., Selvapandiyan, M., Kandasamy, M., Pugazhenthiran, N., Kumar, S.K., Muneeswaran, T. and Quero, F. (2021). Green synthesis of zinc oxide nanoparticles using *Ficus carica* leaf extract and their bactericidal and photocatalytic performance evaluation. *Chem. Phys. Lett.* 783, 139040.

Ashwini, J., Aswathy, T.R., Rahul, A.B., Thara, G.M. and Nair, A.S. (2021). Synthesis and characterization of zinc oxide nanoparticles using *Acacia caesia* bark extract and its photocatalytic and antimicrobial activities. *Catalysts.* 11, 1507.

Badri, A., Slimi, S., Guergueb, M., Kahri, H. and Mateos, X. (2021). Green synthesis of copper oxide nanoparticles using Prickly Pear peel fruit extract: Characterization and catalytic activity. *Inorg. Chem. Commun.* 134, 109027.

Carlucci, C., Degennaro, L., and Luisi, R. (2019). Titanium dioxide as a catalyst in biodiesel production. *Catalysts*, 9, 75.

Castillo, J., Pérez-López, R., Caraballo, M.A., Nieto, J.M., Martins, M., Costa, M.C., Olías, M., Cerón, J.C. and Tucoulou, R. (2012). Biologically-induced precipitation of sphalerite-wurtzite nanoparticles by sulfate-reducing bacteria: Implications for acid mine drainage treatment. *Sci. Total Environ.* 423, 176–184.

Chahardoli, A., Qalekhani, F., Shokoohinia, Y. and Fattahi, A. (2020). Biological and catalytic activities of green synthesized silver nanoparticles from the leaf infusion of *Dracocephalum kotschyi*. *Glob. Chall.* 5(2), 2000018.

Chen, Y., Chen, Q., Zhao, H., Dang, J., Jin, R., Zhao, W. and Li, Y. (2020). Wheat Straws and corn straws as adsorbents for the removal of Cr(VI) and Cr(III) from aqueous solution: Kinetics, isotherm, and mechanism. *ACS Omega.* 5, 6003–6009.

Cheng, N., Wang, B., Wu, P., Lee, X., Xing, Y., Chen, M. and Gao B. (2021). Adsorption of emerging contaminants from water and wastewater by modified biochar: A review. *Environ. Pollut.* 273, 116448.

Colipai, C., Southam, G., Oyarzún, P., González, D., Díaz, V., Contreras, B. and Nancucheo, I. (2018). Synthesis of copper sulfide nanoparticles using biogenic $H_2S$ produced by a low-pH sulfidogenic bioreactor. *Minerals.* 8, 35.

da Costa, J.P., Girão, A.V., Lourenço, J.P., Monteiro, O.C., Trindade, T. and Costa, M.C. (2013). Green synthesis of covellite nanocrystals using biologically generated sulfide: Potential for bioremediation systems. *J. Environ. Manage.* 128, 226e232.

Das, P., Dutta, T., Manna, S., Loganathan, S. and Basak, P. (2022). Facile green synthesis of non-genotoxic, non-hemolytic organometallic silver nanoparticles using extract of crushed, wasted, and spent *Humulus lupulus* (hops): Characterization, anti-bacterial, and anti-cancer studies. *Environ. Res.* 204, 111962.

Dhanker, R., Hussain, T., Tyagi, P., Singh, K.J. and Kamble, S.S. (2021). The emerging trend of bioengineering approaches for microbial nanomaterial synthesis and its applications. *Front. Microbiol.* 12, 638003.

Dikshit, P.K., Kumar, J., Das, A.K., Sadhu, S., Sharma, S., Singh, S., Gupta, P.K. and Kim, B.S. (2021). Green synthesis of metallic nanoparticles: Applications and limitations. *Catalysts*, 11, 902.

Dreno, B., Alexis, A., Chuberre, B. and Marinovich, M. (2019). Safety of titanium dioxide nanoparticles in cosmetics. *JEADV.* 33(Supplement 7), 34–46.

Eltaweil, A.S., Fawzy, M., Hosny, M., Abd El-Monaem, E.M., Tamer, T.M. and Omer, A.M. (2022). Green synthesis of platinum nanoparticles using Atriplex halimus leaves for potential antimicrobial, antioxidant, and catalytic applications. *Arab. J. Chem.* 15, 103517.

Fagier, M.A. (2021). Plant-mediated biosynthesis and photocatalysis activities of zinc oxide nanoparticles: A prospect towards dyes mineralization. *J. Nanotechnol.* 2021, ID 6629180.

Falih, A., Ahmed, N.M. and Rashid, M. (2022). Green synthesis of zinc oxide nanoparticles by fresh and dry alhagi plant. *Mater. Today: Proc.* 49(8), 3624–3629.

Francis, S., Joseph, S., Koshy, E.P. and Mathew, B., (2017). Green synthesis and characterization of gold and silver nanoparticles using *Mussaenda glabrata* leaf extract and their environmental applications to dye degradation. *Environ. Sci. Pollut. Res.* 24(21), 17347–17357.

Hashemi, Z., Mizwari, Z.M., Mohammadi-Aghdam, S., Mortazavi-Derazkola, S. and Ebrahimzadeh, M.A. (2022). Sustainable green synthesis of silver nanoparticles using *Sambucus ebulus* phenolic extract (AgNPs@SEE): Optimization and assessment of photocatalytic degradation of methyl orange and their in vitro antibacterial and anti-cancer activity. *Arab. J. Chem.* 15, 103525.

Hojjati-Najafabadi, A., Davar, F., Enteshari, Z. and Hosseini-Koupaei, M. (2021). Antibacterial and photocatalytic behaviour of green synthesis of Zn0.95Ag0.05O nanoparticles using herbal medicine extract. *Ceram. Int.* 47, 31617–31624.

Hussain, F., Mustafa, G., Zia, R., Faiq, A., Matloob, M., Shah, H.-ur-R., Raza, A. and Ali Irfan, J. (2018). Constructed Wetlands and their role in remediation of industrial effluents via plant-microbe interaction: A mini review. *J. Bioremediat. Biodegrad.* 9, 447.

Hussain, I., Singh, N.B., Singh, A., Singh, H. and Singh, S.C. (2016). Green synthesis of nanoparticles and its potential application. *Biotechnol. Lett.* 38, 545–560.

Ibrahim, S., Ahmad, Z., Manzoor, M.Z., Mujahid, M., Faheem, Z. and Adnan, A. (2021). Optimization for biogenic microbial synthesis of silver nanoparticles through response surface methodology, characterization, their antimicrobial, antioxidant, and catalytic potential. *Sci. Rep.* 11, 770.

Jaast, S. and Grewal, A. (2021). Green synthesis of silver nanoparticles, characterization and evaluation of their photocatalytic dye degradation activity. *CRGSC*. 4, 100195.

Jadoun, S., Arif, R., Jangid, N.K. and Meena, R.K. (2020). Green synthesis of nanoparticles using plant extracts: A review. *Environ. Chem. Lett.* 19, 355–374.

Jalab, J., Abdelwahed, W., Kitaz, A. and Al-Kayali R. (2021). Green synthesis of silver nanoparticles using aqueous extract of *Acacia cyanophylla* and its antibacterial activity. *Heliyon*. 7, e08033.

Joshi, J., Dandia, A. and Kum, S. (2017). Titanium oxide nanoparticles as valuable catalyst in organic synthesis: A review. *Mini-Rev. Org. Chem.* 14(3), 227–236.

Kanniah, P., Chelliah, P., Thangapandi, J.R. Gnanadhas, G., Mahendran, V. and Robert M. (2021). Green synthesis of antibacterial and cytotoxic silver nanoparticles by *Piper nigrum* seed extract and development of antibacterial silver based chitosan nanocomposite. *Int. J. Biol. Macromol.* 189, 18–33.

Kim, B., Song, W.C., Park, S.Y. and Park, G. (2021). Green synthesis of silver and gold nanoparticles via *Sargassum serratifolium* extract for catalytic reduction of organic dyes. *Catalysts*. 11, 347.

Kimber, R.L., Bagshaw, H., Smith, K., Buchanan, D.M., Coker, V.S., Cavet, J.S. and Lloyda, J.R. (2020). Biomineralization of $Cu_2S$ nanoparticles by *Geobacter sulfurreducens*. *Appl. Environ. Microbiol.* 86, e00967-20.

Krakowiak, R., Musial, J., Bakun, P., Spychała, M., Czarczynska-Goslinska, B., Mlynarczyk, D.T., Koczorowski, T., Sobotta, L., Stanisz, B., and Goslinski, T. (2021). Titanium dioxide-based photocatalysts for degradation of emerging contaminants including pharmaceutical pollutants. *Appl. Sci.* 11, 8674.

Kumar, M. and Pakshirajan, K. (2020). Novel insights into mechanism of biometal recovery from wastewater by sulfate reduction and its application in pollutant removal. *Environ. Technol. Innov.* 17, 100542.

Le Pape, P., Battaglia-Brunet, F., Parmentier, M., Joulian, C., Gassaud, C., Fernandez-Rojo, L., Guigner, J.-M., Ikogou, M., Stetten, L., Olivi, L., Casiot, C. and Morin, G. (2017). Complete removal of arsenic and zinc from a heavily contaminated acid mine drainage via an indigenous SRB consortium. *J. Hazard. Mater.* 321, 764–72.

Li, Y., Li, F., Yang, Y., Ge, B., and Meng, F. (2021). Research and application progress of ligninbased composite membrane. *J. Polym. Eng.* 41(4), 245–258.

Li, Z., Yuanyuan, D.X. and Koehler, G.S. (2015). Surface-functionalized porous lignin for fast and efficient lead removal from aqueous solution. *ACS Appl. Mater. Interfaces.* 7(27), 15000–15009.

Lim, S.H., Ahn, E.-Y. and Park, Y. (2016). Green synthesis and catalytic activity of gold nanoparticles synthesized by *Artemisia capillaris* water extract. *Nanoscale Res. Lett.* 11, 474.

Mahmoud, A.E.D., AlQahtani, K.M., Alfaij, S.O., AlQahtani, S.F. and Alsamhan, F.A. (2021). Green copper oxide nanoparticles for lead, nickel, and cadmium removal from contaminated water. *Sci. Rep.* 11, 12547.

Maluckov, B.S. (2014). Ti-based biomaterials-properties and production. *Optoelectron. Adv. Mat.* 8(5–6), 545–550.

Maluckov, B.S. (2015). Bioassisted phytomining of gold. *JOM.* 67(5), 1075–1078.

Maluckov, B.S. (2017). The catalytic role of *Acidithiobacillus ferrooxidans* for metals extraction from mining - metallurgical resource. *Biodivers. Int. J.* 1(3), 109–119.

Maluckov, B.S. (2020). Understanding electric potential in processes of (bio)corrosion and (bio)leaching. In: Reimer A (ed) *Horizons in World Physics*, vol 302. Nova Science Pub Inc., New York, pp 1–27.

Maluckov, B.S. (2021). Biorecovery of nanogold and nanogold compounds from gold-containing ores and industrial wastes. *Appl. Microbiol. Biot.* 105, 3471–3484.

Maluckov, B.S. and Mitrić, M.N. (2018). Electrochemical behavior of pyrite in sulfuric acid in presence of amino acids belonging to the amino acid sequence of rusticyanin. *Bioelectrochemistry.* 123, 112–118.

Maluckov, B.S., Dimitrijević, M., Kovačević, R. and Mladenović, S. (2017). The electrochemical behavior of chalcopyrite in sulfuric acid in the presence of cysteine. *Rev. Roum. Chim.* 62, 809–814.

Melkamu, W.W. and Bitew, L.T. (2021). Green synthesis of silver nanoparticles using *Hagenia abyssinica* (*Bruce*) *J.F. Gmel* plant leaf extract and their antibacterial and anti-oxidant activities. *Heliyon.* 7, e08459.

Meng, X., Yan, S., Wu, W., Zheng, G. and Zhou, L. (2017). Heterogeneous Fenton-like degradation of phenanthrene catalyzed by schwertmannite biosynthesized using *Acidithiobacillus ferrooxidans. RSC Adv.* 7, 21638.

Moraes, L.C., Figueiredo, R.C., Ribeiro-Andrade, R., Pontes-Silva, A.V., Arantes, M.L., Giani, A. and Figueredo C.C. (2021). High diversity of microalgae as a tool for the synthesis of different silver nanoparticles: A species-specifc green synthesis. *Colloid Interface Sci. Commun.* 42, 100420.

Moustafa, M.T. (2017). Removal of pathogenic bacteria from wastewater using silver nanoparticles synthesized by two fungal species. *Water Sci.* 31, 164–176.

Nabi, G., Majid, A., Riaz, A., Alharbi, T., Kamran, M.A. and Al-Habardi, M. (2021). Green synthesis of spherical $TiO_2$ nanoparticles using Citrus Limetta extract: Excellent photocatalytic water decontamination agent for RhB dye. *Inorg. Chem. Commun.* 129, 108618.

Nadeem, A.S., Ali, J.S., Latif, M., Rizvi, Z.F., Naz, S., Mannan, A. and Zia, M. (2021). Green synthesis and characterization of Fe, Cu and Mg oxide nanoparticles using *Clematis orientalis* leaf extract: Salt concentration modulates physiological and biological properties. *Mater. Chem. Phys.* 271, 124900.

Neupane, N.P., Kushwaha, A.K., Karn, A.K., Khalilullah, H., Khan, M.M.U., Kaushik, A. and Verma, A. (2022). Anti-bacterial effcacy of bio-fabricated silver nanoparticles of aerial part of *Moringa oleifera lam*: Rapid green synthesis, *In-Vitro* and *In-Silico* screening. *Biocat. Agric. Biotechnol.* 39, 102229.

Norrrahim, M.N.F., Kasim, N.A.M., Knight, V.F., Misenan, M.S.M., Janudin, N., Shah, N. Kasim, N.A.A., Yusoff, W.Y.W., Noor, S.A.M., Jamal, S.H., Ong K.K. and Yunus, W.M.Z.W. (2021). Nanocellulose: A bioadsorbent for chemical contaminant remediation. *RSC Adv.* 11, 7347.

Obayomi, K.S., Oluwadiya, A.E., Lau, S.Y., Dada, A.O., Akubuo-Casmir, D., Adelani-Akande, T.A., Fazle Bari, A.S.M., Temidayo, S.O. and Rahman, M.M. (2021). Biosynthesis of *Tithonia diversifolia* leaf mediated zinc oxide nanoparticles loaded with flamboyant pods (*Delonix regia*) for the treatment of methylene blue wastewater. *Arab. J. Chem.* 14, 103363.

Oves, M., Rauf, M.A., Aslam, M., Qari, H.A., Sonbol, H., Ahmad, I., Zaman, G.S. and Saeed, M. (2022). Green synthesis of silver nanoparticles by *Conocarpus Lancifolius* plant extract and their antimicrobial and anticancer activities. *Saudi J. Biol. Sci.* 29, 460–471.

Park, J.K., Rupa, E.J., Arif, M.H., Li, J.F. Anandapadmanaban, G., Kang, J.P., Kim, M., Ahn, J.C., Akter, R., Yang, D.C. and Kang S.C. (2021). Synthesis of zinc oxide nanoparticles from *Gynostemma pentaphyllum* extracts and assessment of photocatalytic properties through malachite green dye decolorization under UV illumination-A Green Approach. *Optik: Int. J. Light Electron Opt.* 239, 166249.

Patino-Ruiz, D., Sanchez-Botero, L., Tejeda-Benitez, L., Hinestroza, J. and Herrera, A. (2020). Green synthesis of iron oxide nanoparticles using *Cymbopogon citratus* extract and sodium carbonate salt: Nanotoxicological considerations forpotential environmental applications. *Environ. Nanotechn. Monit. Manag.* 14, 100377.

Punnoose, M.S., Bijimol, D. and Mathew B. (2021). Microwave assisted green synthesis of gold nanoparticles for catalytic degradation of environmental pollutants. *Environ. Nanotechn. Monit. Manag.* 16, 100525.

Radini, I.A., Hasan, N., Malik, M.A and Khan Z. (2018). Biosynthesis of iron nanoparticles using Trigonella foenum-graecum seed extract for photocatalytic methyl orange dye degradation and antibacterial applications. *J. Photochem. Photobiol. B: Biol.* 183, 154–163.

Raffa, C.M., Chiampo, F. and Shanthakumar, S. (2021). Remediation of metal/metalloid polluted soils: A short review. *Appl. Sci.* 11, 4134.

Rodríguez-Felix, F., Lopez-Cota, A.G., Moreno-Vasquez, M.J. Graciano-Verdugo, A.Z., Quintero-Reyes, I.E., Del-Toro-Sanchez, C.L. and Tapia-Hernandez, J.A. (2021). Sustainable green synthesis of silver nanoparticles using safflower (*Carthamus tinctorius* L.) waste extract and its antibacterial activity. *Heliyon.* 7, e06923.

Saied, E., Eid, A.M., El-Din Hassan, S., Salem, S.S., Radwan, A.A., Halawa, M., Saleh, F.M., Saad, H.A., Saied, E.M. and Fouda, A. (2021). The catalytic activity of biosynthesized magnesium oxide nanoparticles (MgO-NPs) for inhibiting the growth of pathogenic microbes, tanning effluent treatment, and chromium ion removal. *Catalysts.* 11, 821.

Saif, M.S., Zafar, A. Waqas, M., Hassan, S.G., Haq, Au, Tariq, T., Batool, S., Irshad, M., Hasan, M. and Shu, X. (2021). Phyto-reflexive zinc oxide nano-flowers synthesis: An advanced photocatalytic degradation and infectious therapy. *J. Mater. Res. Technol.* 13, 2375–2391.

Saitoh, N., Fujimori, R., Nakatani, M., Yoshihara, D., Nomura, T. and Konishi, Y. (2018). Microbial recovery of gold from neutral and acidic solutions by the baker's yeast *Saccharomyces cerevisiae. Hydrometallurgy.* 181, 29–34.

Salvadori, M.R., Lepre, L.F., Ando, R.A., Oller do Nascimento, C.A. and Corre^a, B. (2013). Biosynthesis and uptake of copper nanoparticles by dead biomass of hypocrea lixii isolated from the metal mine in the Brazilian Amazon region. *PLOS ONE.* 8(11), e80519.

Sarwar, N., Humayoun, U.B., Kumar, M., Zaidi, S.F.A.,Yoo, J.H., Ali, N., Jeong, D.I., Lee, J.H. and Yoon, D.H. (2021). Citric acid mediated green synthesis of copper nanoparticles using cinnamon bark extract and its multifaceted applications. *J. Clean. Prod.* 292, 125974.

Schaffie, M. and Hosseini, M.R. (2014). Biological process for synthesis of semiconductor copper sulfide nanoparticle from mine wastewaters. *J. Environ. Chem. Eng.* 2, 386–391.

Silva, P.M.P., Lucheta, A.R., Bitencourt, J.A.P., do Carmo, A.L.V., Cuevas, I.P.Ñ., Siqueira, J.O., de Oliveira, G.C. and Alves, J.O. (2019). Covellite (CuS) production from a real acid mine drainage treated with biogenic $H_2S$. *Metals.* 9, 206.

Singh, A., Singh, N.B., Afzal, S., Singh, T. and Hussain, I. (2018). Zinc oxide nanoparticles: A review of their biological synthesis, antimicrobial activity, uptake, translocation and biotransformation in plants. *J. Mater. Sci.* 53(1), 185–201.

Singh, S., Kumar, A. and Gupta, H. (2020). Activated banana peel carbon: A potential adsorbent for Rhodamine B decontamination from aqueous system. *Appl. Water Sci.* 10, 185.

Sørmo, E., Silvani, L., Bjerkli, N., Hagemann, N., Zimmerman, A.R., Hale, S.E., Hansen, C.B., Hartnik, T. and Cornelissen, G. (2021). Stabilization of PFAS-contaminated soil with activated biochar. *Sci. Total Environ.* 763, 144034.

Sravanthi, K., Ayodhya, D. and Yadgiri Swamy, P. (2018). Green synthesis, characterization of biomaterial-supported zero-valent iron nanoparticles for contaminated water treatment. *J. Anal. Sci.Technol.* 9, 3.

Srećković, N.Z., Nedić, Z.P., Liberti, D., Monti, D.M., Mihailović, N.R., Katanic Stanković, J.S., Dimitrijević, S. and Mihailović, V.B., (2021). Application potential of biogenically synthesized silver nanoparticles using *Lythrum salicaria* L. extracts as pharmaceuticals and catalysts for organic pollutant degradation. *RSC Adv.* 11, 35585–35599.

Srinath, B.S., Namratha, K. and Byrappa, K. (2017). Eco-friendly synthesis of gold nanoparticles by gold mine bacteria *Brevibacillus formosus* and their antibacterial and biocompatible studies. *IOSR J. Pharm.* 7, 53–60.

Sudhakar, C., Poonkothai, M., Selvankmuar, T., Selvam, K., Rajivgandhi, G., Siddiqi, M.Z., Alharbi, N.S., Kadaikunnan, S. and Vijayakumar, N. (2021). Biomimetic synthesis of iron oxide nanoparticles using *Canthium coromandelicum* leaf extract and its antibacterial and catalytic degradation of Janus green. *Inorg. Chem. Commun.* 133, 108977.

Sukweenadhi, J., Setiawan, K.I., Avanti, C., Kartini, K. Rupa, E.J. and Yang, D.-C. (2021). Scale-up of green synthesis and characterization of silver nanoparticles using ethanol extract of *Plantago major* L. leaf and its antibacterial potential. *South African J. Chem. Eng.* 38, 1–8.

Tailor, G., Yadav, B.L., Chaudhary, J., Joshi, M. and Suvalka, C. (2020). Green synthesis of silver nanoparticles using *Ocimum canum* and their anti-bacterial activity. *Biochem. Biophys. Rep.* 24, 100848.

Tanwar, N., Dhiman, V., Kumar, S. and Kondal N. (2022). Plant extract mediated ZnO-NPs as photocatalyst for dye degradation: An overview. *Materials Today: Proceedings.* 48, 1401–1406.

Tasić, V., Kovačević, R., Maluckov, B., Apostolovski-Trujić, T., Matić, B., Cocić, M. and Šteharnik, M. (2017). The content of As and heavy metals in TSP and PM10 near copper smelter in Bor, Serbia. *Water Air Soil Pollut.* 228–230.

Tasić, V., Maluckov, B., Kovačević, R., Apostolovski-Trujić, T., Šteharnik, M. and Stanković, S. (2014). Analysis of $SO_2$ concentrations in the urban areas near copper mining and smelting complex Bor, Serbia. *Chem. Eng. Trans.* 42, 103–108.

Thi, H.P.N., Thi, K.T.P., Thi, L.N., Nguyen, T.T., Nguyen, P.T.M., Vu, T.T., Le, H.T., Dang, T.-D., Huynh, D.C., Mai, H.T., La, D.D., Chang, S.W. and Nguyen, D.D. (2022). Green synthesis of an Ag nanoparticle-decorated grapheme nanoplatelet nanocomposite by using *Cleistocalyx operculatus* leaf extract for antibacterial applications. *Nano-Structures & Nano-Objects.* 29, 100810.

Tian, D., Hu, J., Bao, J., Chandra, R.P., Saddler, J.N. and Lu, C. (2017). Lignin valorization: Lignin nanoparticles as high-value bio-additive for multifunctional nanocomposites. *Biotechnol Biofuels.* 10, 192.

Vidhu, V.K. and Philip, D. (2014). Catalytic degradation of organic dyes using biosynthesized silver nanoparticles. *Micron.* 56, 54–62.

Vijayan, R., Joseph, S. and Mathew, B. (2019). Green synthesis of silver nanoparticles using *Nervalia zeylanica* leaf extract and evaluation of their antioxidant, catalytic, and antimicrobial potentials. *Particul. Sci. Technol.* 37 (7), 809–819.

Vijilvani, C., Bindhu, M.R., Frincy, F.C., AlSalhi, M.S., Sabitha, S., Saravanakumar, K., Devanesan, S., Umadevi, M., Aljaafreh, M.J. and Atif, M. (2020). Antimicrobial and catalytic activities of biosynthesized go silver andpalladium nanoparticles from *Solanum nigurum* leaves. *J. Photochem.Photobiol., B: Biol.* 202, 111713.

Vitor, G., Palma, T.C., Vieira, B., Lourenço, J.P., Barros, R.J. and Costa, M.C. (2015). Start-up, adjustment and long-term performance of a two-stage bioremediation process, treating real acid mine drainage, coupled with biosynthesis of ZnS nanoparticles and $ZnS/TiO_2$ nanocomposites. *Miner. Eng.* 75, 85–93.

Yan, F. and Reible, D. (2015). Electro-bioremediation of contaminated sediment by electrode enhanced capping. *J. Environ. Manage.* 155, 154–161.

Younes, M., Aquilina, G., Castle, L., Engel, K-H., Fowler, P., Frutos Fernandez, M.J., Furst, P., Gundert-Remy, U., Gurtler, R., Husøy, T., Manco, M., Mennes, W., Moldeus, P., Passamonti, S., Shah, R., Waalkens-Berendsen, I., Wolfle, D., Corsini, E., Cubadda, F., De Groot, D., Fitz Gerald, R., Gunnare, S., Gutleb, A.C., Mast, J., Mortensen, A., Oomen, A., Piersma, A., Plichta, V., Ulbrich, B, Van, Loveren, H, Benford, D, Bignami, M, Bolognesi, C, Crebelli, R, Dusinska, M, Marcon, F., Nielsen, E., Schlatter, J., Vleminckx, C., Barmaz, S., Carfi, M., Civitella, C., Giarola, A., Rincon, A.M., Serafimova, R., Smeraldi, C., Tarazona, J., Tard, A. and Wright, M. (2021). EFSA FAF panel (EFSA Panel on Food Additives and Flavourings): Scientific opinion onthe safety assessment of titanium dioxide (E171) as a food additive. *EFSA J.* 19(5), 6585.

Zamora-Ledezma, C., Negrete-Bolagay, D., Figueroa, F., Zamora-Ledezma, E., Ni, M., Alexis, F. and Guerrero, V.H. (2021). Heavy metal water pollution: A fresh look about hazards, novel and conventional remediation methods. *Environ. Technol. Innov.* 22, 101504.

Zhao, C., Wang, B., Theng, B.K.G., Wu, P., Liu, F., Wang, S., Lee, X., Chen, M., Li, L. and Zhang, X. (2021). Formation and mechanisms of nano-metal oxide-biochar composites for pollutants removal: A review. *Sci. Total Environ.*767, 145305.

Zhu, W., Hu, C., Ren, Y., Lu, Y., Song, Y., Ji, Y., Han, C. and He, J. (2021). Green synthesis of zinc oxide nanoparticles using *Cinnamomum camphora* (L.) Presl leaf extracts and its antifungal activity. *J. Environ. Chem. Eng.* 9, 106659.

# 4 Hybrid Bio-nano Techniques

## *Novel Approach for the Removal of Emerging Contaminants from the Environment*

*Sonam Paliya, Mehak Puri, Ashootosh Mandpe, Divyesh Bhisikar, M. Suresh Kumar, and Sunil Kumar*

**CONTENTS**



DOI: 10.1201/9781003270959-4

## 4.1 INTRODUCTION

Growing population and industrialization have promoted economic development, thereby harming the environment by releasing a variety of pollutants (Edgar et al., 2020). The Stockholm Convention targets the identification and aims to restrict the release of new unidentified organic pollutants. The occurrence of trace or low concentration organic compounds are called "emerging contaminants (ECs)." The predominant challenge is the determination, quantification, and eradication of low concentration contaminants from various environmental matrices (Gogoi et al., 2018). Lack of information on the fate, sources (nonpoint), and harmful impacts on the biota and environment pose additional constraints on the mitigation of the emerging contaminants. Steroid hormones, such as estradiol, estriol, and estrone, cause reproductive defects in fishes and amphibians, while engineered nanomaterials cause oxidative stress in fishes and inhibit photosynthesis in algae (Naidu et al., 2016).

Conventional wastewater treatment plants are ineffective in removing such low concentration compounds. Hence, sophisticated and sensitive treatment options, which are environmentally and economically viable, are the need of the hour. Bio-nano techniques are one of the most effective, cost-efficient green treatment options in eradicating ECs from complex environmental matrices (Chen et al., 2021).

## 4.2 EMERGING CONTAMINANTS

### 4.2.1 Sources and Fate of Emerging Contaminants in Different Environmental Compartments

Contamination by emerging chemicals in the environment is mainly the result of anthropogenic activities (Gomes et al., 2017; Paliya et al., 2021b). Usually, the widespread pollution from chemicals occurs through varied and innumerable point and nonpoint sources. Wastewater treatment plants (WWTPs) are the main contributors of contaminants in the environment. Stormwater drains, agricultural runoff, vehicular exhaustion, electronic-waste disposal and recycling sites, and sludge from semi-treated wastewater treatment plants are some examples of sources of chemicals in environmental matrices (Lofrano et al., 2020). Mixed effluent waste from personal care products, such as shampoos, soaps, toothpaste, etc.; metabolized products from the discharge of human feces and urine; hospital effluent discharges; and landfill leaching are other important sources (Figure 4.1). Sources and occurrences of ECs in different environmental matrices depend on climatic conditions and characteristics of environmental matrices in which the chemical is bonded (Rout et al., 2021). Wastewater treatment plants are designed to treat the organic load, and the low range contaminants are left out and discharged into various compartments of the environment (Mohapatra et al., 2016). Pharmaceuticals, such as ciprofloxacin and cetirizine, were found to be present in groundwater up to 14 and 28 mg/L, respectively, while 0.01–2.5 mg/L and 0.005–0.53 mg/L concentrations of the respective drugs were detected in river water in India (Gani and Kazmi, 2016). Synthetic perfumes and personal care products have been detected from 3690–7330 ng/L in influent to

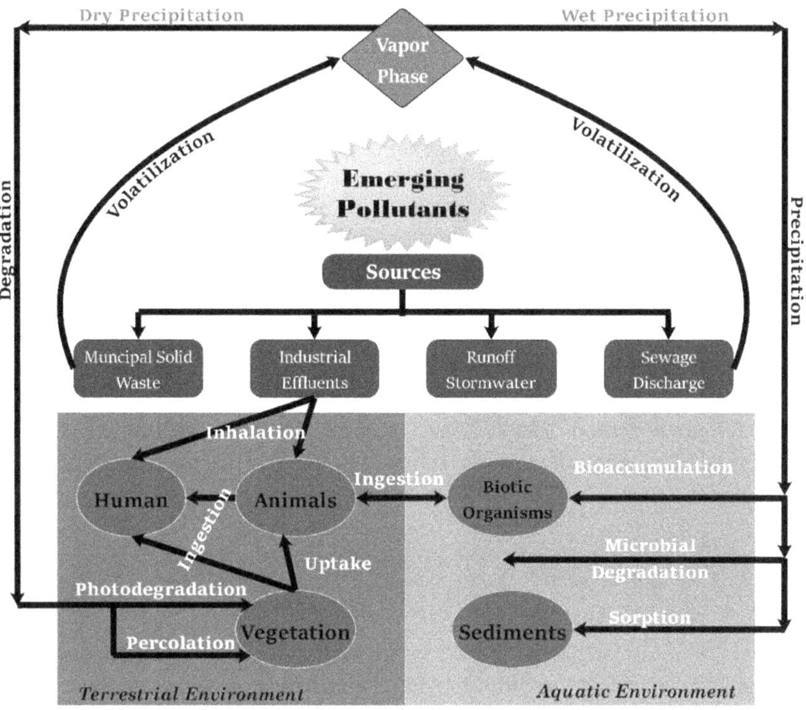


**FIGURE 4.1** Representation of the possible sources and fate of emerging contaminants in the environment.

960–2960 ng/L in the effluent in sewage wastewater, with 150–16700 ng/L concentrations in surface water (Lee et al., 2010). The migration of emerging contaminants in the environment follows multiple diverse pathways depending on the physicochemical nature of the contaminant and the usage pattern of the individual or mixture of chemicals (Gwenzi et al., 2018). Dilution, percolation, leaching, filtration, sorption-desorption, and evaporation are a few known phenomena deciding the fate of the chemical contaminants in varied environmental compartments. The mechanism of the fate of emerging contaminants depends on the volatility, persistence or pseudo-persistence, and polarity of the compounds. The octanol-water partitioning coefficient ($K_{OW}$) plays a vital role in determining the fate of chemicals either in a liquid or solid matrix. The general overview is if log $K_{OW} < 2.5$, the compound has high hydrophilic mobility; $2.5 < \log K_{OW} < 4$ the compound has medium mobility; and log $K_{OW} > 4$ indicates low mobility or high retention in the soil matrix (Rogers, 1996). For example, carbamazepine, an antiepileptic, has low affinity to river water than galaxolide, a synthetic musk. Ion-exchange capacity and organic matter influence the adsorption of the contaminant in soils and sediments (Wilkinson et al., 2017). Compounds such as triclosan, erythromycin, and bezafibrate were detected in sludge ranging from 354 to 15,600, 6 to 79, and 17 to 64 ng/g dry weight, respectively (Rout et al., 2021). Volatilization is the main mechanism in the transportation of the emerging chemicals in the air. The volatile contaminants from the surface or released from the emissions of the industries and incinerators volatilize in the atmosphere. The major contributor of chemicals in the atmosphere is anthropogenic activities. Polybrominated diphenyl ethers (PBDEs) were found in the vapor phase in varied urban regions ranging from 0.004 to 340 pg/m$^{3.}$ The contaminant, Bisphenol A (BPA), was investigated for spatial and temporal presence and was found to be in concentration of 1–17,400 pg/m$^3$ in aerosols collected from urban, rural, marine, and polar sites (Barroso et al., 2019). Dibutyl phthalate (DBP), di(2-Ethylhexyl) phthalate (DEHP), and nonylphenol (NP) were detected in indoor air with 2300 ng/m$^3$, 1000 ng/m$^3$, and 680 ng/m$^3$, respectively; while in the outdoor air, the concentration detected for respective compounds were 55 ng/m$^3$, 45 ng/m$^3$, and 30.5 ng/m$^3$ (Wilkinson et al., 2017). Precipitation leads to the deposition of the contaminants on the land or in water and further percolates deep into groundwater (Vasilachi et al., 2021). Each phenomenon is interlinked in the determination of the pathway of the contaminants; still, more detailed investigation on the fate of the emerging contaminants in different environmental compartments is needed.

## 4.2.2 Risks Associated with Emerging Contaminants

### 4.2.2.1 Ecological Toxicity

Emerging contaminants (ECs) are regarded as one of the low toxic concentrations (nanogram/liter) substances with a potential to harm the ecosystem and humans at large. Genotoxicity, hormonal disruptions, and immune toxicity are some ill effects caused by the ECs on organisms (Vasilachi et al., 2021). ECs are capable of eliciting alterations in biological responses in both target and nontarget species, even when present in trace levels in the environment. Ecological toxicity depends on the exposed organisms, nature and concentration of the contaminant, duration of the exposure, and bioavailability (Wilkinson et al., 2015). Ecotoxicity calculations are majorly based on the extrapolation of the high doses accounting for the acute toxicity of the bioindicator species, mainly algae, Daphnia, and fish and neglecting the chronic toxicity and the non-monotonic dose–response relationship with the variable toxic effects on different species (Beshaa et al., 2017; Wilkinson et al., 2015). Hayes et al. (2003) report 100 ng/L (or 0.1 ppb) of atrazine causing testicular oogenesis in *Rana pipiens* (American Leopard frog) (Hayes et al., 2003). Diclofenac, a common analgesic with 5 µg/L of aquatic concentration, caused renal lesions and gill alterations in rainbow trout after exposure for 28 days, but a higher concentration exposure of diclofenac (1000–2000 µg/L) on zebrafish showed no effect on its reproduction. Hence, it can be concluded that toxicity varies from species to species with respect to the contaminant dose and exposure on an individual species (Hallare et al., 2004; Schwaiger et al., 2004). Continuous discharge of chemicals takes place, percolating and transporting to soil and in turn affecting the microflora and fauna of the environment. Lemos et al. (2009) reported contamination of agricultural soil with 1000 mg/Kg dw of bisphenol A. The 1000 mg/Kg dw contaminant concentration exposure to adult isopods *Porcellio scaber* led to endocrine disruption by molting disturbances (Lemos et al., 2009). EC contamination affects the growth of microflora, quality of soil, and biomass of species. Domene et al. (2009) investigated toxicity of nonylphenol (NP) and nonylphenol polyethoxylates in earthworms (*E. andrei*), soil invertebrate species (*Folsomia candida*), and plants (*Lolium perenne and Brassica rapa*) on loamy soil. The toxicity ($EC_{50}$) for plants, invertebrates, and earthworms for NP was found to be above 1000 mg/Kg, 64–226 mg/Kg, and 240–523 mg/Kg, respectively. Similarly, toxicity (EC50) results for nonylphenol polyethoxylates were found as over 10,000 mg/Kg for plants, 356–1876 mg/Kg for invertebrates, and 784 to >3000 mg/Kg for earthworms. The toxicity results observed ten folds higher concentration in case of plants, five times in invertebrate species, and three times in earthworms for nonylphenol polyethoxylates (Domene et al., 2009). Several environmental toxicity studies have been reported for a single compound and in the case of mixture, limitations occur. According to the Warne and Hawker funnel hypothesis (Warne and Hawker, 1995), mixtures contain a large number of chemicals with a possibility of exhibiting a baseline mechanism. Hence, the risk assessment and toxicological studies must account for the concentration of the contaminant load. This further explains the different modes of action in mixtures of compounds, with either of them behaving synergistically or antagonistically (Thomaidi et al., 2016). Critical investigations and further inclusion of the parameters for the appropriate assessment of the ecotoxicity studies must be initiated. Table 4.1 summarizes the environmental toxicity caused by the presence of emerging contaminants.

#### 4.2.2.2 Impacts on Human Health

Emerging contaminants are mostly found in daily routine chemicals such as personal and beauty care products, disinfectants, pharmaceuticals, UV filters, and many more. The ubiquitous presence of the compounds in the environment via transportation, filtration, percolation, and volatilization has its ill effects on the environment and further impacts human health (Gogoi et al., 2018). The nano concentration ECs has been shown to disrupt target cells and other organs at the molecular and genetic levels (Ullah et al., 2018). Perfluorinated compounds (PFCs) such as perfluoro octane sulfonate (PFO) cause liver adenomas, cancer, Leydig cell adenomas, and pancreatic adenocarcinoma (Chang et al., 2014). Personal care products (PCPs) primarily damage the reproductive and hypothalamus-pituitary-thyroid axis, further leading to infertility, a decrease in fecundity, and reproductive complications (Krause et al., 2012; Shrestha et al., 2015). Paliya et al. (2021a) reviewed the neurotoxic, genotoxic, and reproductive–endocrine disruptions by polybrominated diphenyl ethers (PBDEs) in humans, leading to alterations in the thyroid system and harming the brain and developmental growth in humans (Paliya et al., 2021a). Nanomaterial ECs are known to exhibit carcinogenic and developmental toxicity in humans. Even lung cancer and reproductive disorders have been reported (Bilal et al., 2019). Transgenerational, sexually dimorphic neuroendocrine disorders, type II diabetes, obesity, and neural and cardiovascular damage are collective human health effects by the emerging contaminants (Vasilachi et al., 2021). Low concentration individual compounds have proven to have ill effects on humans. The assessment and quantification of risk and impact by the mixture of these compounds is yet to be well established due to the lack of strong corelation of determination of effects by ECs on the humans by monotonic dose response curve and acute toxicity results (Riva et al., 2018). The mechanism of toxicity in humans has been difficult to establish due to the scarcity

**TABLE 4.1**
**Effects of Emerging Contaminants at Variable Concentrations on Biota**

| Target contaminant | Environmental matrix | Contaminant concentration | Effects on biota | References |
|---|---|---|---|---|
| Diclofenac (DCF) | Municipal wastewater | 10 ng/L | Impairment of osmoregulation in green crab (*Carcinus maenas*) | (Ensano et al., 2017) |
| Bisphenol A (BPA) | Drinking water | 1.4–2.7 ng/mL | Non-alcoholic fatty liver disease in humans | (Verstraete et al., 2018) |
| BPA | Wastewater | 131 mg/L | Endocrine health effects on human and animals | (Christofilopoulos et al., 2019) |
| Ciprofloxacin (CIP) | | 1.21 mg/L | | |
| Sulfamethoxazole (SMX) | | 3.6 mg/L | Reproductive damage in human, terrestrial organisms, and aquatic biota | |
| Ibuprofen (IBU) | Wastewater | 0.02–1 mg/L | Lemna gibba L hinder the growth and photosynthesis | (Baccio et al., 2017) |
| Diclofenac Carbamazepine Amoxicillin | Wastewater | 10 mg/L | Chronic toxicological and endocrinological impacts on the terrestrial biota | (Naddeo et al., 2020) |
| Paracetamol | Wastewater | 0.3–10 mg/L | Endocrine disruptive properties in both animals and human | (García-Mateos et al., 2015) |
| Ibuprofen (IBU) Ketoprofen | Aqueous medium | 25–150 mg/L | Toxic effect on wildlife and humans | (Fröhlich, Foletto, and Dotto, 2019) |
| Carbamazepine | Water | 10 mg/L | | (Delgado et al., 2019) |
| Sulfamethoxazole | Aqueous medium | 5 mg/L | Human (hepatic cancer and alteration of genetic traits) | (Pamphile et al., 2019) |
| Antipyrine | Aqueous medium | 50 mg/L | - | (Bedia et al., 2018) |
| Tetracycline | Water | 100 mg/L | Antimicrobial resistance and endocrine disruption in human and aquatic biota | (Choi et al., 2020) |
| Sulfamethoxazole | Water | - | Thyroid disruption in male zebra fish | (Kwon et al., 2016) |
| Naproxen | Water | 0.5 mg/L | Decrease of survival of juvenile medaka fish and increase of transcription of erβ2 gene | (Kwak et al., 2018) |
| 17β-estradiol (E2) | Water | 10 mg/L | Ratio of E2 and testosterone significantly increased in H295R cells | (Kwak et al., 2018) |
| Caffeine | Aqueous medium | 50 mg/L | Reproductive disorder in humans | (Anastopoulos, Katsouromalli, and Pashalidis, 2020) |

of epidemiological and toxicological data for the multiple compounds (Gwenzi et al., 2018). However, non-monotonic dose responses (NMDRs) have recently been developed for the determination of the toxicological effects from ECs. The drawback is the extrapolation of the effects during the determination of low concentration of ECs which distorts the correct estimation of the impacts (Riva et al., 2018). Table 4.1 summarizes the studies at variable concentrations impacting the human health.

## 4.3 TECHNOLOGICAL OPTIONS FOR THE TREATMENT OF EMERGING CONTAMINANTS

### 4.3.1 Conventional Methods for the Removal of Emerging Contaminants

Wastewater treatment plants are usually designed to remove organics and inorganics and decrease the microbial load to meet the standard water quality norms. ECs are persistent and unregulated compounds, and conventional wastewater treatment plants are not designed for the removal of those chemicals (Roccaro, 2018). The conventional treatment options comprise of both physicochemical and biological techniques. The selection of treatment options depends on the type of wastewater generated, its characteristics, and concentration. Cost, feasibility, environmental impacts, and formation of potential toxic by-products are also important during the selection of waste removal techniques (Crini and Lichtfouse, 2018). This section reviews various physicochemical and biological methods for the complete or partial removal of ECs from effluent.

### 4.3.2 Physicochemical Methods

The quality of water is dependent on the type of treatment applied. Physicochemical methods are common wastewater treatment techniques that tend to remove the inorganic and organic loads effectively. Sedimentation, coagulation-flocculation, precipitation, ion exchange, adsorption, membrane filtrations are physicochemical options employed in wastewater treatment.

1. Sedimentation: This process takes place in the primary settling tanks for 2–4 hours keeping the tank undisturbed. The heavier particles (sludge) settle at the bottom due to gravitational force, and the supernatant passes for further treatment (Gangaraju et al., 2021).
2. Coagulation-flocculation: This process helps to remove the turbidity and the organic load from the wastewater. Commonly employed coagulants are hydroxides of iron and aluminium operating at 8 and 4.5 pH, respectively. Polyferricsulfate (PFS) and polyacrylamide (PAM) are used as flocculants.

$$Al_2(SO_4)_3 + 6H_2O = 2Al(OH)_3 + 3H_2O$$

$$FeCl_3 + 6H_2O = Fe(OH)_3 + 3HCl$$

3. Precipitation: Chemical means are used to bind the heavy metals, phosphorous compounds using hydroxides of sodium and calcium, to insoluble form. The insoluble precipitates are further removed by sedimentation or the filtration process.

   Ion exchange. A highly selective method that causes the exchange of resins between the resin (solid media) and the electrolytic solution (liquid media) without changing the structural integrity of the target pollutant. <10ppm or >100ppm metal concentrations can be removed using ion exchange process, such as using the weakly basic macroporous-type anion exchange resin.
4. Adsorption and membrane filtration: Adsorption is a highly feasible, low-cost option to remove pollutants from wastewater through large micro-mesoporous volumes, due to the greater surface area of the adsorbent. Activated carbon is a widely used adsorbent. Membrane filtration has an edge over other physicochemical techniques in effectively removing the small size or low concentration micropollutants. Ultrafiltration (UF) removes the molecules ranging from 1000 to 100,000 Da molecular weight and pore size of 5–20 nm retaining the high molecular weight pollutants against the pressure gradient and further treating the lower weight pollutants. Chitosan and alginate UFs remove the micropollutants with 90–100% removal efficiency. Nanofiltration (NF) is also one of the types of membrane filtration which removes contaminants ranging from 0.0001 to 0.001 μm. It filters the monovalent ions while rejecting the divalent and multivalent ions. Reverse osmosis (RO) is an effective option for eliminating pollutants from water, with wide pressure operations from 4.5 to 15 and a pH range from 3 to 11. The pressure pumps generate high pressure against the semi-permeable membranes rejecting 90–99% of the pollutants (Crini and Lichtfouse, 2018).

### 4.3.3 Biological Approach

Biological treatment options are economical, simple, and well accepted by the public for the removal of emerging contaminants. The process of microorganisms (either pure or mixed cultures) completely degrading the recalcitrant group of compounds is termed "biodegradation." Sometimes they even transform from more toxic to less toxic components or form intermediates in a process called biotransformation; other times they mineralize to carbon dioxide and water (Crini and Lichtfouse, 2020). Biodegradation is dependent on various factors, such as contaminant load, physicochemical properties of contaminant, biomass characteristics, and operating conditions. Higher sludge retention time (SRT) favors the growth of microorganisms with low concentrations of the compound, while the higher the hydraulic retention time (HRT), the more degradation of the recalcitrant compounds due to increased reaction time (Norvill et al., 2016). Commonly used biological methods are activated sludge process (ASP), constructed wetlands (CW), and membrane bioreactor systems. The activated sludge process involves the uptake of the chemicals by the group of microorganisms (bacteria, algae, protozoa, fungi) which are suspended in the environment at environmental conditions (Rajasulochana and Preethy, 2016). For example, methoxy triclosan is degraded to 81% and biotransformed to 56.5% using the ASP process (Xiong et al., 2017). Constructed wetlands are regarded as one of the most cost-effective and ecologically accepted tertiary treatment techniques for the removal of ECs. Extensive studies have been conducted by various researchers on the elimination of ECs from the wastewater using constructed wetlands. Matamoros et al. (2017) studied the removal of 16 ECs using 12 full-scale horizontal-flow constructed wetlands (HFCWs) (18,000 $m^2$ in total). The removal efficiency varied from 0–92% (average 43 %) (Matamoros et al., 2017). Membrane bioreactor systems (MBRs) have an edge over ASP and CW systems due to its mechanism i.e., size exclusion and adsorption of the contaminant over the biofilm layer through electrostatic interactions, further degrading the pollutant. Majorly polar contaminants are more efficiently eliminated than the non-polar ones. For example, Bisphenol A removal was reported 75% in polysulfone made UF-MBR while ethinylestradiol, sulfamethoxazole, and acetaminophen showed removal efficiencies from 90% to 99% at pilot scale MBR (Rodriguez et al., 2017). Similar studies have been reported in which pharmaceuticals have been reduced more effectively from the wastewater using the MBR technique than ASP. Diclofenac was reduced to 87.4% in MBR compared with 50% in ASP (Tiwari et al., 2017). Table 4.2 addresses various conventional treatment options employed for the management of ECs.

## 4.4 BIO-NANO TECHNIQUES

Biotechnology and nanotechnology are two of the most promising technologies of the 21st century. Nanotechnology is described as the design, development, and implementation of materials and technologies with a minimum functional make-up on the nanoscale scale. In general, nanotechnology is concerned with the development of materials, electronics, or other structures having at least one dimension ranging from 1 to 100 nanometres. Biotechnology, on the other hand, is concerned with the metabolic and other physiological processes of biological entities, such as bacteria. The combination of these two technologies, namely bionanotechnology, has the potential to play a critical role in the development and implementation of several valuable instruments in the study of life (National Nanotechnology Initiative, 2021; Sawicki et al., 2019).

Nanotechnology encompasses a wide range of activities, from extending traditional device physics to establishing whole new ways based on molecular self-assembly, from developing novel nanoscale materials to researching if we can directly manipulate matter on/in the atomic scale/level. While biotechnology is concerned with the metabolic and other physiological processes of biological entities, such as microorganisms, nanotechnology, when combined with biotechnology, has the potential to create and deploy a wide range of helpful instruments in the study of life. This idea entails the application of fields of science as diverse as surface science, organic chemistry, molecular biology, semiconductor physics, microfabrication, etc (Fakruddin et al., 2012).

**TABLE 4.2**
**Physicochemical and Biological Treatment Options for the Treatment of Emerging Contaminants**

| Contaminant | Contaminant concentration | Treatment option | Dose during the treatment | Removal efficiency (%) | References |
|---|---|---|---|---|---|
| 17β-Estradiol (E2) | 1 mg/L | Aerobic treatment | - | >95 | (Liu, Kanjo, and Mizutani, 2008) |
| Triclosan | - | Coagulation-flocculation | - | 89 | (Yang et al., 2017) |
| Butyl benzyl, Diethyl and Dimethyl phthalate | 100 mg/L | Biodegradation | - | 100 | (Hwang, Choi, and Song, 2008) |
| Triclosan (TCS) | 5 mg/L | Biodegradation | - | 100 | (Mulla et al., 2016) |
| Bisphenol A (BPA), E2 | - | $H_2O_2$/UV | 200 mg/L | >99 | (Esplugas et al., 2007) |
| 17Alpha-ethnylestradiol | - | Ozonation | 0.2 and 0.5 mg/L | 97 | |
| Galaxolide (HHCB) | 1 mg/L | ASP | | 100 | (Feijoo, Moreira, and Lema, 2011) |
| DCF, CBZ, TCS, E1 | - | Ozonation | - | 70 | (Sui et al., 2011) |
| BPA | - | Anaerobic | - | 18-91 | (Chouhan, Yadav, and Prakash, Jay, swati, Singh, 2013) |
| Bisphenol A (BPA), Caffeine (CFN), IBP, E2, E1, NPX, Nonylphenol (NP) | - | Conventional treatment options | - | >70 | (Grandclement et al., 2017) |
| Carbamazepine (CBZ) | - | MBR-NF | - | 93 | (Beshaa et al., 2017) |
| Ibuprofen, Acetaminophen | - | MBR | - | 99 | |
| Sulpiride, Amisulpride, Lamotrigine | - | ASP | - | 85 | (Franka et al., 2016) |
| Diethyl phthalate | 170 mg/L | Biodegradation | - | >99 | (Y. Wang et al., 2018) |
| Diclofenac | 200mg/L | Adsorption | 168.04 mg/g | 82 | (de Souza dos Santos et al., 2020) |
| Acetaminophen | - | Adsorption | - | 94.1 | (Jung, Oh, and Yoon, 2015) |
| Naproxen | 10 mg/L | Adsorption | 1 g/L | 95 | (Sekulić et al., 2018) |
| Tri-allate | - | MBR | - | >99.9 | (Racar et al., 2020) |
| Oxcarbazepine | - | AOP | - | 80 | (J. Wang and Wang, 2017) |

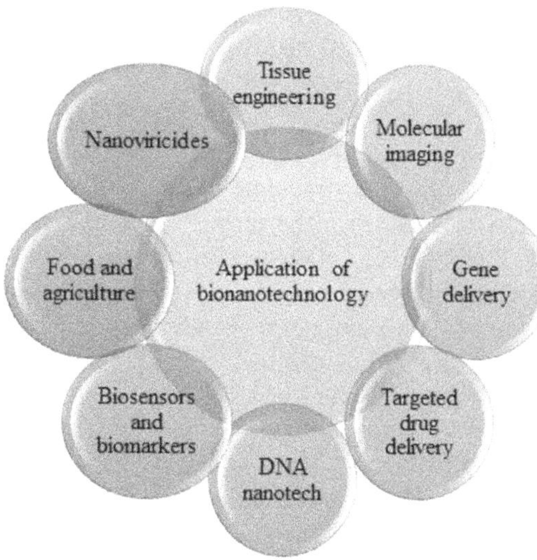


**FIGURE 4.2** Application of bionanotechnology in different areas.

Bionanotechnology is a unique synthesis of biotechnology and nanotechnology that allows traditional microtechnology to be combined with a molecular biological approach in the real world (Figure 4.2). By replicating or integrating biological systems, or by developing microscopic instruments to analyze or regulate numerous aspects of a biological system on a molecular scale, atomic or molecular grade machines can be built. By merging cutting-edge applications of information technology and nanotechnology with modern biological challenges, biotechnology may simplify several avenues of life sciences. This technology has the ability to straddle the line between biology, physics, and chemistry, as well as influence our existing notions and understanding towards green technology (Alshora et al., 2016).

Examples of bionanotechnological study include mechanical properties of materials, such as cell interaction with surfaces, nanopatterns and nanoparticles; electrical and optical effects, such as electrical stimulation, energy storage, absorption, luminescence, and fluorescence; and computing, via chemical "wetware" computers and DNA computing (Papazoglou and Parthasarathy, 2007).

Bionanotechnology is the intersection of biotechnology and nanotechnology, encompassing nanotechnology's use in the life sciences (Figure 4.2). This field of nanotechnology is currently widely employed in "Nanobiotechnology – Definition and Applications" (2021):

1. New molecular imaging techniques, which are used early to identify disease and study the consequences of treatments.
2. Quantitative analytical methods that show how a cell functions at the molecular level.
3. A physical model of the cell as a machine that can help us comprehend the disease's process and efficiently attack it.

4. Improved *ex vivo* procedures and existing laboratory techniques.
5. More effective drug delivery methods.

## 4.5 NANOTECHNOLOGY AND BIONANOTECHNOLOGY

Engineering and manufacturing at nanoscale scales with atomic accuracy is what nanotechnology is characterised as. The terms "molecular nanotechnology" and "molecular bionanotechnology" are often used interchangeably (Goodsell, 2004).

Bionanotechnology is a subset of nanotechnology in which the biological world serves as the source of inspiration and/or the final product. Nano-biomimetics (atomic-level engineering and manufacturing guided by biological precedent) or classical nanotechnology applied to biological and biomedical demands (Goodsell, 2004).

To get a sense of the relative size of a nanometer, compare some ordinary things with certain biological building blocks using a nano-ruler; micelles, glucose, nanoparticles, and hemoglobin are a few examples listed in Table 4.3 (Papazoglou and Parthasarathy, 2007). Human hair, for example, is 50,000 nanometers thick, but a glucose molecule is less than 1 nanometer thick. It's incredible that our metabolic functions are powered by a molecule 50,000 times smaller than a single strand of human hair. Table 4.3 compares bionanotechnology entities to everyday objects to provide a clear image of what it takes to be nano and to understand how small they are. Other nano creatures can be discovered in other places (Crandall, 1997).

**TABLE 4.3**
**Nanoparticles vs Macro World Matter**

| S. NO | Nano-ruler | | | |
|---|---|---|---|---|
| 1. | Glucose molecule diameter: 1 nm **1:** | Thickness of human hair 0.050 mm **50,000** | Thickness of human hair 0.050 mm **1:** | A man of 8.3 feet height **50, 000** |
| 2. | Gold nanoparticle of diameter: 8 mm **1:** | Apple of diameter: 8 cm **10,000,000** | Apple of diameter 8 cm **1:** | 65% of diameter of earth 0.65* 12,576 km = 8,170 km **10,000,000** |
| 3. | Micelle diameter: 13 cm **1:** | Soap bubble 1.3 cm **10,000,000** | Soap bubble of 1.3 cm **1:** | Diameter of earth 12,576 km |
| 4. | Quantum dot diameter 20 nm **1:** | Diameter of a cent 1.9 cm **1:** | Diameter of a cent 1.9 cm **1:** | 55% of diameter of moon 1,9050 km **10,000,000** |
| 5. | Hemoglobin diameter: 6.5 mm **1,000,000** | Riffle bullet of diameter: 6.5 mm **1,000,000** | Riffle bullet of diameter: 6.5 mm **1:** | A land of diameter 6.5 km (three times as big as Vatican City) **1,000,000** |

## 4.6 EFFECTS OF NANOPARTICLES ON THE ENVIRONMENT AND THE FOOD CHAIN

Consider the food-chain hierarchy depicted in Figure 4.3, which places man at the top. Many vital parts of the food chain are shown here, such as aquatic life, air, soil, and other elements in which living organisms of various levels flourish. There's a good chance that NPs will pollute the food chain and biosphere.

Pilot research by Oberdorster (2004) reveals that C60 may have negative effects on the brains of the fish who were exposed to it. The research findings showed a faster rate of fat oxidation in the fish brain, despite the fact that the results were at lower concentrations, and the study also suggests that NPs may have long-term detrimental impacts on aquatic life (Moore, 2006; Oberdorster, 2004). Furthermore, because the food-chain includes thousands of microbiological creatures, seaweeds, plants, and other aquatic components, contamination of the environment takes less time to harm people.

Nanoparticles have the potential to damage the environment due to their long-term persistence, bioaccumulation, and toxicity. This is why, although being non-toxic, inert nanoparticles may cause environmental harm due to their persistence or bioaccumulation potential.

Aside from toxicology research in people and animals, other elements of NP contamination, such as the physical and chemical characteristics of NPs exposed to diverse environmental circumstances, their capacity to react with the environment, their stability, their active lifetime, and their agglomeration capacities (Tsuji et al., 2006), among others, must be addressed.

Soil and water pollution caused by NPs poses a considerable risk to the environment and human health since it affects vegetables, fruits, and plants, among other things. Fertilizers containing NPs, which are supposed to enrich the soil, may have

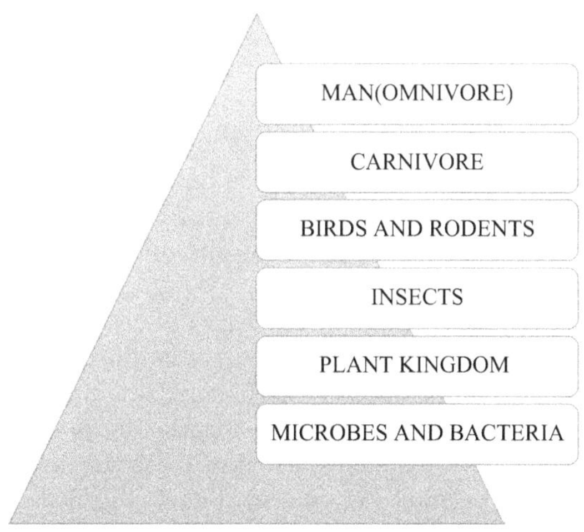


**FIGURE 4.3** A food chain diagram with man at the top of the hierarchy.

negative consequences as well. The poisoning of ground water by iron nanoparticles, for example, is significant (Zhang, 2003). As a result, environmental interactions as well as long-term harmful impacts must be investigated in soil studies and materials used as fertilizers and in remediation. The longer nanoparticles are allowed to collect undetected in the environment, the quicker their quantities may rise to dangerous levels (Park, 2007). For example, iron nanoparticles may move up to 20 meters in the groundwater table while being active for 4–8 weeks (Zhang, 2003), providing a potential hazard to the food chain members at any level.

## 4.7 POTENTIAL HAZARDS AND TOXICITY OF NANOMATERIAL

Nanoparticles have the same potential risks as particulate matter because of their extreme microscopic dimension, which provides a distinct benefit (Li et al., 2007). These particles have the potential to induce respiratory, cardiovascular, and gastrointestinal system diseases (Nijhara and Balakrishnan, 2006). Carbon nanotubes have the potential to produce a variety of lung diseases, including epithelioid granuloma, interstitial inflammation, peribronchial inflammation, and lung necrosis when injected intratracheally in mice. Carbon nanotube toxicity was reported to be higher than carbon black and quartz toxicity (Lam, 2003).

Nanomaterials have been proven to enter the human body through a variety of routes. Accidental or unintentional contact during manufacture or use is most likely to occur through the lungs, from which a fast translocation to other key organs via circulation is feasible. Nanoparticles have been shown to have the capacity to operate as a gene vector at the cellular level. Nanoparticles can enter the central nervous system via olfactory axons or systemic circulation via the olfactory bulb. The olfactory pathway has been found to accumulate carbon and manganese nanoparticles in the olfactory bulb in monkeys and rats. This demonstrates that nanoparticle-mediated administration may one day provide an alternative pathway for bypassing the blood-brain barrier. This can, however, result in inflammatory reactions/responses in the brain, which must be assessed (Sawicki et al., 2019).

Radomski et al. (2005) discovered that nanotubes exhibit pro-aggregatory effects on platelets *in vitro* and accelerate vascular thrombosis in rats. It was also shown that fullerenes could not cause platelet aggregation. As a result, fullerenes may be a safer option for building nanoparticle-based medication delivery systems than nanotubes (Medina et al., 2007). The risk and challenges associated with use of bionanotechnology in the development of different products are illustrated in Figure 4.4.

Nanoparticle toxicity can be extended to the gastrointestinal tract, causing inflammatory bowel illness. Nanoparticle toxicity might be linked to their propensity to trigger the release of pro-inflammatory mediators, resulting in an inflammatory response and organ damage. If consumed, nanoparticles can enter the bloodstream and travel to many organs and systems, potentially causing poisoning (Chen et al., 2006). These have been examined *in vitro* in animal models, but extrapolating the effects to the human system is problematic. Further research and prudence are required before they may be used in common masses.

Challenges in developing bionanotechnology based products

- Safety issues
- Commercialization
- Disposal of nanowaste
- Regulatory guidelines
- Social responsibility
- Ethical considerations
- Risk assesment
- Environmental issues

**FIGURE 4.4** Challenges associated with the use of bionanotechnology in development of different products.

## 4.8 FUTURE PERSPECTIVE OF THE BIO-NANO TECHNIQUES

The potential ramifications of bionanotechnology are hotly debated. There is great potential for its use in a variety of innovative materials and gadgets that could be valuable in fields such as medicine, electronics, biomaterials, and energy generation. Nonetheless, as with any new technology, this method poses a number of challenges, including concerns about toxicity of nanoparticles and environmental impacts (Buzea et al., 2007), as well as their possible implications on world economies and conjecture about numerous apocalyptic scenarios. These issues have sparked a discussion among advocacy organizations and governments over whether nanobiotechnology requires statutory control.

Despite certain disagreements, this technology provides tremendous optimism for the future. It might spark new ideas by playing a key role in a variety of medicinal applications, including medication delivery and gene therapy, as well as molecular imaging, biomarkers, and biosensors. Target-specific medication therapy and approaches for early identification and treatment of illnesses are two of these applications that are now the focus of study (Sahoo and Labhasetwar, 2003). Both in clinical diagnostics and R&D, two sorts of medical applications are currently surfacing: applications for imaging, such as quantum dot technology, have already been licensed, and applications for monitoring cellular processes in the tissue will be forthcoming shortly; the development of extremely precise and sensitive methods of detecting nucleic acids and proteins is the second major kind of application (Thomas, 2006). We will see the products evaluated at bench scale making their way into commercialization between 2015 and 2020. Applications for sparse cell isolation and molecular filtering should be available by then. Some medication delivery systems should be available for purchase or be in advanced clinical studies. For example, Nano Systems and American Pharmaceutical Partners have created drug delivery systems and are currently investigating the encapsulation of Taxol, a cancer medication, in a nano polymer called paclitaxel. Most medical gadgets and medicines are still a decade or more away from commercialization. As a result, manipulating medication targets as well as implanting devices necessitates a complicated technical infrastructure, such as nanotechnology, as well as extensive regulatory management (Hamad-Schifferli et al., 2002).

The discharge of recalcitrant or pseudo-persistent chemicals is increasing day by day in the environment. Low concentration contaminants are the real challenge, with long- and short-term harmful impacts on biota and humans. Existing conventional wastewater treatment technologies are able to decrease the organic and inorganic load in the wastewater but are unable to treat the emerging contaminants effectively. Certain biological and advanced oxidation processes are successful in eliminating the ECs with 90% removal efficiency. However, these processes have their own set of constraints, such as membrane fouling, the large amount of sludge production, use of chemicals for treatment, economic and on-field feasibility, energy consumption, and operation maintenance of the process. Therefore, nano techniques have emerged to solve the above possible disadvantages for the treatment of ECs. These techniques are highly selective, with large surface-area-to-volume ratio with high adsorption capability, and reversible with economic feasibility. Hence, combining the two eco-friendly technologies give rise to bio-nano techniques. The technology offers many advantages, such as facilitating the EC removal by enhancing microbial growth; inducing the remediating enzymes of microbes for the target compounds; helping to improve the solubility and absorption of the contaminant over the bio-nano materials, thereby further degrading the ECs effectively from soil and water matrices.

Concluding the future perspectives of the bio-nano techniques, the major findings and suggestions are as follows:

1. Though the technique is more efficient for the removal of different ECs, the structural integrity of the membrane and the microbes has to be further explored.
2. Toxicity evaluation of the microorganisms used at variable concentration using various classes of ECs is required.
3. Improvements in terms of reusability, material use in terms of cost and its functionalities should be computed.
4. Studies are required to investigate the efficiency, microbial and nanomaterial properties, and characteristics for the treatment of a mixture of ECs.

## 4.9 CONCLUSION

In light of future environmental and health safety legislation, especially for wastewater recovery and recycling, trace contaminants must be removed. Conventional wastewater techniques are still prevalent in the eradication of the ECs, hence the techniques must be replaced with more technological and economically feasible methods. Bio-nano techniques have wide application in the removal of ECs but are restricted to batch scale. Hence, the lab-to-field option must be explored, creating a synergistic approach including the nanomaterials and the biological organism for treatment of the natural environment. Researchers must be motivated to design more biodegradable and cost-effective nanomaterials and hybrid technologies, thereby increasing the awareness among the government agencies and public for the need of a strict regulatory framework for the type of nanomaterials and the presence of the emerging contaminants in the environment.

## REFERENCES


Alshora, D.H., Ibrahim, M.A., and Alanazi, F.K. (2016). *Nanotechnology from Particle Size Reduction to Enhancing Aqueous Solubility. Surface Chemistry of Nanobiomaterials: Applications of Nanobiomaterials.* William Andrew Publishing, pp. 163–191. https://doi.org/10.1016/B978-0-323-42861-3.00006-6.

Anastopoulos, I., Katsouromalli, A., and Pashalidis, I. (2020). Oxidized Biochar Obtained from Pine Needles as a Novel Adsorbent to Remove Caffeine from Aqueous Solutions. *J Mol Liq.* 304, 112661. https://doi.org/10.1016/j.molliq.2020.112661.

Baccio, D. Di., Pietrini, F., Bertolotto, P., Perez, S., Barcelo, D., Zacchini, M., and Donati, E. (2017). Response of Lemna Gibba L. to High and Environmentally Relevant Concentrations of Ibuprofen: Removal, Metabolism and Morpho-Physiological Traits for Biomonitoring of Emerging Contaminants. *Sci Total Environ.* https://.doi.org/10.1016/j.scitotenv.2016.12.191.

Barroso, P. J., Santos, J. L., Martin, J., Aparicio, I., and Alonso, E. (2019). Emerging Contaminants in the Atmosphere: Analysis, Occurrence and Future Challenges. *Crit Rev Env Sci Tec.* 49(2), 1–68. https://doi.org/10.1080/10643389.2018.1540761.

Bedia, J., Belver, C., Ponce, S., Rodriguez, J., and Rodriguez, J.J. (2018). Adsorption of Antipyrine by Activated Carbons from FeCl3-Activation of Tara Gum. *Chem Eng J.* 333(1), 58–65. https://doi.org/10.1016/j.cej.2017.09.161.

Beshaa, A.T., Gebreyohannes, A.Yo., Tufa, R. A., Bekele, D.N., Curcio, E., and Giorno, L. (2017). Removal of Emerging Micropollutants by Activated Sludge Process and Membrane Bioreactors and the Effects of Micropollutants on Membrane Fouling: A Review. *J Environ Chem Eng.* 5(3), 2395–2414. https://doi.org/10.1016/j.jece.2017.04.027.

Bilal, M., Adeel, M., Rasheed, T., Zhao, Y., and Iqbal, H.M.N. (2019). Emerging Contaminants of High Concern and Their Enzyme-Assisted Biodegradation: A Review. *Environ Int.* 124, 336–53. https://doi.org/10.1016/j.envint.2019.01.011.

Buzea, C., Pacheco, I.I., and Robbie, K., (2007). Nanomaterials and Nanoparticles: Sources and Toxicity. *Biointerphases.* 2, MR17–MR71. https://doi.org/10.1116/1.2815690.

Chang, Ellen T., Adami, H., Boffetta, P., Cole, P., Starr, T. B., and Mandel, J.S. (2014). A Critical Review of Perfluorooctanoate and Perfluorooctanesulfonate Exposure and Cancer Risk in Humans. *Crit Rev Toxicol.* 8444(S1), 1–81. https://doi.org/10.3109/10408444.2014.905767.

Chen, H., Shepsko, C., and Sengupta, A.K. (2021). Use of a Novel Bio-Nano-IX Process to Remove SeO $42^{-}$ or Se (VI) from Contaminated Water in the Presence of Competing Sulfate (SO $42^{-}$ ). *Water.* 1, 1859–67. https://doi.org/10.1021/acsestwater.1c00126.

Chen, Z., Meng, H., Xing, G., Chen, C., Zhao, Y., Jia, G., Wang, T., Yuan, H., Ye, C., Zhao, F., Chai, Z., Zhu, C., Fang, X., Ma, B., and Wan, L. (2006). Acute Toxicological Effects of Copper Nanoparticles in vivo. *Toxicol Lett.* 163, 109–120. https://doi.org/10.1016/j.toxlet.2005.10.003.

Choi, Y.K., Choi, T.R., Gurav, R., Bhatia, S.K., Park, YL., Kim, H.J., Kan, E., and Yang, Y.H. (2020). Adsorption Behavior of Tetracycline onto Spirulina Sp. (Microalgae)-Derived Biochars Produced at Different Temperatures. *Sci Total Environ.* 710, 136282. https://doi.org/10.1016/j.scitotenv.2019.136282.

Chouhan, S., Yadav, S.K., Prakash, J., Swati, S, and Surya P. (2013). Effect of Bisphenol A on Human Health and Its Degradation by Microorganisms: A Review. *Annual Microbiol.* 64, 13–21. https://doi.org/10.1007/s13213-013-0649-2.

Christofilopoulos, S., Kaliakatsos, A., Triantafyllous, K., Gounaki, I., Venieri, D., and Kalogerakis, Ns. (2019). Evaluation of a Constructed Wetland for Wastewater Treatment: Addressing Emerging Organic Contaminants and Antibiotic Resistant Bacteria. *New Biotechnol.* 52, 94–103. https://doi.org/10.1016/j.nbt.2019.05.006.

Crandall, B.C. (1997). Nanotechnology: Molecular Speculation on Global Abundance. 34, 3331. https://doi.org/10.5860/choice.34-3331.

de Souza dos Santos, Grazielle E., Ide, A.H., Duarte, J.L.S., McKay, G., Silva, Antonio O. S., and Meili, L. (2020). Adsorption of Anti-Inflammatory Drug Diclofenac by MgAl/ Layered Double Hydroxide Supported on Syagrus Coronata Biochar. *Powder Technol.* 364, 229–40. https://doi.org/10.1016/j.powtec.2020.01.083.

Delgado, N., Capparelli, A., Navarro, A., and Marino, D. (2019). Pharmaceutical Emerging Pollutants Removal from Water Using Powdered Activated Carbon: Study of Kinetics and Adsorption Equilibrium. *J Environ Manage.* 236, 301–8. https://doi.org/10.1016/j.jenvman.2019.01.116.

Díaz-Garduño, B., Pintado-Herrera, M.G., Biel-Maseo, M., Rueda-Marquez, J.J., Lara-Martin, P.A., Perales, J.A., Manzano, M.A., Garrido-Perez, C., and Martin-Diaz, M.L. (2017). Environmental Risk Assessment of Effluents as a Whole Emerging Contaminant: Efficiency of Alternative Tertiary Treatments for Wastewater Depuration. *Water Res.*, 119, 136–149. https://doi.org/10.1016/j.watres.2017.04.021.

Domene, X., Ramirez, W., Sola, L., Alcaniz, J. M., and Andres, P. (2009). Soil Pollution by Nonylphenol and Nonylphenol Ethoxylates and Their Effects to Plants and Invertebrates. *J Soils Sediments.* 9, 555–67. https://doi.org/10.1007/s11368-009-0117-6.

Edgar, V.N., Molina-guerrero, Carlos E., Pena-Castro, J.M., Fernandez-Luqueno, F., and Rosa-Alvarez, Ma Guadalupe de la. (2020). Use of Nanotechnology for the Bioremediation of Contaminants: A Review. *Processes.* 8(826), 1–17. https://doi.org/10.3390/pr8070826.

Ensano, B.B., Borea, L., Naddeo, V., Belgiorno, V., Luna, Mark Daniel G.D., and Ballesteros, F.C. (2017). Removal of Pharmaceuticals from Wastewater by Intermittent Electrocoagulation. *Water-SUI.* 9(2), 85. https://doi.org/10.3390/w9020085.

Esplugas, S., Bila, Daniele M., Krause, Luiz Gustavo T., and Dezotti, M. (2007). Ozonation and Advanced Oxidation Technologies to Remove Endocrine Disrupting Chemicals (EDCs) and Pharmaceuticals and Personal Care Products (PPCPs) in Water Effluents. *J Hazard Mater.* 149, 631–42. https://doi.org/10.1016/j.jhazmat.2007.07.073.

Fakruddin, M., Hossain, Z., and Afroz, H. (2012). Prospects and Applications of Nanobiotechnology: A Medical Perspective. *J Nanobiotechnol.* 10, 31. https://doi.org/10.1186/1477-3155-10-31.

Feijoo, G., Moreira, M T., and Lema, J M. (2011). Degradation of Selected Pharmaceutical and Personal Care Products (PPCPs) by White-Rot Fungi. *World J Microb Biotechnol.* 27, 1839–46. https://doi.org/10.1007/s11274-010-0642-x.

Franka, A., Seitz, W., Prasse, C., Lucke, T., Schulz, W., and Ternes, T. (2016). Occurrence and Fate of Amisulpride, Sulpiride, and Lamotrigine in Municipal Wastewater Treatment Plants with Biological Treatment and Ozonation. *J Hazard Mater.* 320, 204–15. https://doi.org/10.1016/j.jhazmat.2016.08.022.

Fröhlich, A.C., Foletto, Edson L., and Dotto, Guilherme L. (2019). Preparation and Characterization of NiFe2O4/Activated Carbon Composite as Potential Magnetic Adsorbent for Removal of Ibuprofen and Ketoprofen Pharmaceuticals from Aqueous Solutions. *J Clean Prod.* 229, 828–37. https://doi.org/10.1016/j.jclepro.2019.05.037.

Gani, Khalid Muzamil., and Kazmi, Absar Ahmad. (2016). Contamination of Emerging Contaminants in Indian Aquatic Sources: First Overview of the Situation. *J Hazard Toxic Radioact Waste.* 1–12. https://ascelibrary.org/doi/abs/10.1061/(ASCE)HZ.2153-5515.0000348.

García-Mateos, F J., Ruiz-Rosas, R., Marques, M.D., Cotoruelo, L.M., Rodriguez-Mirasol, J., and Cordero, T. (2015). Removal of Paracetamol on Biomass-Derived Activated Carbon: Modeling the Fixed Bed Breakthrough Curves Using Batch Adsorption Experiments. *Chem Eng J.* 279, 18–30. https://doi.org/10.1016/j.cej.2015.04.144.

Gedda, Gangaraju., Balakrishna, Kolli, Devi, Randhi Uma, and Shah, Kinjal J. (2021). Chapter 1 *Introduction to Conventional Wastewater Treatment Technologies: Limitations and Recent Advances.* https://doi.org/10.21741/9781644901144-1.

Gogoi, A., Mazumder, Pl., Tyagi, Vinay K., Chaminda, G.G. T., An, Alicia K., and Kumar, M. (2018). Occurrence and Fate of Emerging Contaminants in Water Environment: A Review. *Groundw Sustain Dev.* 6, 169–80. https://doi.org/10.1016/j.gsd.2017.12.009.

Gomes, A. R., Justino, C., Rocha-Santos, T., Fretias, Ana C., Duarte, Armando C., and Pereira, R. (2017). Review of the Ecotoxicological Effects of Emerging Contaminants on Soil Biota. *J Environ Sci Health A.* 0, 1–16. https://doi.org/10.1080/10934529.2017.1328946.

Goodsell, D.S., (2004). *Bionanotechnology.* John Wiley & Sons, Inc. https://doi.org/10.1002/0471469572.

Grandclement, C., Seyssiecq, Ie., Piram, A., Wong-Wah-Chung, P., Vanot, G., Tiliacos, Nicolas., and Doumenq, P. (2017). From the Conventional Biological Wastewater Treatment to Hybrid Processes, the evaluation of organic micropollutant removal. *Water Res.* 111, 297–317. https://doi.org/10.1016/j.watres.2017.01.005.

Grégorio, Crini., and Lichtfouse, Eric. (2018). Advantages and Disadvantages of Techniques Used for Wastewater Treatment. *Environ Chem Lett.* 17(1), 145–55. https://doi.org/10.1007/s10311-018-0785-9.

Gwenzi, Willis., Mangori, Lynda., Danha, Concilia., Chaukura, Nhamo., Dunjana, Nothando., and Sanganyado, Edmond. (2018). Sources, Behaviour, and Environmental and Human Health Risks of High- Technology Rare Earth Elements as Emerging Contaminants. *Sci Total Environ.* 636, 299–313. https://doi.org/10.1016/j.scitotenv.2018.04.235.

Hallare, A V., Kohler, H.R., and Triebskorn, R. (2004). Developmental Toxicity and Stress Protein Responses in Zebrafish Embryos after Exposure to Diclofenac and Its Solvent, DMSO. *Chemosphere.* 56, 659–66. https://doi.org/10.1016/j.chemosphere.2004.04.007.

Hamad-Schifferli, K., Schwartz, J.J., Santos, A.T., Zhang, S., and Jacobson, J.M. (2002). Remote Electronic Control of DNA Hybridization Through Inductive Coupling to an Attached Metal Nanocrystal Antenna. *Nature.* 415, 152 155. https://doi.org/10.1038/415152a.

Hayes, T., Haston, K., Tsui, M., Hoang, A., Haeffele, C., and Vonk, A. (2003). Atrazine-Induced Hermaphroditism at 0.1 Ppb in American Leopard Frogs. *Environ Health Persp.* 111(4), 568–75. https://doi.org/10.1289%2Fehp.5932.

Hwang, Soon-Seok., Choi, Hyoung Tae., and Song, H.G. (2008). Biodegradation of Endocrine-Disrupting Phthalates by Pleurotus Ostreatus. *J Microbiol Biotechn.* 18(4), 767–72. https://www.jmb.or.kr/submission/Journal/018/JMB018-04-24.pdf.

Jung, Chanil., Oh, Jeill., and Yoon, Yeomin. (2015). Removal of Acetaminophen and Naproxen by Combined Coagulation and Adsorption Using Biochar: Influence of Combined Sewer Overflow Components. *Environ Sci Pollut R.* 22(13), 10058–69. https://doi.org/10.1007/s11356-015-4191-6.

Krause, M., Klit, A., Jensen, M B., Soeborg, T., Frederiksen, H., Schlumpf, M., Lichtensteiger, W., Skakkebaek, N.E., and Drzewiecki, K.T. (2012). Sunscreens: Are They Beneficial for Health? An Overview of Endocrine Disrupting Properties of UV-Filters. *Int J Androl.* 35(191), 424–36. https://doi.org/10.1111/j.1365-2605.2012.01280.x.

Kwak, K., Ji, K., Kho, Y., Kim, P., Lee, Jaean, Ryu, Jg., and Choi, K. (2018). Chronic Toxicity and Endocrine Disruption of Naproxen in Freshwater Water Fleas and Fish, and Steroidogenic Alteration Using H295R Cell Assay. *Chemosphere.* 204, 156–62. https://doi.org/10.1016/j.chemosphere.2018.04.035.

Kwon, Bareum, Kho, Younglim, Kim, Pan-gyi, and Ji, K. (2016). Thyroid Endocrine Disruption in Male Zebrafish Following Exposure to Binary Mixture of Bisphenol AF and Sulfamethoxazole. *Environ Toxicol Phar.* 48, 168–74. https://doi.org/10.1016/j.etap.2016.10.018.

Lam, C.-W. (2003). Pulmonary Toxicity of Single-Wall Carbon Nanotubes in Mice 7 and 90 Days After Intratracheal Instillation. *Toxicol Sci.* 77, 126–134. https://doi.org/10.1093/toxsci/kfg243.

Lee, I., Lee, Sung-hee, and Oh, J. (2010). Occurrence and Fate of Synthetic Musk Compounds in Water Environment. *Water Res.* 44(1), 214–22. https://doi.org/10.1016/j.watres.2009.08.049.

Lemos, Marco F L., Gestel, Cornelis A.M.V., and Soares, M.V.M.A. (2009). Endocrine Disruption in a Terrestrial Isopod under Exposure to Bisphenol A and Vinclozolin. *J Soil Sediments.* 9, 492–500. https://doi.org/10.1007/s11368-009-0104-y.

Li, Z., Hulderman, T., Salmen, R., Chapman, R., Leonard, S.S., Young, S.-H., Shvedova, A., Luster, M.I., and Simeonova, P.P. (2007). Cardiovascular Effects of Pulmonary Exposure to Single-Wall Carbon Nanotubes. *Environ Health Perspect.* 115, 377–382. https://doi.org/10.1289/ehp.9688.

Liu, Z., Kanjo, Y., and Mizutani, S. (2008). Removal Mechanisms for Endocrine Disrupting Compounds (EDCs) in Wastewater Treatment: Physical Means, Biodegradation, and Chemical Advanced Oxidation: A Review. *Sci Total Environ.* 407(2), 731–48. https://doi.org/10.1016/j.scitotenv.2008.08.039.

Lofrano, Giusy et al. (2020). Visible Light Active Structured Photocatalysts for the Removal of Emerging Contaminants Occurrence and Potential Risks of Emerging Contaminants in Water. *Elsevier Inc.* https://doi.org/10.1016/B978-0-12-818334-2.00001-8.

Matamoros, Vr., Rodríguez, Yolanda., and Bayona, Josep M. (2017). Mitigation of Emerging Contaminants by Full-Scale Horizontal Flow Constructed Wetlands Fed with Secondary Treated Wastewater. *Ecol Eng.* 99, 222–27. https://doi.org/10.1016/j.ecoleng.2016.11.054.

Medina, C., Santos-Martinez, M.J., Radomski, A., Corrigan, O.I., and Radomski, M.W. (2007). Nanoparticles: pharmacological and toxicological significance. *Br J Pharmacol.* 150, 552–558. https://doi.org/10.1038/sj.bjp.0707130.

Mohapatra, D P., Cledon, M., Brar, S K., and Surampalli, R Y. (2016). Application of Wastewater and Biosolids in Soil: Occurrence and Fate of Emerging Contaminants. *Water Air Soil Poll.* 227, 77, 1–14. https://doi.org/10.1007/s11270-016-2768-4.

Moore, M.N. (2006). Do nanoparticles present ecotoxicological risks for the health of the aquatic environment? *Environ Int.* 32, 967–976. https://doi.org/10.1016/j.envint.2006.06.014.

Mulla, Sikandar I., Wang, Han., Sun, Qian., Hu, Anyi., and Yu, Chang-Ping. (2016). *Characterization of Triclosan Metabolism in Sphingomonas Sp.* Nature Publishing Group. 1–11. https://doi.org/10.1038/srep21965.

Naddeo, V., Secondes, Mona Freda N., Borea, Laura., Hasan, Shadi W., Jr., Florencio, Ballesteros, and Belgiorno, V. (2020). Removal of Contaminants of Emerging Concern from Real Wastewater by an Innovative Hybrid Membrane Process – UltraSound, Adsorption, and Membrane Ultrafiltration (USAMe®). *Ultrason Sonochem.* 68, 105237. https://doi.org/10.1016/j.ultsonch.2020.105237.

Naidu, R., Espana, Victor Andreas, Liu, Yanju, and Jit, J. (2016). Emerging Contaminants in the Environment: Risk-Based Analysis for Better Management. *Chemosphere.* 154, 350–57. https://doi.org/10.1016/j.chemosphere.2016.03.068.

Nanobiotechnology: Definition and Applications [WWW Document]. (2021). URL https://www.nanowerk.com/nanobiotechnology.php (accessed 10.25.21).

National Nanotechnology Initiative. (2021). What is the NNI? | National Nanotechnology Initiative [WWW Document]. URL https://www.nano.gov/nanotech-101/what/definition (accessed 10.25.21).

Nijhara, R., and Balakrishnan, K. (2006). Bringing Nanomedicines to Market: Regulatory Challenges, Opportunities, and Uncertainties. *Nanomed Nanotechnol Biol Med.* 2, 127–136. https://doi.org/10.1016/j.nano.2006.04.005.

Norvill, Zane N., Shilton, A., and Guieysse, B. (2016). Emerging Contaminant Degradation and Removal in Algal Wastewater Treatment Ponds: Identifying the Research Gaps. *J Hazard Mater.* 313, 291–303. https://doi.org/10.1016/j.jhazmat.2016.03.085.

Oberdörster, E. (2004). Manufactured Nanomaterials (Fullerenes, C 60) Induce Oxidative Stress in the Brain of Juvenile Largemouth Bass. *Environ Health Perspect.* 112, 1058–1062. https://doi.org/10.1289/ehp.7021.

Paliya, S., Mandpe, A., Bombaywala, S., Kumar, M.S., Kumar, S., and Morya, V.K. (2021a). Polybrominated diphenyl ethers in the environment: A wake-up call for concerted action in India. *Environ. Sci. Pollut. Res.,* https://doi.org/10.1007/s11356-021-15204-7

Paliya, S., Mandpe, A., Kumar, M.S., and Kumar, S., (2021b). Aerobic degradation of decabrominated diphenyl ether through a novel bacterium isolated from municipal waste dumping site: Identification, degradation and metabolic pathway. *Bioresour. Technol.,* 333, 125–208. https://doi.org/10.1016/j.biortech.2021.125208.

Pamphile, N., Xuejiao, L., Guangwei, Y., and Yin, W. (2019). Synthesis of a Novel Core-shell-structure Activated Carbon Material and its Application in Sulfamethoxazole Adsorption. *J Hazard Mater.* 368, 602–12. https://doi.org/10.1016/j.jhazmat.2019.01.093.

Papazoglou, E.S., and Parthasarathy, A. (2007). BioNanotechnology. *Synth Lect Biomed Eng.* https://doi.org/10.2200/S00051ED1V01Y200610BME007.

Park, B. (2007). Chapter 1. Current and Future Applications of Nanotechnology. Nanotechnology: Consequences for Human Health and the Environment. *The Royal Society of Chemistry*, 24, 1–18. https://doi.org/10.1039/9781847557766-00001.

Racar, M., Dolar, D., Karadakic, K., Cavarovic, N., Glumac, N., Asperger, D., and Kosutic, K. (2020). Challenges of Municipal Wastewater Reclamation for Irrigation by MBR and NF/RO: Physico-Chemical and Microbiological Parameters, and Emerging Contaminants. *Sci Total Environ.* 137959. https://doi.org/10.1016/j.scitotenv.2020.137959.

Radomski, A., Jurasz, P., Alonso-Escolano, D., Drews, M., Morandi, M., Malinski, T., and Radomski, M.W. (2005). Nanoparticle-induced platelet aggregation and vascular thrombosis. *Br J Pharmacol.* 146, 882–893. https://doi.org/10.1038/sj.bjp.0706386.

Rajasulochana, P., and Preethy, V. (2016). Comparison on Efficiency of Various Techniques in Treatment of Waste and Sewage Water: A Comprehensive Review. *Resource-Efficient Technologies.* https://doi.org/10.1016/j.reffit.2016.09.004.

Riva, F., Castiglioni, S., Fattore, E., Manenti, A., Davoli, E., and Zuccato, E. (2018). Monitoring Emerging Contaminants in the Drinking Water of Milan and Assessment of the Human Risk. *Int J Hyg Envir Heal.* 0–1. https://doi.org/10.1016/j.ijheh.2018.01.008.

Roccaro, P. (2018). Treatment Processes for Municipal Wastewater Reclamation: The Challenges of Emerging Contaminants and Direct Potable Reuse. *Curr Opin Environ Sci Health.* https://doi.org/10.1016/j.coesh.2018.02.003.

Rodriguez, Oscar, Peralta-hernandez, Juan Manuel, Goonetilleke, Ashantha, and Bandala, Erick R. (2017). Treatment Technologies for Emerging Contaminants in Water: A Review. *Chem Eng J.* https://doi.org/10.1016/j.cej.2017.04.106.

Rogers, Howard R. (1996). Sources, Behaviour and Fate of Organic Contaminants during Sewage Treatment and in Sewage Sludges. *Sci Total Environ.* 185, 3–26. https://doi.org/10.1016/0048-9697(96)05039-5.

Rout, Prangya R., Zhang, Tian C., Bhunia, P., and Surampalli, Rao Y. (2021). Treatment Technologies for Emerging Contaminants in Wastewater Treatment Plants: A Review. *Sci Total Environ.* 753, 141990. https://doi.org/10.1016/j.scitotenv.2020.141990.

Sahoo, S.K., and Labhasetwar, V. (2003). Nanotech Approaches to Drug Delivery and Imaging. *Drug Discov Today.* 8, 1112–1120. https://doi.org/10.1016/S1359-6446(03)02903-9.

Sawicki, K., Czajka, M., Matysiak-Kucharek, M., Fal, B., Drop, B., Męczyńska-Wielgosz, S., Sikorska, K., Kruszewski, M., and Kapka-Skrzypczak, L. (2019). Toxicity of Metallic Nanoparticles in the Central Nervous System. *Nanotechnol Rev.* 8, 175–200. https://doi.org/10.1515/ntrev-2019-0017.

Schwaiger, J., Ferling, H., Mallow, U., Wintermayr, H., and Negele, R D. (2004). Toxic Effects of the Non-Steroidal Anti-Inflammatory Drug Diclofenac Part I: Histopathological Alterations and Bioaccumulation in Rainbow Trout. *Aquat Toxicol.* 68, 141–50. https://doi.org/10.1016/j.aquatox.2004.03.014.

Sekulić, M.T., Pap, S., Stojanovic, Z., Boskovic, N., Radonic, J., and Knudsen, T.S. (2018). Efficient Removal of Priority, Hazardous Priority and Emerging Pollutants with Prunus Armeniaca Functionalized Biochar from Aqueous Wastes: Experimental Optimization and Modeling. *Sci Total Environ.* 614, 736–50. https://doi.org/10.1016/j.scitotenv.2017.09.082.

Shrestha, S., Bloom, M.S., Yucel, R., Seegal, R.F., Wu, Q., Kannan, K., Rej, R., and Fitzgerald, E.F. (2015). Perfluoroalkyl Substances and Thyroid Function in Older Adults. *Environ Int.* 75, 206–14. https://doi.org/10.1016/j.envint.2014.11.018.

Sui, Qian, and Huang, Jun, Deng, Shubo, Chen, Weiwei, and Yu, Gang. (2011). Seasonal Variation in the Occurrence and Removal of Pharmaceuticals and Personal Care Products in Different Biological Wastewater Treatment Processes. *Environ Sci Technol.* 3341–48. https://doi.org/10.1021/es200248d.

Thomaidi, Vasiliki S., Stasinakis, Athanasios S., Borova, Viola L., and Thomaidis, Nikolaos S. (2016). Assessing the Risk Associated with the Presence of Emerging Organic Contaminants in Sludge-Amended Soil: A Country-Level Analysis. *Sci Total Environ.* 548–549, 280–88. https://doi.org/10.1016/j.scitotenv.2016.01.043.

Thomas, J. (2006). An Introduction to Nanotechnology: The Next Small Big Thing. *Development.* 49, 39–46. https://doi.org/10.1057/palgrave.development.1100315.

Tiwari, Bhagyashree, Sellamuthu, Balasubramanian, Ouarda, Yassine, Drogui, Patrick, Tyagi, Rajeshwar D., and Buelna, Gerardo. (2017). Review on Fate and Mechanism of Removal of Pharmaceutical Pollutants from Wastewater Using Biological Approach. *Bioresource Technol.* 224, 1–12. https://doi.org/10.1016/j.biortech.2016.11.042.

Tsuji, J.S., Maynard, A.D., Howard, P.C., James, J.T., Lam, C., Warheit, D.B., and Santamaria, A.B. (2006). Research Strategies for Safety Evaluation of Nanomaterials, Part IV: Risk Assessment of Nanoparticles. *Toxicol Sci.* 89, 42–50. https://doi.org/10.1093/toxsci/kfi339.

Ullah, S., Zuberi, A., Alagawany, M., Farag, Mayada R., Dadar, Maryam., Karthik, K., Tiwari, R., Dhama, K., and Iqbal, Hafiz M.N. (2018). Cypermethrin Induced Toxicities in Fish and Adverse Health Outcomes: Its Prevention and Control Measure Adaptation. *J Environ Manage.* 206, 863–71. https://doi.org/10.1016/j.jenvman.2017.11.076.

Vasilachi, Ionela C., Asiminicesei, Dana M., Fertu, D.I., and Gavrilescu, M. (2021). Occurrence and Fate of Emerging Pollutants in Water Environment and Options for Their Removal. *Water.* 13(181), 1–34. https://doi.org/10.3390/w13020181.

Verstraete, Sofia G., Wojcicki, Janet M., Perito, Emily R., and Rosenthal, Philip. (2018). Bisphenol a Increases Risk for Presumed Non-Alcoholic Fatty Liver Disease in Hispanic Adolescents in NHANES 2003–2010. *Environ Health.* 1–8. https://doi.org/10.1186/s12940-018-0356-3.

Wang, J., and Wang, S. (2017). Activation of Persulfate (PS) and Peroxymonosulfate (PMS) and Application for the Degradation of Emerging Contaminants. *Chem Eng J.* https://doi.org/10.1016/j.cej.2017.11.059.

Wang, Y., Liu, H., Peng, Y., Tong, L., Feng, L., and Ma, K. (2018). New Pathways for the Biodegradation of Diethyl Phthalate by Sphingobium Yanoikuyae SHJ. *Process Biochem.* https://doi.org/10.1016/j.procbio.2018.05.010.

Warne, Michael St. J., and Hawker, Darryl W. (1995). The Number of Components in a Mixture Determines Whether Synergistic and Antagonistic or Additive Toxicity Predominate: The Funnel Hypothesis. *Ecotox Environ Safe.* 31, 23–28. https://doi.org/10.1006/eesa.1995.1039.

Wilkinson, J., Hooda, Peter S., Barker, J., Barton, S., and Swinden, J. (2017). Occurrence, Fate and Transformation of Emerging Contaminants in Water: An Overarching Review of the Field. *Environmental Pollution.* 231, 954–70. https://doi.org/10.1016/j.envpol.2017.08.032.

Wilkinson, J. L., Hooda, P. S., Barker, J., Barton, S., and Swinden, J. (2015). Ecotoxic Pharmaceuticals, Personal Care Products and other Emerging Contaminants: A Review of Environmental, Receptor-mediated, Developmental, and Epigenetic Toxicity with Discussion of Proposed Toxicity to Humans. *Crit Rev Env Sci Tec.* 3389. https://doi.org/10.1080/10643389.2015.1096876.

Xiong, J., Kurade, M.B., and Jeon, B.H. (2017). Can microalgae remove pharmaceutical contaminants from water? *Trends Biotechnol.* 1–15. https://doi.org/10.1016/j.tibtech.2017.09.003.

Yang, Y., Ok, Y.S., Kim, K.-H., Kwon, E.E., and Tsang, Y.F. (2017). Occurrences and Removal of Pharmaceuticals and Personal Care Products (PPCPs) in Drinking Water and Water / Sewage Treatment Plants: A Review. *Sci Total Environ.* 596–597, 303–20. https://doi.org/10.1016/j.scitotenv.2017.04.102.

Zhang, W. (2003). Nanoscale Iron Particles for Environmental Remediation: An Overview. *J Nanoparticle Res.* 5, 323–332. https://doi.org/10.1023/A:1025520116015.

# 5 Eco-design and Modification Study of Bionanoparticles and Life Cycle Assessment

## *Keys to Sustainability*

*Naveen Dwivedi, Shubha Dwivedi, and Deepa Sharma*

## CONTENTS



## 5.1 INTRODUCTION

Human thoughts and imagination frequently give rise to new knowledge and skills. Nanotechnology is a boon for the 21st century frontier. Nanotechnology is a branch of technology that encompasses understanding and control of matter at dimensions between 1 and 100 nm, where exceptional phenomena enable novel applications. In 1925, the Nobel Prize Laureate in chemistry, Richard Zsigmondy, first proposed the

DOI: 10.1201/9781003270959-5

concept of a "nanometer." The term "nanometer" was coined by him and clearly refers to particle size characterization and he was the first to quantity the size of particles, such as gold colloids, using a microscope. The prefix 'nano' comes from Greek, a prefix meaning "dwarf" or something very small and depicts one thousand millionth of a meter ($10^{-9}$ m).

Nanotechnology advancements are present in almost every field of science, and nanotechnology makes life easier and more convenient in this age. Nanotechnology epitomizes an increasing research area, which includes structures, devices, and systems with innovative properties and functions, due to the organization of their atoms on the 1–100 nm scale. The field became the focus of an emerging public awareness and debate in the early 2000s; and in turn, it was the early period of the production of commercial applications of nanotechnology. Nanotechnology contributes to practically every field of science, including, climate change, water remediation, agriculture, physics, chemistry, materials sciences, biology, computer science, and engineering. Today, nanotechnology influences everyday human life. The key benefits are many and miscellaneous, although, because of extensive human exposure to nanoparticles, there is a substantial worry about the potential health and environmental dangers. These worries led to the development of additional scientific disciplines, including nanotoxicology and ecotoxicology.

Nanotechnology is a new innovative and promising area that suggests unique ideas for the growth and expansion of such tools which could be used in various sectors like medicine, environmental science, security, agriculture, healthcare, information technology, food safety, energy, transportation, and sustainable development. The ability to convert nanoscale concepts to useful applications by detecting, calculating, manipulating, gathering, regulatory framework, and engineering at the nanometer scale has allowed nanotechnology. The National Nanotechnology Initiative (NNI) in the United States asserts that nanotechnology is "science, engineering, and technology conducted at the nanoscale (1 to 100 nm)." This chapter provides a detailed overview of the eco-designing and modification process of bionanoparticles/bionanocomposites with possible applications and a life cycle assessment.

### 5.1.1 The Foundational Principle behind the Eco-design Concept of Making Nanoparticles

Nanotechnology has been an emerging field of research for the last several decades. Meanwhile, during the 1959 American Physical Society meeting at Caltech, a state-of-the-art lecture on nanotechnology was given by the 1965 Nobel Prize Laureate in physics, Richard P. Feynman, entitled "There's Plenty of Room at the Bottom" (Feynman, 1960). This talk had more impact and triggered more various and revolutionary developments than had ever been made in the field of nanotechnology until that time. Modern nanotechnology is the brainchild of his vision. In his lecture, he said "The marvellous biological system." Further, he explained that cells in biological system can be exceptionally small. Cells are very small, but they are very dynamic, active, and vigorous; assembled various substances through biosynthetic pathways, they move around; they wriggle; and they do all kinds of amazing, stunning, and marvellous things at a very small scale. Also, they work as powerhouses: just like transformers that store energy, cells store genetic material in codon form which consists of genetic information. From here, the foundation stone has been laid down for future researchers to work at the miniature level to develop the "*Science of Miniature*" or "*Science of Small*," where all the energy is stored at one small point but has the potential to work on a large and economically sound spectrum. Feynman said that biological phenomenon has huge potential by which chemical forces are used in repetitious fashion to produce all kinds of weird effects. This novel idea established that Feynman's hypotheses have been confirmed and correct; and for this reason, he is considered the father of modern nanotechnology. After 15 years, the Japanese scientist, Norio Taniguchi, was the first to use and define the term "nanotechnology" in 1974: "nanotechnology mainly comprises of the processing of separation, consolidation, and deformation of materials by one atom or one molecule." On January 21, 2000, during a speech at Caltech, President Bill Clinton encouraged financial assistance for research in the field of nanotechnology (Lok, 2010). Three years later, President George W. Bush clearly stated that the 21st century is known as for nanotechnological advancements and signed into law the Nanotechnology Research and Development Act (21st Century Nanotechnology Research, 2019).

## 5.2 ECO-DESIGNING OF BIONANOPARTICLES AND BIONANOCOMPOSITES

After Feynman, many researchers started working on new technology that worked on the nano scale. A storm came in the field of nanotechnology, and various nanoparticles have been synthesized which have huge potential in several of the fields stated above. Just as any boon or miracle comes with some challenges, nanotechnology is one of them. Excessive use of nanoparticles in any dimension of life generates nanotoxicity, which imparts harmful effects on the environment and humans. This leads to the evolution of the new analytical approach which is well equipped with mathematical tools to analyze the intensity of harm on the environment and humans. This tool is termed "life cycle analysis." Nanoparticles synthesized by the physical or chemical process have some toxic side effects on environment. They leach out and create potential hazards to living beings. Researchers suggest two approaches for the synthesis of nanoparticles: the top-down approach and the bottom-up approach, which differ in cost, time, and quality. The top-down approach is basically the breaking down of bulk material to obtain nano-sized particles. The bottom-up approach, or self-assembly approach, builds nanostructures from the bottom upward, atom-by-atom or molecule-by-molecule, using physical and chemical methods which are in the nanoscale range (1 nm to 100 nm). K. Eric Drexler (1986), a famous nanotechnologist of his time, published the first book on nanotechnology titled "Engines of Creation: The Coming Era of Nanotechnology," which provided a detail history of nanotechnology, since the beginning of this science, and also laid down the foundation of a new theory, "molecular engineering," as an additional input in nanotechnology research. After five years, in 1991, Drexler, Peterson, and Pergamit jointly published another book titled "Unbounding the Future: The Nanotechnology

Revolution" in which they coined new terms, like "nanobots" and "assemblers" for nano-processes in medical applications. Soon after, the term and practice of "nanomedicine" followed (Drexler et al., 1991; Victor et al. 2012).

There is so much new terminology related to nanotechnology, attracting the attention of researchers and society, in general, which are listed in Table 5.1.

With the help of various manufacturing pathways, engineered bio-nanostructures have been synthesized for various applications. The use of engineered bio-nanostructures often like a double sword weapon in sustainability perspective, though they are advantageous in treatment of societal problems, still consist some disadvantages in terms of eco-toxicity. Keeping in the view, the eco-designed concept has been gaining momentum day-by-day. In subsequent section a detailed study has been discussed. In the present scenario, bionanocomposite/bionanostructures are becoming very important because of the extraordinary properties and safety by design approach.

**TABLE 5.1**
**Terminology for Various Branches Emerged from Nanotechnology**

| New Terminology Related With Nanotechnology | Specific Area |
|---|---|
| *Bionanotechnolgy* | Synthesis of nanostructures from bacteria, deals with the problem-solving of environmental issues with the help of natural/green manufacturing of nanostructures. |
| *Phytonanotechnology* | Deals with plant system, manufacturing of nanostructures from plant species. |
| *Phyconanotechnology* | Synthesis of nanostructures from algal species and deals with problem-solving of environmental issues. |
| *Myconanotechnology* | Synthesis of nanostructures from fungal species and deals with problem-solving of environmental issues. The term "myconanotechnology" is a novel word coined in 2009 by Rai M. from India. |
| *Nanoinformatics* | The term "nanoinformatics" was officially documented after an initial Workshop on Nanoinformatics Strategies, June 12–13, 2007, Arlington, VA, USA.<br>Deals with the assembling, sharing, envisaging, modeling, and evaluation of significant nanoscale level data and information. |
| *Nano-oncology* | Deals with cancer and tumors. |
| *DNA nanotechnology* | Deals with molecular biology concepts. The theoretical basis for DNA nanotechnology was first laid out by Nadrian Seeman in 1982. |
| *Nanopharmaceuticals* | Drug delivery and gene therapy approaches. |
| *Green nanotechnology* | Environment and to produce more efficient and cost-effective energy, such as generating less pollution during the manufacture of materials, producing solar cells that generate electricity at a competitive cost, and cleaning up organic chemicals polluting groundwater. |
| *Nanomedicine* | Application of nanobiotechnology to medicine. |

### 5.2.1 Eco-design Model

A new concept, "Safety by Design (SbD)," is the basis for designing eco-friendly nanoparticles. It is actually the design consideration should be kept in mind during the design and modification of some important nanoparticle features. The "Safety by Design" concept is highly applicable to the nanotechnology/nanotoxicology sector and widely applicable in the drug delivery and development sector (Hjorth et al., 2017a). The basic principle behind the "Safety by Design" concept is to avoid those undesirable properties of structured nanomaterials, which makes them hazardous for the environment and human health, in the process of nanomaterial design. Further, this principle advocates that only those properties should be incorporated as design parameters during product development which will maintain the nanomaterials' effectiveness, security, and safety. Life cycle assessment of nanomaterial should provide suitable ecotoxicity data in terms of exposure and effects. Such knowledge should be used to select only those properties of nanomaterials which will assure their eco-friendly nature and sustainable application in the future.

The goal is an eco-design model for the synthesis of structured nanomaterials in such a way that they will be a boon in the era of nanotechnological advancement. An environmental application obtained from an eco-toxicological testing approach will allow the selection of the best green, eco-friendly, and ecologically sustainable nanomaterials and will significantly remove/check any potential side effects in order to ensure no toxicological risk for natural ecosystems, natural habitat, and human health. A thorough green, eco-safe analytical assessment approach is proposed which possesses the following significant features and characteristics:

- Collects information on the behavior of nanomaterial in environmental condition which to be remediated in terms of physicochemical transformations occurring.
- Identifies the fate and reactivity potential of nanoparticles which might affect biological targets.
- Develops a detailed mechanism-based assessment of ecotoxicity from single model species up to ecosystem level (from microcosm to mesocosm and *in situ* studies).

Standardized methodologies able to measure the nanomaterial on the basis of effectiveness, environmental safety, and economic sustainability within the context of existing environmental regulations are therefore immediately needed. All these aspects will certainly support patenting and pilot applications of new engineered nanomaterials (ENMs) developed based on eco-safety by design approach. "Safety by Design" is actually a reverse approach in which the endpoint is the starting point for development of eco-friendly nanoparticles as shown in Figure 5.1. The following approaches should be kept in mind while designing the eco-design nanomaterial (Ilaria et al., 2018):

- Upstream approach should be incorporated in eco-safety design parameters.
- Ending point serves as the initiation point of the process.

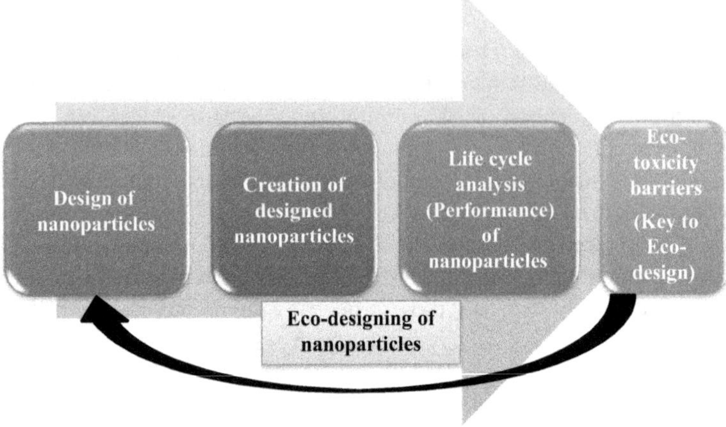


**FIGURE 5.1** Reverse approach of eco-designing of nanoparticles.

- Analytical and high output approach.
- Consistent and reliable validation protocols for stakeholders.

The usage of naturally obtained materials, like biopolymers, agro-based materials, phyto-extracts, microorganisms, etc., offers various advantages of eco-friendly environment and biocompatibility for various environments, food packaging, and medicinal and pharmaceutical applications, while toxic chemicals are not consumed in the production process. The synthesis and assembly of eco-designed nanoparticles by biological methods are the keys to the development of clean, nontoxic, and environmentally suitable measures, probably involving a wide range of living organisms, like bacteria, fungi, and even plants.

The three key ingredients, involved in the formation of eco-designed nanoparticles should be assessed from an eco-chemistry viewpoint:

- Solvent medium used for synthesis.
- Green neutral reducing agents.
- Nonhazardous materials for the stabilization of the nanoparticles.

From here, the concept of green-socio nano industries is developed as shown in Figure 5.2.

Nowadays, the synthesis of eco-designed nanoparticles using plants as the source has drawn the attention of more researchers, as it's an easy single-step process – and the branch of such study is termed "phyto-nanotechnology." Plant-arbitrated biosynthesis is a very informal and cost-effective procedure for the production of nanoparticles. As we know, scientists have proved that synthesizing nanoparticles from plants is a superior method, as it is free from various kinds of toxic elements, since they are in nanoscale; and secondly, the plant readily provides them with natural capping agents, e.g., the synthesis of gold, silver, copper, and zinc nanoparticles and

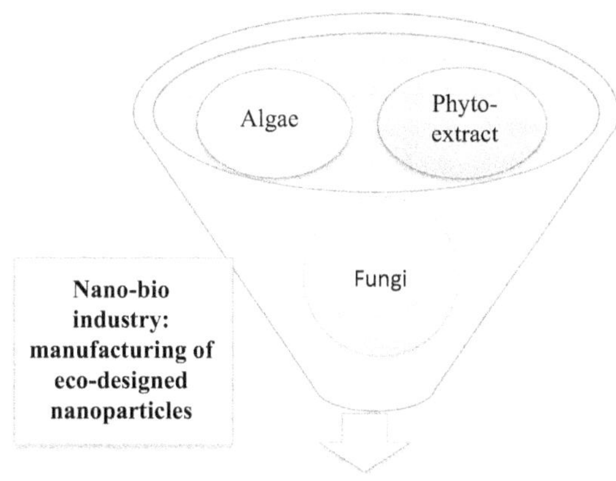


**FIGURE 5.2** Nano-bioindustry: manufacturing of eco-designed nanoparticles.

**TABLE 5.2**
**Names of Nanoparticles with Their Plant Sources**

| Name of Nanoparticles | Plant Source | Reference |
|---|---|---|
| Silver nanoparticles | *Azadirachta indica* | Shankar et al., 2004 |
| | *Phyllanthin* extract | Kasthuri et al., 2009 |
| | *Mentha piperita* | Parashar et al., 2009 |
| | *Ocimum sanctum* | Mallikarjuna et al., 2011 |
| | *Marigold* flower | Kaur et al., 2011; Bansal et al., 2017 |
| Gold nanoparticles | *Garcinia mangostana* | Lee et al., 2016 |
| | *Acanthella elongate* | Inbakandan et al., 2010 |
| | *Avena sativa* | Armendariz et al., 2004 |
| Copper nanoparticles | *Punica granatum* | Kaur et al., 2018; Kaur et al., 2016 |
| Old-iron and silver iron core–shell nanoparticles | *Punica granatum* | Kaur et al., 2018; Kaur et al.,2016 |

their oxides, using various plant extracts, like *Andrographis paniculate* (Shankar et al., 2004), extracts of aloe vera plant, the sun-dried leaf extract of *Cinnamomum camphora*, and *Azadiracta indica* (Shankar et al., 2004). A brief of different types of nanoparticles with their plant source are presented in Table 5.2.

Today, the synthesis of nanoparticles, nanofilm, nanofiber, etc from bacteria has gained massive interest to the researcher due to its extensive application. Various researchers report the formation of extracellular and intracellular metal nanoparticles by bacteria, like *Pseudomonas stutzeri*, *Escherichia coli*, *Pseudomonas aeruginosa*, *Plectonema boryanum*, *Salmonella typhus*, *Staphylococcus currens*, *Vibrio cholera*, etc (Klaus et al., 1999). A brief of different types of nanoparticles with their bacterial sources are presented in Table 5.3.

**TABLE 5.3**
**Different Types of Nanoparticles with Their Bacterial Source**

| Name of Nanoparticles | Bacterial Source | Reference |
|---|---|---|
| Silver and gold nanoparticles | *Brevibacterium casei* | Kalishwaralal et al., 2010 |
| Silver and gold nanoparticles | *Escherichia coli* | Du et al., 2007; Gurunathan et al., 2009 |
| Silver nanoparticles | *Corynebacterium* sp. | Zhang et al., 2005 |
| Silver nanoparticles | *Lactic acid bacteria* | Sintubin et al., 2009 |
| Silver nanoparticles | *Staphylococcus aureus* | Nanda and Saravanan, 2009 |
| Silver nanoparticles | *Bacillus cereus* | Babu and Gunasekaran, 2009 |
| Silver nanoparticles | *Morganella* sp. | Parikh et al., 2008 |
| Mercury nanoparticles | *Enterobacter* sp. | Sinha and Khare, 2011 |
| Cadmium nanoparticles | *Desulfobacteraceae* sp | Labrenz et al., 2000 |
| Gold nanoparticles | *Salmonella enterica* | Mortazavi et al., 2017 |
| Gold nanoparticles | *Ureibacillus thermosphaericus* | Juibari et al., 2011 |

"Myconanotechnology" is the term used for the synthesis of nanostructures from fungi. Increasing interest in this process is largely due to the rich diversity of fungal species and their unique characteristics and metal-binding affinity. Synthesis of nanomaterials, like nanoparticles, nanofilm, nanofiber, etc from fungi is of enormous interest to researchers due to its properties, like reducing toxicity, toleration of toxic contaminants, and increasing higher bioaccumulation. Also, its simple downstream processing and ease of use make this biomass very attractive. Biosynthesis of silver nanoparticles by *Aspergillus niger*, *Fusarium solani*, and *Aspergillus oryzae* can be used to produce silver nanocrystals. The prologue of silver ions to *Fusarium oxysporum* leads to the synthesis of stable Ag hydrosols. The nanoparticle can be synthesized by taking a culture of fungi as a medium, containing an appropriate substrate for particular enzymes, which will activate the production of nanoparticles from fungus by the reduction of silver. Bansal et al., 2017 reported that fungus *Aspergillus terreus* mycosynthesized silver nanoparticles using silver nitrate solution as a precursor with the culture supernatants. Gold nanoparticles (AuNPs) have been reported as a widespread research tool in various fields of healthcare, medicine, and agriculture due to their biocompatibility and stability. Fungi-mediated methods for the synthesis of gold nanoparticles were suggested as the two main precursors $HAuCl_4$ and AuCl which dissociate to $Au^{3+}$ ions and $Au^+$, respectively (Kitching et al., 2015). Bansal et al. (2017a) reported that zirconia nanoparticles were synthesized by the fungus *Fusarium oxysporum*. *Aspergillus fumigatus* also produced zinc oxide nanoparticles (Raliya and Tarafdar, 2013). It was also reported by many researchers that gold nanoparticles could also be mycosynthesized from many fungal species, like *Rhizopus oryzae*, *Aspergillus niger*, *Aureobasidium pullulans*, *Fusarium semitectum*,

*Fusarium oxysporum*, *Penicillium brevicompactum*, *Neurospora crassa* (Das et al., 2009; Xie et al., 2007; Zhang et al., 2011; Sawle et al., 2008; Mishra et al., 2011; Castro-Longoria et al., 2011).

Yeast cells also are potential generators of eco-designed nanoparticles during the process of fermentation. In this method, the extracellular production of nanoparticles could be beneficial in the fermentation industry. Production of nanoparticles from yeast cells in nanotechnology, however, is a somewhat new approach. A large quantity of synthesis of nanoparticles is possible by using a very simple method of downstream processing (Kowshik et al., 2002). Scientists have isolated silver-tolerant yeast strain MKY3, which was inoculated with aqueous silver nitrate resulting in the formation of nano-size 2–5 nm silver nanoparticles. Dameron showed that cadmium nanoparticles can be biosynthesized by using *Candida glabrata* and *Schizosaccharomyce*. The silver and gold nanoparticles can be biosynthesized from a yeast strain that was isolated from acid mine drainage. Sulphide nanoparticles were synthesized intracellularly by *S. pombe* for the fabrication of diode cadmium (Kowshik et al., 2002).

### 5.2.2 Reformation of Bionanoparticles and Biocomposites According to Need

Molecular material can be manipulated to enhance the properties which will depend on the type of element chosen. How much quantity needs to be taken, how they would be distributed, the texture of the material, composition of the molecular material, and the structural interface between components. By taking care of these properties, we can design bionanocomposites to meet the growing need of various sectors like the agricultural field, information technology, health sector, transportation, food and safety, aerospace and aviation, homeland security, packaging industry, imaging, textile industry, medicine, energy, and environmental sciences for sustainable development. Researchers have discovered that if they add epoxy composite to graphene it may result in a stronger and stiffer material than current epoxy composites. This property could result in the production of components with a higher strength to weight ratio e.g., windmill blades. Researchers showed that when nanotube-polymer-based bionanocomposite is used as a kind of scaffold, replacement bone grows faster, but they have yet to discover the exact mechanism of how these bionanocomposites increase bone growth.

Bionanocomposite conductive paper was made by clubbing nanocomposite of cellulose materials and nanotubes. It is a very simple, easy, and nontoxic method that can be designed just by dipping this conductive paper into an appropriate electrolyte and having a flexible battery at the ready. Researchers are working to join both magnetic nanoparticles and fluorescent nanoparticles into bionanocomposites which would result in a material having both the property of magnetism as well as fluorescence. This nanocomposite could be used, for instance, to make tumors more visible during MRIs, which are routinely performed before surgery. The fluorescent property of the nanocomposite will help the surgeon see where tumors are during surgery.

## 5.3 SIGNIFICANT CHALLENGES AND INFORMATION GAPS

Challenges, in the form of various physical and chemical properties, and the rate of production of bionanocomposites, need to be addressed. Scientists are doing research to find unique fillers and films by making use of unexplored natural resources which can be integrated into the nanobiocomposites giving them undoubtedly unique properties which can be designed and applied according to the needs of society. One of the biggest challenges faced in one of the applications is intensifying polymer conductivity which generates energy through controlled photo-generation. Researchers are doing research to find new resources, tools, and technology so that they can design unique bionanocomposite. They are also working to develop unique composite materials which they can standardize for their stiffness and strength so that the exact ratio of additives could be applied to the polymeric matrix which would be coupled to the appropriate filler, thereby increasing the quality of the products. So, the most important issue in the biosynthesis of bionanocomposites is the unique design so that society can take maximum advantage of that product.

Even though eco-designed nanomaterials and products have exposed amazing and exciting results in all dimensions of application in water treatment, nano-medicine pharmaceuticals, drug delivery, etc., as well as changing the mindset of researchers and scientists, still, there are certain crucial issues that require attention in order to provide solutions for commercial applications.

- Energy contributions.
- Varying cost and market.
- Carbon emission, economy.
- Climate change and global warming.
- Sustainability and the Energy Information Administration (EIA).
- Regaining and regeneration.
- Wellbeing and nanotoxicity.

## 5.4 LIFE CYCLE ASSESSMENT OF BIONANOPARTICLES/ NANOCOMPOSITES

The life cycle assessment (LCA) approach to bionanotechnology and bio-nanoproducts can deliver valuable information about the foremost environmental impacts and benefits of this emerging technology. Potential LCA approaches are needed and experimental data on characteristics and toxicity of bionanoparticles coming from research projects should be included in LCA methodologies. Altered exposure and fate modeling are required in order to have complete results on the environmental performance of nanoproducts during all life cycle stages.

At the present time, nanomaterial and bionanotechnology-based products are widely used in different applications. This tendency has brought argument due to the lack of data on the potential impact of nanotoxicity on human health and the environment. Though LCA is considered the best approach to assess the environmental behavior of emerging technologies, currently the potential impacts of the released nanomaterial associated with human and environmental health are not yet included in LCA databases, and several uncertainties and data gaps exist. Figure 5.3 shows the various areas of applicability of life cycle analysis. The main objective of applying LCA approaches in nanotechnology-based products is to derive characterization factors for bionanomaterials using existing models as a starting point (Bjørn and Hauschild 2018). Inventory and impact data could be included in LCA databases and impact methods to allow performing comprehensive LCA studies on nanotechnology-based products and their applications, comparing them to conventional materials, and providing the basis for future regulations and policies in this field.

Life cycle assessment (LCA) can deliver a more thoughtful approach for the major environmental issues and guarantee the environmental sustainability of nanomaterials. LCA is an all-inclusive approach to the evaluation of environmental impacts of a product throughout its complete life cycle by recognizing the materials used and energy and emissions released to the environment – all vital in evaluating the potential impacts of nanomaterial releases, as shown in Figure 5.4. LCA is a worldwide standardized accepted methodology, based on the International Organization for Standardization (ISO) 14040 series (ISO 2006; 2006; 2006; 2006d), encompassing four phases, as follows:

1. Goal and scope.
2. Life cycle inventory.
3. Life cycle impact assessment.
4. Life cycle interpretation.

This methodology was established to assess the environmental impact of yields, and the processes linked with these products. Consequently, the current state-of-the-art LCA application in bionanotechnology needs to be explored to gain insights into the current practices of LCA application in bionanotechnology and its future outlook. Due to the fast technological advancements in nanotechnology, the environmental toxicity pathways of the nanomaterial require further investigation from an LCA viewpoint.

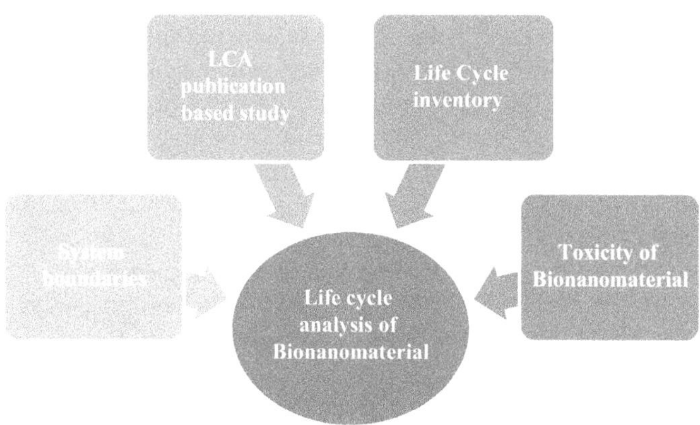


**FIGURE 5.3** Major areas of the life cycle assessment.

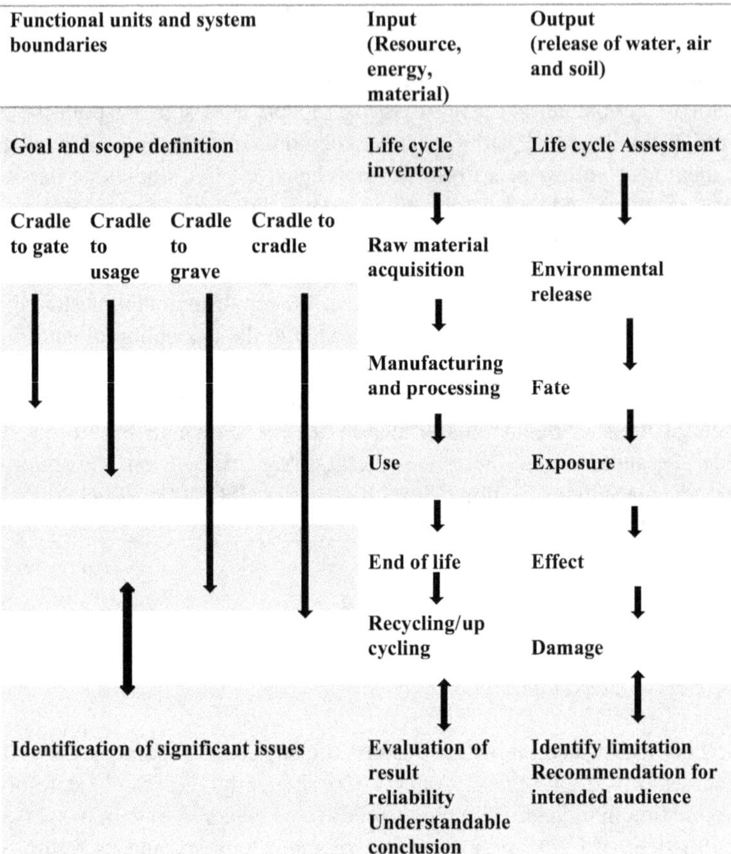


**FIGURE 5.4** Life cycle assessment framework.

A review has been carried out on the inclusion of the LCA study, which shows that the most research so far was carried out in the year 2020. Data were searched on online engines, like Google Scholar, ResearchGate, academia, etc., and found that initially, 182 studies were there on LCA; though, the numbers were reduced to 126 studies, seeing only studies published in scientific indexed journals. To ensure the LCA and nanomaterial relevancy, only the literature with a focus on the nanomaterial synthesis pathways, concerning the potential environmental impacts and relevant case studies, were found. Studies in the last 19 years (2001–2020) were considered, to explore the trends of the LCA approach to nanomaterial (Nurul et al., 2021).

Figure 5.5 shows the geographic provenance of published papers. It shows that 37 studies were from Europe, and 28 studies were from North America, which represents about 95% of the published papers. In the year 2020, Europe published the most articles (8 articles out of 11). Moreover, the earliest article published on the LCA of nanoparticles was also from Europe in 2001, which indicated early investigation of the environmental effects of nanoparticles in this area. Figure 5.6 shows the country-wise statistics of published articles based on LCA studies on nanoparticles.

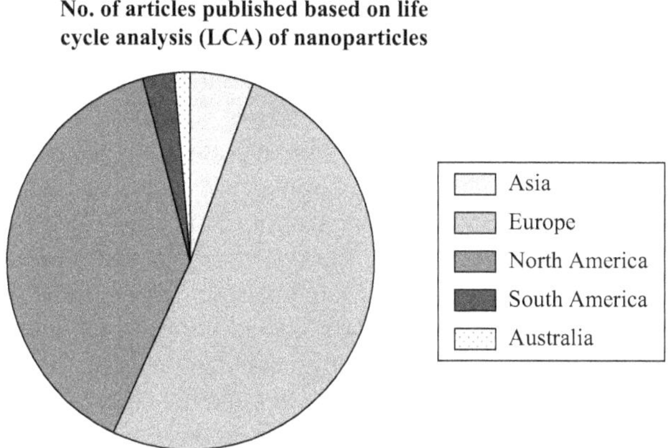


**FIGURE 5.5** Number of articles published worldwide based on LCA of nanoparticles.

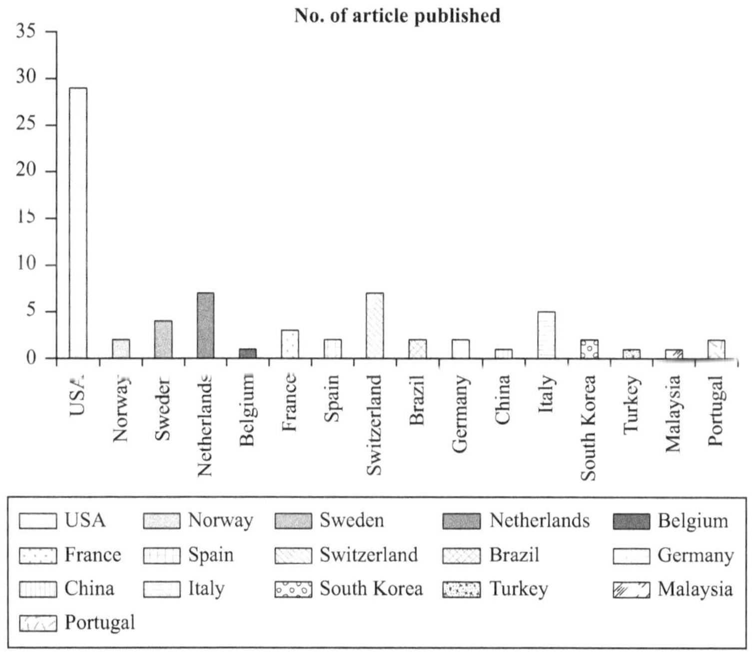


**FIGURE 5.6** Number of articles published country-wise based on LCA of nanoparticles.

### 5.4.1 System Boundary

As shown in Figure 5.4, the system boundary describes all progressions that contribute to the life cycle of nanomaterial, processes, and activities (Pallas et al., 2019). Due to the complex nature of cradle-to-cradle approach, it is not commonly studied, and grounded with the circular economy concept a higher level of reutilization of constituents is required. This approach suggests that the end product of one cycle should be the raw material of another new process. Due to the structural complexity of physical properties and chemical compositions of nanoparticles, process intermediate interactions of nanoparticles may vary throughout the life cycle, especially at the disposal stage, making them challenging to repurpose into something new, since nanomaterials (NM) properties can be unpredictable (Chappell et al., 2017; Bartolozzi et al., 2019; Salieri et al., 2018). Nanoparticle disposal management is also required for the safeguarding of the environment (Pati et al., 2014). Bartolozzi et al. (2019) specified that during the study of life cycle analysis, use of the final product at end-of-life stages should be included. It should also be kept in mind that the cradle-to-grave system boundary approach is considered during the final disposal of the nanomaterial product. In his study, Bartolozzi asserted that midpoint indicators only reveal bearings somewhere between the releases and the endpoint of the nanomaterial life cycle, while end-of-life stages are well-defined at the level of the protection areas *viz* the environment, human health, and natural resources.

### 5.4.2 Life Cycle Inventory

As shown in Figure 5.4, the life cycle inventory (LCI) is the repository of all the inputs required for any process operation of nanoparticle synthesis. The LCI segment is also known as the data collection phase and is central to any LCA study. It works as the backbone of any life cycle analysis process. This phase is the most labor-intensive, rigorous, and time-consuming phase in an LCA, due to the requirement of detailed data input of all the operational processes. The inventory is based on the collection of comprehensive and consistent primary, secondary, and generic data, which contains clear descriptions of applied assumptions, advantages and drawbacks, and transparency and credibility criteria, etc. The input for inventory includes raw materials, industrial processes, energy, resources, transport, basic materials, waste disposal, energy (renewable and nonrenewable), and water, while outputs are the products and by-products: emissions to air, water, and soil (Moloi et al., 2021; Miseljic and Olsen, 2014; Mullen and Morris, 2021).

### 5.4.3 Life Cycle Impact Assessment

Life cycle impact assessment (LCIA) is the process of conversion of life cycle inventory data from a life cycle assessment into a set of potential impacts in a customized and conventional way. This permits experts and decision-makers to better understand the harm caused by resource use and emissions. For example, in one process of methane generation, a process that emits 10 g of carbon dioxide emissions and 1 g of methane emissions gives a partial understanding of climate change.

The LCIA method is the debated research. Due to the complex nature of the LCIA, methodologies have been established to simplify and optimize the LCIA process. Two approaches are to be considered; that is, a midpoint approach and an endpoint approach. LCIA approach framework is shown in Figure 5.7. The endpoints characterize the end of the chain and areas of protection – those things we want to protect and conserve for our wellbeing and for the wellbeing of the Earth. Let's understand midpoints with an example first: when an emission is created, it

| Input for LCIA | Midpoint | Output | Endpoint |
|---|---|---|---|
| Raw materials<br>Industrial processes<br>Energy resources → | Toxicity<br>Climate change<br>Acidification<br>Eco-toxicity<br>Land use<br>Water use<br>Eutrophication<br>Particulate matter formation | → Air Emission<br>Water contamination<br>Soil fertility check<br>Environmental stress | → Environment protection<br>Natural resource conservation<br>Human health<br>Energy conservation |

**FIGURE 5.7** LCIA approach framework with the midpoint and endpoint approaches.

undergoes a certain physical/chemical change in the environment and forms a steady state deliberation. Frequently, the models used to calculate the midpoints measure these steady-state deliberations that form the basis of the midpoint characterization factors. The endpoint methods assess the actual impairment caused by an emission in terms of species lost or human life years lost or lived disabled.

LCIA usually has multiple steps, which consist of classification, characterization, normalization, and valuation. *Classification* deals with the impact categories and its selection on the basis of impact of product. It is also a data sorting exercise for making an inventory. *Characterization* involves the application of premium factors or equivalence to unite all relevant substances within each impact category. This step provides a mode of straight comparison of the LCI results within each category. *Normalization* establishes a common reference to enable comparison of different environmental impacts. To attain this goal, a reference quantity is used to make the data "dimensionless." The final step, *valuation*, is used for the assessment of the relative importance of the potential environmental impacts identified in the previous steps by assigning them weight. It involves defining the intensity of impacts in relation to others. The concluding goal is to obtain an exclusive result. *Grouping* influences characterize the impact categories resulting from characterization and normalization according to certain fixed conditions. *Integration* allows weight to the impact categories to achieve a single index weighting. Three categories are used to evaluate the environmental impact of a product. In Ecoindicator 99, these categories are ecosystem quality, depletion of nonrenewable resources, and human health. Advantages and limitations of LCIA are presented in Table 5.4.

### 5.4.4 Toxicity of Bio-nanomaterials

There are certain toxicities which are associated with bionanomaterials and basic procedures required to overcome the effect of these nanomaterials. Bionanomaterials/nanomaterials enter the environment through water, soil, and air during various

**TABLE 5.4**
**Advantages and Limitations of LCIA**

| Advantages | Limitations | References |
|---|---|---|
| Allows improved understanding of the potential environmental problems. | Limited research and knowledge, lack of inventory data. | Lee et al., 2020; Cucurachi and Rocha, 2019; Parsons et al., 2015; Lacirignola et al., 2017; Walser and Gottschalk, 2014; Beloin-Saint-Pierre et al., 2017; Igos et al., 2019; Gilbertson et al., 2015; Inshakova and Inshakov, 2017; Jakovljevic et al., 2020 |
| Environmental impacts can be quantified. | The lack of characterization factors. | |
| Investigation held in significant shifts to measure environmental impacts. | Uncertainity of LCA for bionanomaterials. | |
| LCA can be useful to compare and study the human and ecological impacts between two or more rival products/processes. | Lack of correlation between the mathematical models. | |
| It enables decision-makers to make long-term decisions and improvements. | Uncertainties in the inconsistency of laboratory data. | |
| Worked as useful inventory. | Invalid assumptions in the interpretation stage. | |
| Safeguards the environmental sustainability of bionanomaterial by assessing the environmental impacts. | Emissions from various waste management processes. | |

human activities. As a result, this environmental damage attracts increasing concern from all stakeholders. Though there are various advantages associated with nanomaterials, like their small size, high reactivity and great capacity, they could become lethal factors by inducing adverse cellular toxic and harmful effects. Research also shows that nanoparticles can enter organisms during ingestion or inhalation and can translocate within the body to various organs and tissues where they react and create toxicological effects. Although some studies have also addressed the toxicological effects of NPs on animal cells and plant cells, the toxicological studies with magnetic NPs on plants to date is still limited. Research shows, for instance, that zinc-based nanoparticles exert harmful effects on algal species, like algal growth inhibition, oxidative stress, ROS generation, genotoxicity, disruption of membrane integrity, morphological alteration of roots, oxygen consumption, etc. (Hjorth et al., 2017b; Nguyen et al., 2018; Schiwy et al., 2016; Ghosh and Mukherjee, 2017). One research on iron-based nanoparticles revealed similar effects as discussed above when aquatic organisms (bacteria, freshwater microalga, freshwater crustaceans, earthworms, plants, fish) were kept under the influence of structured engineered nanoparticles (Qualhato et al., 2017). In another research on titanium oxide-based nanoparticles which were used for photo-degradation of organic contaminants, harmful effects in bacteria crustaceans, visible in terms of ROS generation, oxidative stress, membrane damage, cell viability reduction, reproduction impairment, tissue alterations and gill histopathology, and neurotoxicity were reported (Mathur et al., 2017; Callaghan and MacCormack, 2017; Yang et al., 2013).

Furthermore, such belongings have been defined over different levels of biological organizations, ranging from plants and algae to aquatic invertebrate and vertebrate species, identifying diverse toxicological endpoints. Such indications highlight the necessity to move toward the life cycle impact analysis for further protection of human health, ecosystems, and natural resources (Zhang et al., 2018; Gottschalk et al., 2013).

## 5.5 FUTURE RECOMMENDATIONS

One recommendation for future work is that process optimization should be required for different operational parameters like temperature, pressure, time, pH, etc. They can play important roles in controlling the shape and morphology of the nanomaterial so that high yield is received in terms of product development. Though widespread research and progressive efforts are being made in the field of green bionanoparticles as innovative eco-designed functional units, we are still far away from a bioeconomy-driven sustainable future planet.

Most importantly, environmental issues should be taken into account before using these materials for any applications like wastewater treatment, food packaging, drug delivery, etc. Even though LCA is powerfully recommended as a tool to assess the sustainability of nanomaterial throughout their life cycle, the research community currently finds several information gaps, which hamper the proper application of LCIA in the field (Hischier and Walser, 2012; Miseljic and Olsen, 2014). Closing these gaps is the pressing need. These gaps consist of basically two broad issues, namely, the difficulty of including the whole life cycle of nanomaterial and the challenge of fully measuring their impacts on human toxicity and ecotoxicity. The life cycle approach should develop in an incremental way to access the LCIA in the broad spectrum (Gavankar et al., 2015). As for future concerns, it is relevant to highlight some areas of urgent priority:

1. All-inclusive and standardized investigational work to include the complete spectrum of nanostructure, properties, application, and suitability.
2. A synchronized terminology/nomenclature system and adherence to the 12 principles of green chemistry developed by Paul Anastas and John Warner (1998).
3. Techno-socio-economic assessment of eco-designed materials using quantitative approaches like life cycle assessment and eco-safety.

## 5.6 CONCLUSION

Eco-designed nanoparticle synthesis has gained attention recently, which is speedily substituting traditional chemical syntheses and is of great interest due to its eco-friendliness, economic value, viability, and wide range of applications in several areas. Recently, numerous types of biological entities which serve a dual role as both the reducing and stabilizing agents have been used in the synthesis of bionanoparticles. The state of the art defined so far proves that LCA is definitely a valuable, appreciable, and powerful tool for the assessment of sustainability parameters as emerging technologies; nevertheless, several procedural issues require resolution in order to deliver an inclusive assessment in the field of nanotechnologies. Certainly, its main challenge will be to overcome the ecotoxicological characterization. LCA marked its presence in setting up the provision of the development of green nanotechnology, which is cleaner and greener in its approach. Life cycle assessment is a valuable tool to implement the design thinking strategy. Satisfactory use of transparent and complete characterization models, during the interpretation phase of LCA, requires extra effort from LCA experts and researchers. and toxicity analyst that they should be discovered more widely for a comprehensive and reliable LCA study. LCA can deliver wide-ranging support for the eco-design of emerging nanomaterials, already significant at the early stage of their development at the lab scale, and can guide the transition to the scale-up phase.

## REFERENCES


21st Century Nanotechnology Research and Development Act. [(Accessed on 22 July 2019)]; Available online: https://www.congress.gov/bill/108th-congress/senate-bill/189.

Anastas, P.T., Warner, J.C. (1998). *Green Chemistry: Theory and Practice*, Oxford University Press: New York, p. 30. By permission of Oxford University Press.

Armendariz, V., Herrera, I., Peralta-Videa, J.R., Jose- Yacaman, M., Troiani, H., Santiago, P., Gardea-Torresdey, J.L. (2004). Size controlled gold nanoparticle formation by *Avena sativa* biomass: Use of plants in nanobiotechnology. *Nanoparticle Res.* 6, 377–382.

Babu, M.M.G., Gunasekaran, P. (2009). Production and structural characterization of crystalline silver nanoparticles from *Bacillus cereus* isolate. *Colloids and Surfaces B: Biointerfaces.* 74, 191–195.

Bansal, P., Kaur, P., Duhan J.S. (2017a). Biogenesis of silver nanoparticles using *Fusarium pallidoroseum* and its potential against human pathogens. *Annals Biol.* 33, 180–185.

Bansal, P., Kaur, P., Surekha, Kumar, A., Duhan, J.S. (2017b). Microwave assisted quick synthesis method of silver nanoparticles using citrus hybrid "Kinnow" and its potential against early blight of tomato. *Research on Crops.* 18(4), 650–655.

Bartolozzi, I., Daddi, T., Punta, C., Fiorati, A., Iraldo, F. (2019). Life cycle assessment of emerging environmental technologies in the early stage of development: A case study on nanostructured materials. *J. Ind. Ecol.* 24, 101–115.

Beloin-Saint-Pierre, D., Turner, D.A., Salieri, B., Haarman, A., Hischier, R. (2017). How suitable is LCA for nanotechnology assessment? Overview of current methodological pitfalls and potential solutions. *Int. J. Life Cycle Assess.* 23, 191–196.

Bjørn, A., Hauschild, M.Z. (2018). Cradle to Cradle and LCA. In *Life Cycle Assessment.* Springer: Cham, Switzerland. pp. 605–631.

Callaghan, N.I., MacCormack, T.J. (2017). Ecophysiological perspectives on engineered nanomaterial toxicity in fish and crustaceans. *Comp. Biochem. Physiol. Part C Toxicol. Pharmacol.* 193, 30–41.

Castro-Longoria, E., Vilchis-Nestor, A.R., Avalos-Borja, M. (2011). Biosynthesis of silver, gold and bimetallic nanoparticles using the filamentous fungus *Neurospora crassa. Colloids Surf B Biointerfaces.* 83(1), 42–8.

Chappell, M.A., Shih, W.S., Bledsoe, J.K., Cox, C., Janzen, D., Gibbons, S., Patel, R., Kennedy, A.J., Brame, J., Brondum, M., (2017). Environmental life cycle assessment for a carbon nanotube-based printed electronic sensor platform. *Adv. Mater.* 1, 345–347.

Cucurachi, S., Rocha, C.F.B. (2019). *Nanotechnology in Eco-efficient Construction, 2nd ed. Materials, Processes and Application: Life- Cycle Assessment of Engineered Nanomaterials; Wood head Publishing Series in Civil and Structural Engineering.* Wood head Publishing: Leiden, The Netherlands, pp. 815–846.

Das, S.K., Das, A.R., Guha, A.K. (2009). Gold nanoparticles: Microbial synthesis and application in water hygiene management. *Langmuir.* 25, 8192–8199.

Drexler, E.K. (1986). *Engines of Creation: The Coming Era of Nanotechnology.* Anchor Press; Garden City.

Drexler, E.K., Peterson, C., Pergamit, G. (1991). *Unbounding the Future: The Nanotechnology Revolution.* William Morrow and Company, Inc.; New York.

Du, L., Jiang, H., Liu, X., (2007). Wang, E. Biosynthesis of gold nanoparticles assisted by *Escherichia coli* DH5α and its application on direct electrochemistry of haemoglobin. *Electrochem. Communicat.* 9, 1165–1170.

Gavankar, S., Suh, S., Keller, A.A. (2015). The role of scale and technology maturity in life cycle assessment of emerging technologies: A case study on carbon nanotubes. *J. Ind. Ecol.* 19, 51–60.

Ghosh, I., Mukherjee, A. (2017). In planta genotoxicity of nZVI: Influence of colloidal stability on uptake, DNA damage, oxidative stress and cell death. *Mutagenesis.* 32, 371–387.

Gilbertson, L.M., Wender, B.A., Zimmerman, J.B., Eckelman, M.J. (2015). Coordinating modeling and experimental research of engineered nanomaterials to improve life cycle assessment studies. *Environ. Sci. Nano.* 2, 669–682.

Gottschalk, F., Sun, T., Nowack, B. (2013). Environmental concentrations of engineered nanomaterials: Review of modeling and analytical studies. *Environ. Pollut.* 18, 1287–300.

Gurunathan, S., Kalishwaralal, K., Vaidyanathan, R., Venkataraman, D., Pandian, S.R.K., Muniyandi, J., Eom, S.H. (2009). Biosynthesis, purification, and characterization of silver nanoparticles using Escherichia coli. *Colloids and Surfaces B: Biointerf.* 74, 328–335.

Hischier, R., Walser, T. (2012). Life cycle assessment of engineered nanomaterials: State of the art and strategies to overcome existing gaps. *Sci. Total Environ.* 425, 271–282.

Hjorth, R., Coutris, C., Nguyen, N.H.A., Sevcu, A., Gallego-Urrea, J.A., Baun, A., Joner, E.J. (2017a). Ecotoxicity testing and environmental risk assessment of iron nanomaterials for sub-surface remediation–Recommendations from the FP7 project Nano Rem. *Chemosphere.* 182, 525–531.

Hjorth, R., Van, H.L., Wickson, F. (2017b). What can nanosafety learn from drug development? The feasibility of "safety by design". *Nanotoxicology.* 11, 305–312.

Igos, E., Benetto, E., Meyer, R., Baustert, P., Othoniel, B. (2019). How to treat uncertainties in life cycle assessment studies? *Int. J. Life Cycle Assess.* 24, 794–807.

Ilaria, C., Andrea, F., Giacomo, G., Irene B., Tiberio, D., Lucio, M., Carlo, P. (2018). Environmentally sustainable and ecosafe polysaccharide-based materials for water nano-treatment: An eco-design study. *Materials.* 11(7), 1228.

Inbakandan, D., Venkatesan, R., Khan S.A. (2010). Biosynthesis of gold nanoparticles utilizing marine sponge *Acanthella elongata* (Dendy, 1905) *Colloids and Surf. B: Biointerfaces.* 81, 634–639.

Inshakova, E., Inshakov, O. (2017). World market for nanomaterials: Structure and trends. *MATEC Web Conf.* 129, 02013.

Jakovljevic, M., Sugahara, T., Timofeyev, Y., Rancic, N. (2020). Predictors of (in)efficiencies of healthcare expenditure among the leading Asian economies: Comparison of OECD and Non-OECD Nations. *Health Policy.* 13, 2261–2280.

Juibari, M.M., Abbasalizadeh, S., Jouzani, G.S., Noruzi, M. (2011). Intensified biosynthesis of silver nanoparticles using a native extremophilic *Ureibacillus thermosphaericus* strain. *Mater. Lett.* 65, 1014–1017.

Kalishwaralal, K., Deepak, V., Ram Kumar, S. Pandian, M. Kottaisamy, S. BarathmaniKanth, B. Kartikeyan, S. Gurunathan. (2010). Biosynthesis of silver and gold nanoparticles using *Brevibacterium casei Colloids Surf. B: Biointerfaces.* 77, 257–262.

Kasthuri, A.K., Kathiravan, A., Rajendiran, N. (2009). Phyllanthin-assisted biosynthesis of silver and gold nanoparticles: A novel biological approach. *J. Nanopart. Res.* 11, 1075–1085.

Kaur, P., Thakur, R., Chaudhary, A. (2011). Biosynthesis of silver nanoparticles using flower extract of *Calendula officinalis* Plant NCACT, Amity, Conf. Proc. pp. 109–110. ISBN 978-81-8424-705-3.

Kaur, P., Thakur, R., Chaudhury, A. (2016). Biogenesis of copper nanoparticles using peel extract of *Punica granatum* and their antimicrobial activity against opportunistic pathogens. *J. Green Chem.: Review Lett.* 9(1), 33–38.

Kaur, P., Malwal, H., Thakur, R., Manuja, A., Chaudhury, A. (2018). Biosynthesis of biocompatible and recyclable silver/iron and gold/iron core-shell nanoparticles for water purification technology. *Biocataly. Agricul. Biotechnol.* 10.1016/j.bcab.2018.03.002.

Kitching, M., Ramani, M., Marsili, E. (2015). Fungal biosynthesis of gold nanoparticles: Mechanism and scale up. *Microbial Biotechnol.* 8, 904–917.

Klaus, T., Joerger, R., Olsson, E., Granqvist, C.G. (1999). Silver based crystalline nanoparticles, microbially fabricated *Proceedings of the National Academy of Sciences of the United States of America.* 96, 13611–13614.

Kowshik, M., Deshmukh, N., Vogel, W., Urban, J., Kulkarni, S.K., Paknikar, K.M. (2002). Microbial synthesis of semiconductor CdS nanoparticles, their characterization, and their use in the fabrication of an ideal diode. *Biotechnol. Bioeng.* 78(5), 583–8.

Labrenz, M., Druschel, G.K., Thomsen-Ebert, T., Gilbert, B., Welch, S.A., Kemner, K.M., Lai, B. (2000). Formation of sphalerite (ZnS) deposits in natural biofilms of sulfate-reducing bacteria. *Science.* 290, 1744–1747.

Lacirignola, M., Blanc, P., Girard, R., Perez-Lopez, P., Blanc, I. (2017). LCA of emerging technologies: Addressing high uncertainty on inputs' variability when performing global sensitivity analysis. *Sci. Total Environ.* 578, 268–280.

Lee, J., Ryu, K.H., Ha, H.Y., Jung, K.D., Lee, J.H. (2020). Techno-economic and environmental evaluation of nano calcium carbonate production utilizing the steel slag. *J. CO2 Util.* 37, 113–121.

Lee, K.X., Kamyar, S., Miyake, M., Yew, Y.P. (2016). Green synthesis of gold nanoparticles using aqueous extract of *Garcinia mangostana* fruit peels. *J. Nanomat.* 2, 1–7.

Lok, C. (2010). Nanotechnology: Small wonders. *Nature.* 467, 18–21. doi: 10.1038/467018a.

Mallikarjuna, K., Narasimha, G., Dillip, G.R., Praveen, B., Shreedhar, B., Lakshmi, C.S., Raju, B.D.P. (2011). Green synthesis of silver nanoparticles using Ocimum leaf extract and their characterization. *Digest J. Nanomater. Biostruct.* 6, 181–186.

Mathur, A., Parashar, A., Chandrasekaran, N., Mukherjee, A. (2017). Nano-TiO2 enhances biofilm formation in a bacterial isolate from activated sludge of a waste water treatment plant. *Int. Biodeterior. Biodegrad.* 116, 17–25.

Miseljic, M., Olsen, S.I. (2014). Life-cycle assessment of engineered nanomaterials: A literature review of assessment status. *J. Nanopart. Res.* 16, 2427.

Mishra, A., Tripathy, S., Wahab, R., Jeong, S.H., Hwang, I., Yang, Y.B., Kim, Y.S., Shin, H.S., Yun, S.I. (2011). Microbial synthesis of gold nanoparticles using the fungus *Penicillium brevicompactum* and their cytotoxic effects against mouse mayo blast cancer C2C12 cells. *Appl. Microbiol. Biotechnol.* 92, 617–630.

Moloi, M.S., Lehutso, R.F., Erasmus, M., Oberholster, P.J., Thwala, M. (2021). Aquatic environment exposure and toxicity of engineered nanomaterials released from nano-enabled products: Current status and data needs. *Nanomaterials.* 11, 2868.

Mortazavi, S.M., Khatami, M., Sharifi, I., Heli, H., Kaykavousi, K., Poor, M.H.S., Nobre, M.A.L. (2017). Bacterial biosynthesis of gold nanoparticles using *Salmonella enterica* subsp. *enterica serovar typhi* isolated from blood and stool specimens of patients. *J. Cluster Sci.* 28, 2997.

Mullen, E., Morris, M. (2021). Green nanofabrication opportunities in the semiconductor industry: A life cycle perspective. *Nanomaterials.* 11, 1085.

Nanda, A., Saravanan, M. (2009). Biosynthesis of silver nanoparticles from *Staphylococcus aureus* and its antimicrobial activity against MRSA and MRSE. *Nanomedicine.* 5, 452–456.

Nguyen, N.H.A., Von, M.N.R., Slaveykova, V.I., Mackenzie, K., Meckenstock, R.U., Thũmmler, S., Bosch, J., Sevcu, A. (2018). Biological effects of four iron-containing nanoremediation materials on the green alga *Chlamydomonas* sp. *Ecotoxicol. Environ. Saf.* 154, 36–44.

Nurul, U.M. Nizam, M.M., Hanafiah, Kok, S.W. (2021). A content review of life cycle assessment of nanomaterials: Current practices, challenges, and future prospects. *Nanomaterials.* 11, 3324.

Pallas, G., Vijver, M.G., Peijnenburg, W.J.G.M., Guinee, J. (2019). Life cycle assessment of emerging technologies at the lab scale: The case of nanowire-based solar cells. *J. Ind. Ecol.* 24, 193–204.

Parashar, V., Parashar, R., Sharma, B., Pandey, A.C. (2009). Parthenium leaf extract mediated synthesis of silver nanoparticles: A novel approach toward weed utilization. *Digest J. Nanomater. Biostruct.* 4, 45–50.

Parikh, R.Y., Singh, S., Prasad, B.L., Patole, M.S., Sastry, M., Shouche Y.S. (2008). Extracellular synthesis of crystalline silver nanoparticles and molecular evidence of silver resistance from *Morganella* sp.: Toward understanding biochemical synthesis mechanism *Chem. BioChem.* 9, 1415–1422.

Parsons, S., Murphy, R., Lee, J., Sims, G. (2015). Uncertainty communication in the environmental life cycle assessment of carbon nanotubes. *Int. J. Nanotechnol.* 12, 620.

Pati, P., Mcginnis, S., Vikesland, P.J. (2014). Life cycle assessment of "green" nanoparticle synthesis methods. *Environ. Eng. Sci.* 31, 410–420.

Qualhato, G., Rocha, T.L., Oliveira, L.C., Silva, D.M., Cardoso, J.R., Koppe, G.C., Sabóia-Morais, S.M.T. (2017). Genotoxic and mutagenic assessment of iron oxide (maghemite-γ-Fe2O3) nanoparticle in the guppy Poecilia reticulata. *Chemosphere.* 183, 305–314.

Rai, M., Yadav, A., Paul, B. and Gade, A. (2009). Myconanotechnology: A new and emerging science. *CAB International. Appl. Mycology.* (eds M. Rai and P.D. Bridge) 258–267.

Raliya, R., Tarafdar, J.C. (2013). ZnO nanoparticle biosynthesis and its effect on phosphorous-mobilizing enzyme secretion and gum contents in Clusterbean (*Cyamopsis tetragonoloba* L.). *Agric. Res.* 2, 48–57.

Richard, P., Feynman. (1960). There's plenty of room at the bottom. *Eng. Sci.* 22, 22–36.

Salieri, B., Turner, D.A., Nowack, B., Hischier, R. (2018). Life cycle assessment of manufactured nanomaterials: Where are we? *Nano Impact.* 10, 108–120.

Sawle, B.D., Salimath, B., Deshpande, R., Bedre, M.D., Prabhakar, B.K., Venkataraman, A. (2008). Biosynthesis and stabilization of Au and Au-Ag alloy nanoparticles by fungus, *Fusarium semitectum. Sci. Technol. Adv. Mater.* 9(3), 1–6.

Schiwy, A., Maes, H.M., Koske, D., Flecken, M., Schmidt, K.R., Schell, H., Tiehm, A., Kamptner, A., Thümmler, S., Stanjek, H. (2016). The ecotoxic potential of a new zero-valent iron nanomaterial, designed for the elimination of halogenated pollutants, and its effect on reductive dechlorinating microbial communities. *Environ. Pollut.* 216, 419–427.

Shankar, S.S., Rai, A., Ahmad, M.S. (2004). Rapid synthesis of Au, Ag and bimetallic Au core-Ag shell nanoparticles using Neem (*Azadirachta indica*) leaf broth. *J. Colloid and Interf. Sc.* 275, 496–502.

Sinha, A., Khare, S.K. (2011). Mercury bioaccumulation and simultaneous nanoparticle synthesis by *Enterobacter* sp. Cells. *Bioresource Technol.* 102, 4281–4284.

Sintubin, L., De, W., Windt, J. Dick, J. Mast, D.V. Ha, W. Verstraete, N. (2009). Boon Lactic acid bacteria as reducing and capping agent for the fast and efficient production of silver nanoparticles. *Appl. Microbiol. Biotechnol.* 84(4), 741–749.

Victor, M., Martin, F., Diana, D.L.I., Raul, E., Cachau, M.G.R., Joyce, A.M., Casimir, K. (2012). Nanoinformatics: A new area of research in nanomedicine. *Int. J. Nanomed.* 7, 3867–3890.

Walser, T., Gottschalk, F. (2014). Stochastic fate analysis of engineered nanoparticles in incineration plants. *J. Clean. Prod.* 80, 241–251.

Xie, J., Lee, J.Y., Wang, D.I.C., Ting, Y.P. (2007). High-yield synthesis of complex gold nanostructures in a fungal system. *J. Phys. Chem.* 111, 16858–16865.

Yang, Y., Zhang, C., Hu, Z. (2013). Impact of metallic and metal oxide nanoparticles on wastewater treatment and anaerobic digestion. *Environ. Sci. Process. Impacts.* 15, 39–48.

Zhang, H., Li, Q., Lu, Y., Sun, D., Lin, X., Deng, X. (2005). Biosorption and bioreduction of diamine silver complex by *Corynebacterium*. *J. Chem. Technol. Biotechnol.* 80, 285–290.

Zhang, W., Lo, I.M.C., Hu, L., Voon, C.P., Lim, B.L., Versaw, W.K. (2018). Environmental risks of nano zerovalent iron for arsenate remediation: impacts on cytosolic levels of inorganic phosphate and MgATP2– in Arabidopsis thaliana. *Environ. Sci. Technol.* 52, 4385–4392.

Zhang, X., He, X., Wang, K., Yang, X. (2011). Different active biomolecules involved in biosynthesis of gold nanoparticles by three fungus species. *J. Biomed. Nanotechnol.* 7, 245–254.

# 6 Bionanocomposite Membranes and Adsorbents for Water and Wastewater Treatment

*Piyush Gupta, Neha Rana, Subhakanta Dash, Namrata Gupta, Monika Singh, and Sumit Kaushik*

**CONTENTS**



DOI: 10.1201/9781003270959-6

## 6.1 INTRODUCTION

Nanotechnology, in the 21st century, is rebuilding the innovative applications in various fields of science and technology with dimensions ranging from 1 to 100 nm. A material's nanostructure is critical in the advancement of novel properties and in controlling structure at nanoscale. A composite material is formed by combining two or more materials with dissimilar properties. In recent years, the term "bionanocomposite" has been coined to the nanocomposites that is comprised of a biological moiety, i.e., biopolymer (naturally occurring polymer) and an inorganic constituent and have at least one measurement on the nanometer scale. Bionanocomposites have become an emerging class of nanostructured hybrid materials and are also referred as nanobiocomposites, green composites, or biohybrids (Darder et al., 2007). These biohybrid materials, like conventional nanocomposites, consisting of synthetic polymers, exhibit enhanced structural and functional properties that are important for a variety of applications. They have a variety of properties, including biodegradability, biocompatibility, thermal stability, and water solubility. Because of their high surface-to-volume ratio and the increased surface reactivity of nano-sized antimicrobial agents, bionanocomposites provide an ideal environment for antimicrobial systems. This has the ability to inactivate microorganisms to a greater extent than their macro- or microscale counterparts. The most important feature of a biomaterial is its ability to be used in medicine and to be biocompatible (i.e., the material would be able to function properly in the human body without causing major side effects while also producing the desired clinical outcome). Bionanocomposites have excellent mechanical properties, biodegradability, and biocompatibility, which makes them perfect green nanocomposites for a variety of pharmaceutical applications such as vaccinations, drug delivery systems, tissue engineering, and wound dressings (Ratner et al., 2004). They can also be used as an eco-friendly, cost-effective, and renewable film material for food packaging, with enhanced antimicrobial activity (Rhim et al., 2013).

Aquatic systems have been polluted as a result of urbanization, industrialization, land cover change, population growth, and climate change and are directly threatened by various human activities. Bacterial contamination of drinking water causes 80% of diseases in countries such as India. Pharmaceutical manufacturing and consumption have grown rapidly in recent years, owing to advances in the medical sciences. The pharmaceutical industry generates a large amount of waste, which may be toxic to biological life. These pollutants may harm aquatic ecosystems and human health by disrupting the endocrine system and promoting the growth of antibiotic-resistant bacteria (superbugs). They've been found in urban and agricultural wastewater, as well as surface water. The removal of various pollutants from water, such as oils, surfactants, heavy metals, pesticides, other industrial and agricultural wastes; dyes or colorants from different chemical industries, including textile, printing, plastics, and paper is a considerable issue around the world. The main problem of many countries is a lack of fresh and pure water due to contamination. Following that, a variety of treatment processes, such as adsorption, coagulation, flocculation, oxidation, ion exchange, membrane separation, and catalytic degradation, have been used to remove pollutants from water. Many scientists and government agencies are focusing on water treatment methods, and to treat wastewater, a combination of an advanced nanotechnology-oriented approach and conventional methods has provided fascinating benefits among other methods. Because of excellent mechanical strength, high surface area, high chemical reactivity, precise binding action, and low cost, nanoscale composite materials have enormous potential to treat water in a variety of ways for the removal of inorganic/organic pollutants and microorganisms from wastewater. Biopolymer nanocomposites or bionanocomposites, metal nanocomposites, metal oxide nanocomposites, polymer nanocomposites, carbon nanocomposites, and membrane nanocomposites all play an active role in water purification. Bionanocomposites perform better than nanoparticles in terms of selectivity, adsorption capacity, greater number of active sites, and stability. The adsorption of numerous pollutants, such as heavy metal cations and dyes from polluted water, using bionanocomposites has received a lot of attention. Advancements in nanoscience and nanotechnology suggest that the use of nanomaterials can greatly reduce many of the current problems related to water quality (Dar et al., 2020). This chapter sheds light on emerging applications of bionanocomposites (as adsorbents and membranes both) in water/wastewater treatment.

## 6.2 BIONANOCOMPOSITES: AN EMERGING APPROACH

Different databases have been searched for literature related to bionanocomposites and the number of publications per year from 2012 to 2021 is shown in Figure 6.1.

Figure 6.2 throws some light on cluster representation of article availability on different keywords used in bionanocomposite searches.

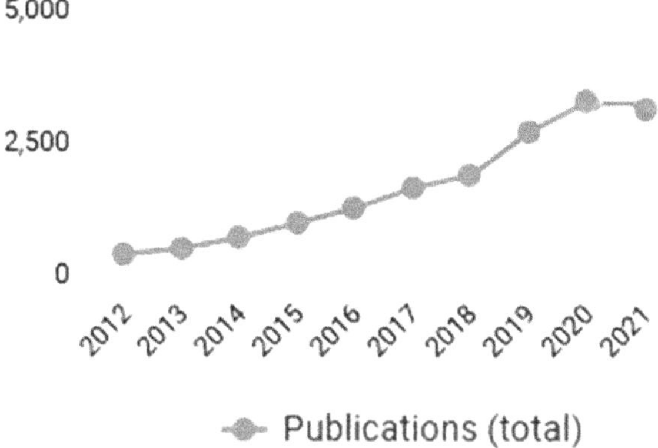


**FIGURE 6.1** Number of publications per year according to literature available in different databases.

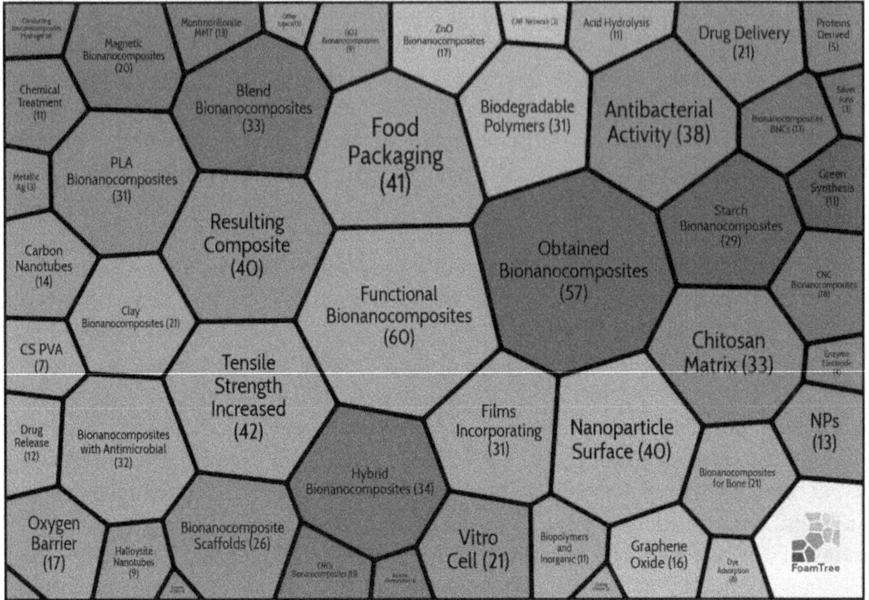


**FIGURE 6.2** Cluster representation of article availability on different keywords used in bionanocomposite search.

### 6.2.1 Classification of Bionanocomposites

Bionanocomposites can be classified into two categories based on the biopolymer used as a principal component (Pande and Sanklecha, 2017):

Polysaccharide-based nanocomposites, like starch, cellulose, lignocellulosic complex (natural fibers), and alginic acid (an edible polysaccharide generally found in brown algae, chitin, chitosan, etc.).

Protein-based nanocomposites (polypeptides of natural origin), like wheat gluten, whey protein, soy protein, corn zein, collagen and gelatin, caseine and caseinate, etc.

### 6.2.2 Preparation of Bionanocomposites

Bionanocomposites can be prepared by various methods (Pande and Sanklecha, 2017) as outlined in the following sections.

#### 6.2.2.1 *In Situ* Intercalative Polymerization

In this process, the nanoparticles are dispersed in a monomer solution (liquid monomer), allowing polymer formation to occur between the intercalated sheets. Polymerization can be accomplished through the use of heat or radiation, the diffusion of a suitable initiator, or the use of an organic initiator or catalyst.

#### 6.2.2.2 Solution Intercalation or Exfoliation-Adsorption

The exfoliation-adsorption method is employed to achieve exfoliation of the layered silicate into single layers using an appropriate solvent. In this system, a biopolymer, such as starch or protein, is mixed with a solvent in which it is completely soluble. Inorganic nanofillers, such as silicate platelets, become swollen in a solvent such as water, chloroform, or toluene. When the biopolymer and swollen nanoparticle solution are mixed, the polymer chains intercalate and displace the solvent within the silicate interlayer. The intercalated structure remains after the solvent is removed, resulting in the formation of a biopolymer/layered silicate bionanocomposite.

#### 6.2.2.3 Template Synthesis

This method uses fundamental sol-gel procedures. This technique is highly adaptable, simple, and straightforward and can be applied to large-scale production. An aqueous solution or gel of water-soluble biopolymers and silicate-building blocks (dispersion of silicate layers) are used to synthesize clay within a polymer matrix in a single-step process to fabricate heterogeneous nanostructured composites. The polymer used serves as a template that provides support for nucleation and growth of silicate crystals.

#### 6.2.2.4 Melt Intercalation

The melt intercalation technique has become the method of choice for producing polymer/layered silicate bionanocomposites. When compared to solution intercalation and *in situ* intercalative polymerization, there are numerous advantages of metal intercalation. The polymer is heated to a specific temperature and mixed with nanoparticles in this process.

### 6.2.3 Characterization of Bionanocomposites

The massive increase in scientific literature produced in the field over the last couple of decades demonstrates the scientific community's growing interest in bionanocomposites. To identify the benefits and performance of these bionanocomposites in practical applications, ranging from the electronics industry to medical diagnostics, their physical properties must first be studied using experimental characterization techniques. Characterization techniques, such as Fourier-transform infrared spectroscopy (FTIR), X-ray diffraction (XRD), SEM, TEM, nuclear magnetic resonance (NMR), thermogravimetric analysis (TGA), differential scanning calorimetry (DSC), and dynamic mechanical thermal analysis (DMTA), are commonly used to investigate structural, morphological, physical, optical, and thermal properties of bionanocomposites. The use of these characterization techniques on bionanocomposites can provide useful information about their physical properties, which can be used in practical applications. Although some characterization techniques, such as XRD, are incapable of fully resolving the properties of bionanocomposites, in such cases, a combination of the characterization techniques listed above can be used to obtain a proper understanding (Abbas et al., 2020). Bionanocomposites are generally characterized by using different methods as outlined below:

- The particles of bionanocomposites are nanoscale in size. It is determined using Motic images of particles taken with a Motic digital microscope or a Malvern Zetasizer (Pande and Sanklecha, 2017).
- Scanning electron microscopy is used to examine the surface morphology of a particle (Pande and Sanklecha, 2017).
- Thermal analysis of bionanocomposites is performed using differential scanning calorimetry. It can determine whether or not the nanoparticles are homogeneously entrapped within the polymer (Pande and Sanklecha, 2017).
- Fourier-transform infrared spectrometer (FTIR) uses the infrared (IR) part of the electromagnetic spectrum as probe radiation to measure the frequencies of the vibrations of chemical bonds between atoms in a given sample. This technique can analyze bionanocomposites such as polychitin nanofibers, CS-ZnO/Ag/halloysite nanotubes, chitin-clay, etc at a faster and more sensitive rate than other IR spectroscopy techniques. FTIR is also a simple method for determining the lower critical solution temperature of bionanocomposites and can be used to confirm the presence of hydroxyl (–OH) and amide ($-NH_2$) groups (Abbas et al., 2020).
- XRD can be used to investigate structural properties of bionanocomposites such as particle size, lattice parameter, dislocation, X-ray density, bulk density, and a variety of other mechanical structures with specific areas, packing factors, strains, and so on. All parameters, however, are affected by the nature of the biomaterials. Combining inorganic nanostructures with biological materials is critical for developing environmentally friendly, cost-effective, and efficient components for catalytic, electrochemical, optical, and magnetic devices (Abbas et al., 2020).

### 6.2.4 Applications of Bionanocomposites

Selvakumar et al. (2015) created an enriched adhesion of talc/ZnO nanocomposites on cotton fabric aided by aloe-vera for biomedical applications – primarily baby diapers.Tanahashi (2010) promoted using polymer nanocomposites as catalysts, gas separation membranes, contact lenses, and bioactive implant materials. Ahmad et al. (2017) concentrated on the medical specialty and cumulative applications of chitosan-based bionanocomposites, such as scaffolds, implants, diagnostics, biomedical devices, tissue engineering, drug and gene delivery, wound healing, and bioimaging. Cherian et al. (2011) created cellulose nanocomposites using nanofibers isolated from pineapple leaf fibers to make a variety of versatile medical implants.

#### 6.2.4.1 Bionanocomposites in Water Treatment

The irresponsible discharge of chemicals such as dyes, pharmaceuticals, pesticides, toxic elements, phenolic compounds, herbicides, personal care products, etc. has a negative impact on water quality and leads to water contamination. These toxins have the potential to be bioaccumulative, persistent, carcinogenic, mutagenic, and toxic to marine species, flora and fauna, and human health. Scientists face significant challenges because water pollution is a growing threat to human health and the environment. Toxin removal from water has become a necessity. Precipitation, coagulation, flocculation, incineration, ion exchange, reverse osmosis, membrane filtration, electrochemistry, photo electrochemistry, advanced oxidation processes (AOPs), adsorption, and biological methods are all used to purify water. Highly advanced nanotechnologies, like bionanocomposites and nanoengineered wastewater treatments could play a significant role in wastewater treatment. Out of all the engineered nanostructured materials, nanocomposites, particularly biopolymer nanocomposites, have received a lot of attention. It is also inexpensive and easily disposable, which may help to reduce pollution. Organic pollutants, inorganic pollutants, and microorganisms/viruses are the three broad categories of wastewater pollutants (Chkirida et al., 2021). The removal of these contaminants from wastewater employing bionanocomposites has been summarized in this section (Figure 6.3).

#### 6.2.4.2 Pathogenic Microorganism Inactivation

Water pollution emerges in a variety of ways, ranging from organic and inorganic pollution to microbiological contamination. The presence of pathogenic microorganisms, such as bacteria, viruses, and parasites, causes the latter. The presence of microorganisms in water is responsible for a variety of infectious diseases that affect the human body. The tendency to control the disinfection of these microorganisms by chlorination, which produces carcinogenic by-products known to be harmful to human health, is no less dangerous. Bionanocomposites have created opportunities for the use and adoption of novel eco-friendly and high-performing materials for antibacterial applications, allowing them to replace traditional techniques.

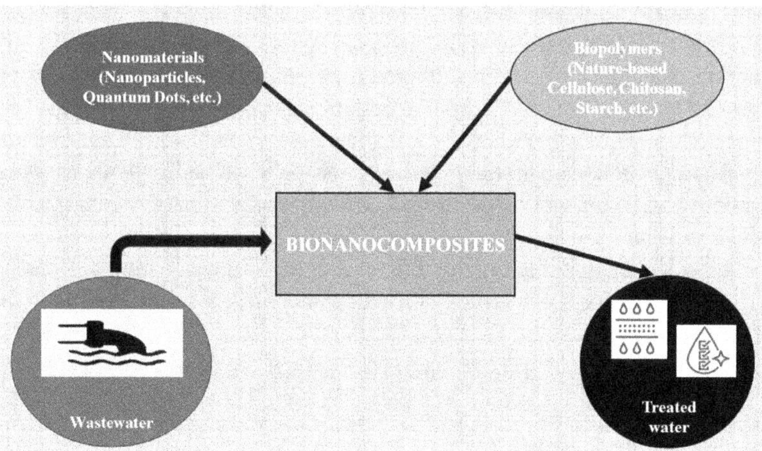


**FIGURE 6.3** Pictorial representations of preparation of bionanocomposites and their application in wastewater treatment.

Ahmad et al. (2019) developed a novel bionanocomposite based on Polyaniline, Cellulose, and Nickel nanoparticles (PANI/C/Ni) through use of chemical oxidative polymerization. The bionanocomposite was designed to support the growth of two pathogenic fungi, *Alternaria alternata* and *Rhizoctonia solani*. The incorporation of nickel nanoparticles into the PANI/C chains significantly improves its fungicidal properties. Indeed, for *A. alternata* and *R. solani*, PANI/C/Ni bionanocomposite inhibited fungal growth more effectively (50 and 42%, respectively) than PANI/C biocomposite. Aside from its antibacterial activity, the bionanocomposite composition is an important factor. This point was emphasized in the study conducted by Nehra et al., (2020). The bionanocomposite demonstrated potent antibacterial activity against *Staphylococcus aureus* and *Escherichia coli* at a minimum concentration of 5 g/ml. The synthesis method involved incorporating Al–Zr metal oxide nanoparticles into humic acid biopolymer through a deep freeze drying process. Xiao et al. (2013) created a nanometer-thick titania/chitosan/Ag-NP bionanocomposite film. The layer-by-layer self-assembly technique was used to coat a uniform titania/chitosan ultrathin biocomposite on cellulose microfibril. The antibacterial activities of bionanocomposites have been tested against gram-negative (*Escherichia coli*) and gram-positive (*Staphylococcus aureus*) bacteria. It was stated that almost all of the inoculated bacteria were disinfected using the bionanocomposite, with increasing antibacterial activities in the order of cellulose/($TiO_2$/CHI)n and cellulose/($TiO_2$/CHI)n/Ag-NP, based on the synergetic effect of cellulose, chitosan, titania, and silver nanoparticles. Chitosan was once again used to coat the surfaces of metallic silver nanoparticles, resulting in a bionanocomposite with high antibacterial activity against *Bacillus subtilis* and *Escherichia coli*. Another advantage of this bionanocomposite is its green, one-pot, and simple synthesis route, which has been shown to maintain antibacterial activity for up to 12 hours (Dutta et al., 2019).

### 6.2.4.3 Organic Pollutants

#### *6.2.4.3.1 COD, BOD, and TOC Removal*

Organic matter encompasses a wide range of organic compounds (over 40 million), making identification difficult, costly, and time-consuming. Three popular parameters are chosen as indicators of organic pollution as an effective alternative for their classification in waste water analysis: chemical oxygen demand (COD), biochemical oxygen demand (BOD), and total organic carbon (TOC). The COD of a water sample is defined as the amount of oxygen required for the oxidation of its oxidizable compounds. The BOD of water, on the other hand, is an approximation of the amount of biochemically degradable organic matter present in a water sample. The amount of oxygen required for the aerobic microorganisms present in the sample to oxidize the organic matter to a stable inorganic form determines it. TOC is another type of organic matter that can be used to determine the level of pollution. TOC determination is critical because it reflects the organic nature of the water sample. The more carbon there is, the more microorganisms grow and, as a result, the more oxygen is consumed. In other words, the higher the COD, BOD, or TOC values of water, the more contaminated it is. As expected, bionanocomposites have been widely used to remove organic matter from contaminated water in the form of COD, BOD, and

TOC. Mostafa et al. (2017) combined nanotechnology and membrane technology to create a new bionanocomposite for wastewater decontamination by entrapping nano zero-valent iron into cellulose acetate membrane. This combination resulted in the development of a reactive membrane with improved mechanical properties, a high surface area, significant resistance to direct fire, and high organic matter removal of 80% for COD and 73% for BOD using the phase inversion technique.

#### 6.2.4.3.2 Phenolic Compound Removal

Phenol (hydroxybenzene) and its derivatives are classified as highly toxic and recalcitrant organic pollutants. These compounds are typically discharged into the environment, after previously serving as starting materials in a variety of chemical industries, as by-products of water chlorination in wastewater treatment plants, or as agricultural resources. Their hazard arises from their high solubility in water, various organic solvents, and oils, as well as the fact that they are carcinogenic, mutagenic, and not very biodegradable. In view of this, wastewater containing phenolic compounds should be treated with caution to prevent their release into the environment. There are several methods for absorbing phenolic compounds, including adsorption, membrane extraction, ozonation, biological processes, electrochemical methods, and liquid–liquid extraction using organic solvents. As a result, phenol and its derivatives have frequently been used as hazardous pollutant models in wastewater treatment, confirming their importance in the field of environmental research. There is a wealth of information available on phenol removal in water decontamination using bionanocomposites. Again, even when dealing with recalcitrant toxic pollutants, these hybrid materials have proven to be extremely effective.

Dinari et al. (2020) used a green method to create a simple recoverable bionanocomposite based on magnetite-Ag/layered double hydroxide/starch for quick reduction of 4-nitrophenol and 2-nitrophenol to aminophenols. The developed bionanocomposite demonstrated improved thermal stability and reusing effectiveness up to the fifth cycle with phenolic compound removal on the order of 90%. Bonardd et al. (2019) conducted a study using chitosan as a biopolymer and gold nanoparticles as reinforcement content. The use of this golden bionanocomposite as a heterogeneous catalyst for the reduction of p-nitrophenol (4-NP) to p-aminophenol has demonstrated significant catalytic performance (4-AP). The addition of gold reinforcements has been found to increase the bionanocomposite's catalytic activity. Yun et al. (2018) used the curing/casting method to create a bionanocomposite film based on three essential components: mung bean starch, PVA, and Zns nanoparticles. In addition to good bisphenol photodegradation, the authors have prioritized improving specific matrix properties such as thermal, mechanical, optical, and water barrier via inorganic nanoparticle addition. Mohammad et al. (2017) have choose to prepare biopolymer–rare earth oxide bionanocomposites for the same purpose. It is about the preparation of cerium oxide $CeO_2$ nanoparticles trimethyl chitosan-loaded bionanocomposite for effective phenol adsorption, as well as its chloro derivatives, 2-chlorophenol and 4-chlorophenol. The combination of the biopolymer and the rare earth oxide created a true synergy, increasing the bionanocomposite's antioxidant and antibacterial capacities while also increasing its adsorption capacity. This latter has gone in the following order: 4-chlorophenol > 2-chlorophenol > phenol. Narayanan and Sakthivel (2011) developed a new green approach for metal bionanocomposite processing by designing a silver bionanocomposite as a heterogeneous catalyst using the fungus *Cylindrocladium floridanum*. The method involves coating fungal mycelia/silver nanoparticles onto the surface of glass beads, where the biological entities catalyze the reduction of toxic p-nitrophenol (4-NP) to p-aminophenol (4-AP).

#### *6.2.4.3.3 Dye Removal*

Dyes are colored chemical substances that are widely used in the textile industry to color fibers. Because of their high, water solubility, light stability, and intense coloration, even at low concentrations, dyes are considered a real challenge to remove from discharged textile wastewater. Aside from being resistant to aerobic digestion and not biologically degradable, dyes are extremely poisonous and even carcinogenic to microbial populations and humans. Dyes are frequently studied in the scientific literature as model molecules to evaluate adsorption and pollution elimination efficiencies of materials. Several studies on the use of bionanocomposites for dye removal have been published due to their wide adaptability, high adsorption capacities, and ease of operation. Kazemi and Javanbakht (2020) developed a bionanocomposite based on chitosan/alginate impregnated with magnetic zeolite in the form of beads for effective cationic dye uptake. The combination of anionic/cationic polysaccharides, zeolite, and magnetic nanoparticles would aid in the removal of dyes from the system. Furthermore, a magnetic separation and regeneration capability of up to four cycles was ensured. Mahmoodi et al. (2020) created a bionanocomposite for Direct Red 23 removal based on nano zeolite and nanocarbon tubes, with a well-ordered enzyme immobilization to provide a high specific area in an attempt to achieve efficient biodegradation. Effectively, enzyme immobilization to a carbon nanotube solid matrix increased the stability and lifetime of bound peptides compared to their native forms.

In general, adsorbents showed a significant decrease in dye removal efficiency when the pH of the solution was changed, particularly in alkaline solutions. Effective adsorbents with a wide pH range are rare to find and study. To address this issue, Ghourbanpour et al. (2019) created an environmentally friendly bionanocomposite based on Fe(III) cross-linked poly (vinyl alcohol) and chitin nanofibers with a high maximum adsorption capacity of 810.4 mg/g for Methyl Orange (MO) that can be maintained almost constant over five cycles. Furthermore, the prepared bionanocomposites can efficiently adsorb MO dye under both alkaline and acidic conditions due to electrostatic interactions between the chitin nanofibers and the adsorbate coupled with chelation between the Fe(III) and methyl orange sulfonic groups. Fan et al. (2018) developed another approach derived from surface-engineering microorganisms for the same purpose: dye removal via bionanocomposites. Because of the combination of biodegradation and physical adsorption, $Fe_3O_4$@MIL-100 core-shell bionanocomposite was found to be more effective for AO10 dye removal than adsorbent alone or free microorganisms. As a result, this strategy highlights the significant potential for dye treatment from industrial effluents.

$TiO_2$ based bionanocomposites have been shown to be competitive for dye removal via a combined effect of adsorption and photocatalysis under UV irradiation. Fathi et al. (2019) proposed embedding $TiO_2$ NPs into sesame protein isolate films for methylene blue (MB) removal. The films had a high dye photodegradation potential as well as a high $O_2$ scavenging activity. $TiO_2$ nanoparticles immobilized in Gum Tragacanth biopolymer to form photocatalyst hydrogels have been shown to be a suitable material for dye removal.

### 6.2.4.4 Inorganic Pollutant Removal

#### *6.2.4.4.1 Oil Removal and Separation*

Free or floating oils are available in a variety of textures, lubricants, grease, fats, and heavy and light hydrocarbons. Oily emulsions released into the environment are stable liquid/liquid systems that obstruct sunlight penetration and oxygen absorption by water bodies. In the absence of adequate separation techniques, this causes an increase in BOD and COD, as well as damage to the natural environment. However, apart from the well-known membrane technology for removing oils, the development of bionanocomposites as new alternatives to draw oils from water has recently been a focus of research. Because of their high porosity and light weight, bionanocomposite aerogels have proven to be excellent absorbents for oil/water separation. Zhuang et al. (2020) reported a tolerant robust superoleophobic aerogel based on $TiO_2$ nanoparticles, reduced graphene oxide (RGO), and alginate. The synthesis of $TiO_2$/RGO/Alginate (TRGA) was a non-toxic, low-cost ionic cross-linking method followed by freeze-drying to produce aerogel. The bionanocomposite combined several functionalities, including 99.96% oil/water separation that could be maintained even after 120 cycles, self-cleaning ability, and excellent recyclability. Yap et al. (2020) created another bionanocomposite that is also decorated with iron oxide magnetic nanoparticles $Fe_3O_4$ for rapid water-oil separation. Based on alginate and reduced graphene, the magnetic bionanocomposite takes the form of foam, which has a highly porous texture that allows water to circulate and pollutants to collect. Because of its unique mixed hydrophobic–hydrophilic nature, the bionanocomposite has demonstrated significant oil absorption of the order of (13–18 g/g) in a very short period of time (30 s). Dai et al. (2019) reported another 3-D porous aerogel for oil/water separation based on $TiO_2$ nanoparticles and sodium alginate. The surface hydrophilicity, in conjunction with microstructure porosity and roughness, has provided the aerogel with a 99.7% oil/water separation capability, as well as improved reusability and antifouling ability. Furthermore, titanium nanoparticles enhance the photocatalytic activity of the aerogel. A plethora of studies have been conducted on the association of biosorbents with magnetic nanoparticles due to their magnetic properties. Indeed, the magnetic behavior of this type of material makes it easily recoverable from aqueous mediums rather than traditional separation techniques, such as centrifugation and filtration. Oil spills typically float on the water's surface for a period of time. Based on these floating characteristics, bionanocomposite adsorbents for this type of handling need to be low-density, easily recoverable, and non-sinkable in order to fully exchange with floating oils and achieve maximum oil removal. Debs et al. (2019) developed a ferromagnetic bionanocomposite (YB-MNP) for oil spill cleanup using a co-precipitation method based on magnetic nanoparticles (MNPs) and yeast biomass. The three oils tested on the bionanocomposite were petroleum (P28API), mixed–used motor oil (MUMO), and new motor oil (NMO). The results showed that the bionanocomposite removed approximately 69% of the oil and was easily recovered due to its superparamagnetic behavior. Lu et al. (2017) developed a bionanocomposite of chitosan grafted magnetic nanoparticles (Cs-MNPs) to evaluate its demulsification performance. After solvothermal synthesis of $Fe_3O_4$ MNPs, a Schiff base reaction was used to coat the surface with amino propyl-functionalized silica (APFS) and aldehyde functionalization with glutaraldehyde for further chitosan (CS) molecular chain grafting. This bionanocomposite's suitability for treating emulsified oil-polluted wastewater is due to its strong magnetic properties and demulsification efficiency under neutral, acidic, and alkaline conditions via electrostatic or hydrophobic interactions. Furthermore, Cs-MNPs bionanocomposite can be reused up to 7 times without significant loss in demulsification performance. De Souza et al. (2010) created another bionanocomposite for oil spill cleanup based on alkyd resin cured with toluene diisocyanate and *in situ* incorporated with magnetic nanoparticles ($\gamma$-$Fe_2O_3$). The new bionanocomposite has two useful features based on its well-proven design of double aromatic/aliphatic nature and magnetic nanoparticles incorporation: good affinity between oil and the composite and easy magnetic recovery.

#### *6.2.4.4.2 Nitrogen and Phosphate Removal*

Nitrogen and phosphate-based compounds are the most common contaminants in drinking water and wastewater. These ions are released into the environment from a variety of sources, including chemical fertilizers, farmland discharges, and so on and are responsible for eutrophication, which occurs when dissolved oxygen is depleted, resulting in the death of aquatic life. Furthermore, at high concentrations, they pose a direct threat to humans by causing diseases and cancers. Biological denitrification and physicochemical processes are the two most commonly used conventional methods for nitrate removal, which require a lot of energy and time to run, are expensive, and require posttreatment for nitrite, nitrate, and ammonia residual wastes. Bionanocomposites have been extensively researched for nitrate and nitrite removal from contaminated water due to their environmentally friendly nature. Zarei et al. (2020) developed an ultra-thin 2D Ag-$TiO_2$/$Al_2O_3$/CS bionanocomposite for photocatalytic nitrate removal using a simple method. The bionanocomposite demonstrated an extra fast and high nitrate removal (74%) in 5 min reaction time in alkaline conditions (pH = 11), excluding any nitrite production as a by-product, due to the synergetic effect of Ag doping $TiO_2$ nanoparticles coupled to $Al_2O_3$ and incorporated into the chitosan biopolymer. He et al. (2018) created a bionanocomposite (Z-Fe/Ni) with Fe/Ni bimetallic nanoparticles incorporated within the inner pores as well as on the outer zeolite surface for simultaneous nitrate reduction and phosphate removal. Ion removal was discovered to involve a variety of mechanisms, including co-precipitation, complexation, and reduction. The results showed that zeolites had a significant dispersal effect all over the surface, promoting the reactivity of the Fe/Ni nanoparticles and improving their reduction performance when compared to non-supported nanoparticles. Qiu et al. (2017) developed an innovative bionanocomposite (Ws-N-La) based on nano-sized La(III) hydroxides anchored within a quaternary amoniated wheat straw within the framework of biomass-supported adsorbent preparation as an economical and effective method for enhanced ion removal (Ws-N). Ws-N-La has been shown to have a higher preferable phosphate adsorption capacity over a wide pH range with no La(III) leaching. Over ten adsorption-desorption cycles, the bionanocomposite demonstrated excellent stability. Song et al. (2016) investigated magnetic amine-crosslinked biopolymer-based corn stalk (MAB-CS) for the same purpose. This study demonstrated the enormous potential of this bionanocomposite for effective nitrate removal, with an adsorption capacity of 102.04 mg/g. Nitrate was removed using a variety of adsorption mechanisms, including ion exchange, surface complex formation, and electrostatic attractions. Furthermore, the bionanocomposite magnetization property indicated that it could be separated and collected from an aqueous medium using an external magnet in less than 10 seconds. Zeolites, like biopolymers, have recently been used as low-cost effective supports for nano-scale particles in water treatment.

#### *6.2.4.4.3 Heavy Metal Removal*

The presence of heavy metals in the environment, even in trace amounts, endangers the fauna, flora, and human life due to their proclivity to accumulate in biological systems. Heavy metal sources are diversifying, and the chemical industrial sectors are the primary source of their production and pollution. Biological, physical, and chemical techniques are used in conventional heavy metal removal processes. There are several established conventional processes for heavy metal uptake from aqueous mediums that range from biological/chemical to physical techniques. Heavy metal ion removal by coagulation/flocculation or expensive membrane filtration techniques has demonstrated high removal efficiencies. They can, however, result in the formation of large particles and concentrated sludge. Other methods, such as electrochemical treatment and oxidation, are quick and practical for removing some metals, but they are generally energy-intensive. Furthermore, biological treatments have been shown to be ineffective. Ion exchange, electrostatic interaction, surface complexation, and physical adsorption are all possible heavy metal removal mechanisms. Bionanocomposite utilization, however, has paved a new, green, and cost-effective path for dealing with heavy metal pollution. Because of their high specific area and abundance of surface functional groups, they are ideal for metal binding through use of adsorption with no harmful by-products. Numerous studies have been published on the adsorption of heavy metals by bionanocomposites, including Hg, Cd, Pb, Cr, Ni, As, and Cr.

Kumar et al. (2020) investigated Hg uptake from simulated wastewater using a bionanocomposite based on chitosan, which is rich in amino and hydroxyl functional groups and embedded with Brassica leaf extract with a maximum adsorption capacity of 50 mg/g for 45 minutes at 60°C. Dinari et al. (2020) recently developed another green bionanocomposite for toxic chromium removal, based on magnetic $Fe_3O_4$ and nickel aluminium layered double hydroxide with guar gum as a polymer. Within

30 minutes, a high chromium adsorption capacity of 101 mg/g was discovered. Bhatt et al. (2019) used chitosan as the matrix and embedded the biopolymer with nano zirconium phosphate to create a unique organic–inorganic bionanocomposite that can be used as a hexavalent chromium effective adsorbent with a maximum of 311.53 mg/g in addition to acting as a catalyst for other oxidative degradation reactions. Lead (Pb) is a cancer-causing element, and scientists have worked to eliminate it from wastewater. Mirza and Ahmad (2018) developed an effective bionanocomposite based on Xanthan gum and montmorillonite for lead uptake from industrial and synthetic wastewater with an adsorption capacity on the order of 187 mg/g within 240 minutes, despite the perilous situation. Furthermore, the bionanocomposite's reusability was demonstrated by a regeneration cycle that could last up to five times. Rais Ahmed et al. (2017) have created yet another outstanding bionanocomposite based on Au-Mica-Alginate, with exceptional Pb adsorption capacities of up to 2000 mg/g. Padilla-Ortega et al. (2016) investigated the removal of cadmium ions using bionanocomposite in the form of macro-porous foams based on chitosan intercalated into vermiculite interlayers. The study found that the bionanocomposite has three times the adsorption capacity of the individual components. Chiew et al. (2016) proposed halloysite/alginate bionanocomposite beads for the uptake of $Pb^{2+}$ from aqueous solutions via a synergism between alginate carboxylate groups and ion exchange. They have achieved an adsorption capacity of up to 325 mg/g, which is roughly four times that of Halloysite nanotubes alone (84 mg/g). Ahmad and Mirza (2015) developed methionine modified bentonite/alginate for the same purpose. The detailed study discovered an adsorption capacity of 217.39 mg/g at 303 K, with the ability to regenerate bionanocomposites up to five times in a row.

#### *6.2.4.4.4 Rare Earth Elements Removal*

Contrary to popular belief, rare earth elements (REEs) are not as limited as their label indicates. These natural elements are abundant in the earth's crust and are commonly found agglomerated in specific types of rocks and minerals. REEs are made up of 17 elements, including scandium, yttrium, and 15 lanthanides (lanthanum, cerium, praseodymium, neodymium, promethium, samarium, europium, gadolinium, terbium, dysprosium, holmium, erbium, thulium, ytterbium, and lutecium). Due to their strategic potential in various fields such as catalysts, magnetics, metallurgy, rechargeable batteries, and superconductors, there has been an increase in demand for REE extraction and separation. Javadian et al. (2020) created a magnetic bionanocomposite by gelating carboxymethyl chitosan, sodium alginate, and magnetic nanoparticles. Adsorption experiments revealed that the uptake efficiency of $Nd^{+3}$, $Tb^{+3}$, and $Dy^{+3}$ was 97.75, 96.83, and 97.85%, respectively. Nkinahamira et al. (2020) created a magnetic bionanocomposite, P-CDP@$Fe_3O_4$, from cross-linked cyclodextrin biopolymer and $Fe_3O_4$ magnetic nanoparticles. Because of the high surface area and porosity of these bionanocomposites, equilibrium was reached in less than 10 minutes, with adsorption capacities of 8.88 mg/g for Nd and 7.76 mg/g for Gd, respectively. These bionanocomposites have additional properties such as easy separation by an external magnetic field, high selectivity towards REEs in the presence of other competing ions, and regeneration for at least five cycles. Iftekhar et al. (2017)

developed a bionanocomposite adsorbent based on cellulose and Zn/Al layered double hydroxide (LDH) for the uptake of Ytterbium, Lanthanum, and Cerium for the same purpose. Cellulose was intercalated into LDH to create a hybrid bionanocomposite capable of rapid REE ion uptake. With capacities of 102.25, 92.51, and 96.25 mg/g for $Y^{3+}$, $La^{3+}$, and $Ce^{3+}$, respectively, it demonstrated good adsorption capacities and selectivity. This bionanocomposite has good reusability for up to five cycles in addition to high adsorption capacities.

#### 6.2.4.5 Bionanocomposites as Adsorbents for Sorption of Various Contaminants

Contaminants like pharmaceutical compounds have certain physicochemical and biological characteristics, as well as functional groups. A number of methodologies have been explored for removing pharmaceuticals from polluted water including bionanocomposites, ozonation, biodegradation, the Fenton process, photocatalysis, adsorption on chitosan beads, adsorption on activated carbon, and removal by micro- and nanoscale iron particles. Nonsteroidal anti-inflammatory drugs (NSAIDs), such as Naproxen, retard several hormones responsible for causing inflammation and pain (Dietrich et al., 2005). Rafati et al. (2016) modified nanoclay with β-cyclodextrin before polymerizing it with polyvinylpyrrolidone (PVP) to add more surfaces for better adsorption. The results revealed that factors such as the initial concentration of naproxen, the amount of adsorbent, and the pH of the solution all had an effect on adsorption. The optimum adsorbent dosage, contact time, initial naproxen concentration, and pH were estimated to be 1 g, 120 minutes, 10 $mgL^{-1}$, and 6, respectively. Adsorption efficiency of 92.2% was also achieved. Based on the findings, it is possible to conclude that the adsorption method using the modified nanocomposite is an efficient, simple, and reliable method for removing naproxen from aqueous solution (Rafati et al., 2016). A new graphite oxide/poly(acrylic acid) grafted chitosan nanocomposite (GO/CSA) was created and used as a biosorbent to remove a pharmaceutical compound (dorzolamide) from contaminated water (Kyzas et al., 2014). Although the use of bionanocomposites in water remediation is in its early stages, these are the most promising materials for use as green and sustainable adsorbents for the safe and clean removal of various water pollutants. As green membranes, bionanocomposites combine the benefits of natural polymers as well as inorganic and organic compounds. A methionine-modified bentonite/alginate (meth-bent/alg) nanocomposite was used as an adsorbent for heavy metal removal such as Pb(II) and Cd(II). This was discovered to be an effective adsorbent for heavy metal remediation up to 98 and 82% (Ahmad and Mirza, 2015). Because of their lack of toxicity, chemical reactivity, and hydrophilicity, bentonites are a better support for composite materials. A simple sorption technique was used to modify the biobased ligand (L-methionine) on bentonite. Because of its amino, carboxylic, and thiol ligand side chains, L-methionine is an intriguing nontoxic biomolecule in the heavy metal trapping field (Faghihian and Nejati-Yazdinejad, 2009). The removal of textile dyes (acid red 88) from aqueous solutions was accomplished using a biosilica/chitosan nanocomposite, and it was discovered that the amount of adsorbed AR88 ($mg\ g^{-1}$) increased with increasing reaction time and adsorbate concentration, as well as decreasing temperature and initial pH (Soltani et al., 2013). Increasing the adsorption dosage from 1 to 3 g $L^{-1}$ resulted in a rapid increase in adsorption, while increasing the adsorption dosage further resulted in an insignificant increase in adsorption (1.66 mg $g^{-1}$). The fast adsorption kinetics following the pseudo-second-order model, which is based on the chemisorption phenomenon, is a significant advantage of this system over other reported adsorbents (e.g., starch and cellulose). In addition, the hydrogel composite is made from biodegradable polymers, such as starch and cellulose. As a result, this starch/cellulose nanowhiskers (CNWs) hydrogel composite demonstrated an exceptional ability to be used in the removal of MB contamination (Gomes et al., 2010). For the removal of MB cationic dye from its aqueous solution, batch adsorption methods were used with chitosan-g-poly (acrylic acid)/montmorillonite (CTS-g PAA/MMT) nanocomposites as adsorbent. The study discovered that as the pH of the nanocomposite increased, so did its adsorption capacity for MB. The CTS-g-PAA/MMT nanocomposite can be considered an effective adsorbent for the removal of MB from aqueous solution (Wang et al., 2008). A novel nanocomposite (h-XG/SiO2) was developed and demonstrated to be an effective adsorbent for the removal of toxic MB and methyl violet (MV) from aqueous solution. This nanocomposite contains hydrolyzed polyacrylamide grafted onto xanthan gum, as well as nanosilica, and has a higher dye adsorbent capacity. The high hydrodynamic volume and specific surface area of the nanocomposite were attributed to the uniform distribution of $SiO_2$ nanoparticles in the polymer matrix of partially hydrolyzed polyacrylamide grafted onto xanthan gum. The adsorption capacity of h-XG/SiO2 was significantly higher than that of other reported cationic dye removal adsorbents, and it also demonstrated good recyclability. As a result, h-XG/SiO2-based nanocomposites may be advantageous adsorbents for the selective removal of cationic dye pollutants from aqueous solution (Ghorai et al., 2014).

Graphene oxide-based composite hydrogels were designed and prepared using chitosan and graphene oxide self-assembly. The catalytic capacity results show that the prepared GO-based composite hydrogels effectively remove two tested dye molecules, MB and Rhodamine B, from contaminated water in good agreement with the pseudo-second-order model (Jiao et al., 2015).

In many areas, groundwater is a significant source of drinking water. Groundwater sources for drinking water supply have been found to have higher levels of dissolved arsenic in several areas, which is mobilized through aquifer As-containing mineral oxidation. Because these are the most common and toxic species in natural waters, inorganic arsenate (As(V)) and arsenite (As(III)) pose the greatest threat to human health. Because of its higher mobility in the environment and faster cellular uptake, arsenite is thought to be more toxic than arsenate. A novel chitosan iron (oxyhydr) oxide composite material for the removal of arsenic from contaminated water supplies has been developed. The iron (oxyhydr) oxide phase has been identified as a nano-sized goethite that is well distributed in the chitosan matrix, giving rise to the term "chitosan goethite bionanocomposite" (CGB). CGB is a promising material for the elimination of arsenic, particularly in developing countries with a wide range of socioeconomic and traditional sanitation and water purification limits (He et al., 2016).

Saifuddin et al. (2011) demonstrated the synthesis of Ag/chitosan bionanocomposites via a melt intercalation process in which a silver nitrate solution and a chitosan solution were combined and microwave-irradiated for pesticide removal. When exposed to irradiation, silver nitrate was reduced, resulting in the formation of Ag nanoparticles. Thus, in column mode, the synthesized Ag/chitosan bionanocomposites removed more atrazine pesticide. Dehaghi et al. (2014) reported the preparation of bionanocomposites with two nanofillers and biopolymers, as well as their application in the removal of dichlorvos. Higher dichlorvos removal by CuO/montmorillonite/chitosan and CuO/montmorillonite/gum ghatti bionanocomposites suggests that biopolymers could play an important role in pesticide adsorption.

Atef et al. (2015) investigated the antibacterial properties of agar/cellulose bionanocomposite films containing a savory essential oil (SEO) and discovered that they were more effective against gram-positive bacteria (*Listeria monocytogenes*, *Staphylococcus aureus*, and *Bacillus cereus*) than gram-negative bacteria (*E. coli*). SEO concentrations of 0.5, 1.0, and 1.5% were incorporated into agar-based nanocomposite films as active packaging during the study to evaluate their physical, mechanical, and antimicrobial properties. The addition of SEO to nanocomposite films reduced their tensile strength and Young's modulus while increasing their percentage of elongation at break. Huang et al. (2000) investigated the effects of soluble soybean polysaccharide (SSPS) films reinforced with nanotitanium dioxide ($TiO_2$-N) on the growth of *E. coli* and *S. aureus*. Without the addition of $TiO_2$-N, SSPS films exhibited no antimicrobial activity and revealed no inhibition zones. The inhibition zone of nanoincorporated SSPS films increased significantly as $TiO_2$-N concentrations increased. This finding implies that SSPS films containing $TiO_2$-N nanoparticles can act as antimicrobial films against microorganisms.

### 6.2.4.6 Bionanocomposites as Membranes in Water Depollution

There are numerous membranes made of polymer materials coated with various biomaterials that have a wide range of applications in decontamination and water treatment. Membranes are designed to remove fine and large particles; however, the process is highly dependent on the pore size of the membrane. Particles of various sizes are used in various processes such as microfiltration, nanofiltration, ultrafiltration, and reverse osmosis. Other processes, such as conventional filtration, filter only small particles that cannot be seen with the naked eye, whereas microfiltration isolates only particles larger than 10 μm, in size, such as microorganisms. The ultrafiltration method is used to separate particles smaller than 1000 Å, such as macromolecules or colloids. Both nanofiltration and reverse osmosis membranes can typically separate ions and salts, but both require pressure to function. After analyzing and comparing various types of membranes, biopolymeric membranes have caught the interest of researchers because they are inexpensive, biodegradable, and safe for humans, animals, and the environment. They discovered that chitosan is the most commonly used material after investigating the difficulties in obtaining pure drinking water and the aforementioned characteristics. Natural and synthetic polymers were studied, and it was found that chitosan-based materials had important properties including biocompatibility, biodegradability, low-cost, nontoxicity as well as antibacterial and antioxidant activities, making them suitable for use in water treatment applications. Chitosan, the most widely used hydrophilic biopolymer, has numerous advantages for use in applications such as biomedicine, biosensors, the food industry, cosmetics, and agriculture, but its most widely evaluated application is water treatment. Chitosan's excellent antimicrobial activity has attracted the interest of those engaged in using it in membranes for industrial desalination. Chitosan functionalized with graphene oxide is an excellent adsorbent for heavy metal removal from wastewater systems. Furthermore, adding an inorganic filler, ZnO, to the chitosan membrane composition improved the adsorptive capacity, fouling performance, and membrane durability. Using this type of filler in the composition of chitosan-based membranes improves their performance significantly. Furthermore, the addition of different nanomaterials as fillers (C nanotubes, ZnO, Ag, Cu, and $TiO_2$) could overcome their limitations by increasing retention, hydrophilicity, thermo-mechanical stability, and swelling, which prevents its use at a larger scale. As a result, this significant achievement would be a significant breakthrough for the water treatment industry. Chitosan's applicability in nanocomposite membranes could overcome its limitations and yield significant results (Dongre et al., 2019; Spoiala et al., 2021).

Tang et al. (2017) combined chitosan (CTS) and polyvinylidene fluoride (PVDF) in a mixture (CTS/PVDF) to create a composite in a single step. The chitosan-based composite polymeric membrane has excellent mechanical strength and good selectivity over Pb when compared to Cd(II). Wang et al. (2016) used immersion precipitation to achieve Cu(II) adsorption on a porous chitosan membrane. Silica was used as a porogen material in the development of a porous chitosan membrane. The polymeric membranes obtained had a reticular structure with pores ranging from 1.9 to 4.6 m and an adsorption capacity of 87.5 mg/g. Habiba et al. (2017) used electrospinning to create an intriguing membrane based on chitosan/polyvinyl alcohol (PVA)/zeolite nanofibrous composite. Toxins can be effectively removed from wastewater using a chitosan/PVA/zeolite membrane. It also establishes that the synthesized composite polymeric membrane had regeneration and reusability capacity, as well as no loss of adsorption capacity. Makaremi et al. (2016) created an intriguing functionalized chitosan-based membrane. They functionalized the chitosan membrane with polyacrylonitrile nanofibers and ZnO nanoparticles to improve its filtration capability, mechanical properties, and antibacterial activity. After testing this type of porous membrane, it was discovered that adding nanofibers and nanoparticles improved Cr(III) adsorption performance. Bibi et al. (2015) investigated the adsorption capacity of a membrane-based chitosan, nanotubes, and PVA with silane for the removal of naphthalene. Nitrophenol is another organic pollutant that must be removed from water through the use of chitosan-based nanocomposite polymeric membranes. To identify and eliminate 4-nitrophenol, Khan et al. (2019) created a chitosan/GO nanocomposite coated with Cu nanoparticles. The results revealed a constant reduction rate, high sensitivity, and accurate detection.

Pinem et al. (2019) investigated the synthesis and characterization of chitosan-silica membranes for the treatment of hotel wastewater containing polyethylene glycol and PVA. The study's goal was to determine the mechanical properties, permeability, and selectivity of membranes, as well as the effect of membrane composition on their performance. Experiments revealed that increasing the amount of polyethylene glycol increased the pore diameter while decreasing the flux rate; however, the rejection value was lower. Adding PVA, on the other hand, shrank the pore diameter and increased the rejection value while decreasing the flux rate. Rosdi et al. (2019) conducted another study to investigate the role of silica in improving chitosan membrane performance in removing Pb(II) ions from water solutions. The results showed that composite polymeric membranes outperformed pure chitosan membranes in terms of lead removal efficiency.

Another study involved the use of chitosan/zeolite composite membranes for the removal of trace metal ions, such as Cr, As, Cd, and Pb from wastewater. For this study, researchers used chitosan in conjunction with zeolites to create composite membranes filled with glutaraldehyde in order to eliminate the previously stated metal cations via the evacuation permeation process (EPP). Finally, the results demonstrated the potential utility of chitosan/zeolite composite membranes in wastewater purification (Truong et al., 2019).

This study was carried out to design novel, environmentally friendly membranes by *in situ* and *ex situ* routes based on bacterial cellulose (BC) as a template for the chitosan (Ch) as functional entity for copper elimination in wastewater. Two paths led to bionanocomposites with distinct characteristics and physicochemical properties. The mechanical behavior in wet state, which is strongly related to crystallinity and water holding capacity, differed greatly depending on the preparation route, despite the fact that the Ch content was very similar: 35 and 37 wt% for *in situ* and *ex situ* membranes, respectively. The morphological characterization suggests that the Ch would be better incorporated into the BC matrix via the *in situ* route. The copper removal capacity of these membranes was investigated, and *in situ* prepared membranes demonstrated the highest values, around 50%, for initial concentrations of 50 and 250 mg L-1. Furthermore, the membranes' reusability was evaluated. For the first time, the entire 3D nano network BC membrane was used to provide physical integrity for chitosan in order to develop eco-friendly membranes with potential applications in heavy metal removal. The potential of combining the bacterial cellulose (BC) nanofibril network as a template and chitosan (Ch) as the active phase of membranes prepared by two different preparation routes, *in situ* and *ex situ*, for the elimination of copper in wastewater is investigated in this study. *In situ* modified BC membranes were created by adding different concentrations of Ch to the culture media during BC biosynthesis, whereas *ex situ* modification was accomplished by immersing BC membranes in Ch-acetic acid solutions. The two paths resulted in bionanocomposites with varying physicochemical properties. The FTIR spectra of *in situ* membranes show new bands indicating interactions between Ch and BC, whereas *ex situ* membranes show characteristic peaks of both cellulose and chitosan but no new peaks. The crystallinity and water holding capacity, which are strongly related to mechanical performance, differed greatly depending on the preparation route. Because of the lower interaction with water molecules, *in situ* prepared membranes outperformed *ex situ* and neat BC membranes in the wet state. This was confirmed by a swelling study after 24 hours, which revealed that the *in situ* prepared membrane had a WHC that was 3.5 times lower than the *ex situ* membrane and 8.5 times lower than the neat BC. SEM and AFM morphological characterization revealed a more homogeneous reticulated structure with the addition of Ch during BC biosynthesis, implying a better incorporation of Ch into the BC matrix via the *in situ* route. BET analysis was used to determine the surface area, pore diameter, and volume of BC/Ch membranes, all of which were lower than the neat BC due to the filling of the pores and compaction by the Ch. Finally, the copper removal efficiency of these novel membranes was evaluated in two solutions with varying copper concentrations and compared to that of traditional glutaraldehyde crosslinked Ch membranes, with *in situ* prepared membranes demonstrating higher removal capacity. Although the prepared membranes had similar compositions (35 and 37% Ch for *in situ* and *ex situ* membranes, respectively), incorporation of the Ch into the BC matrix via the *in situ* route results in a more homogeneous network with less tendency to swell in water and thus a better interaction with the copper ions. Furthermore, the reusability of the membranes was evaluated, and after two cycles, a decrease in removal efficiency of less than 10% was observed. As a result, these findings point to the possibility of developing fully biobased reusable filtering membranes for copper removal from wastewater (Urbina et al., 2018).

Polyvinylidene fluoride (PVDF) membrane is a trademark known and recognized as a polymeric hydrophobic membrane. To address PVDF membrane limitations, such as fouling caused by strong hydrophobic-hydrophobic interactions, a coating trend of the membrane with bionanocomposites has emerged. The PVDF/bionanocomposite membranes have an underwater superoleophobic top surface as well as a hydrophobic substrate. This hydrophilic/oleophobic structure produces a variety of synergistic effects that improve oil removal efficiency by preventing oil adhesion on the membrane surface Indeed, the modified PVDF membrane's hydrophilic surface can effectively react with water to form a hydration layer that is responsible for foulant rejection (Liu et al., 2016). Liu et al. (2016) reported a modified PVDF membrane with chitosan-silica nanoparticles for ultra-low oil adhesion, underwater superoleophobicity, and super hydrophilicity. The membrane coating has resulted in an exceptional water-oil separation capability of>99.0%. Ardeshiri et al. (2019) investigated chitosan and silica coated PVDF via a one-step coating route to prepare an oleophobic/hydrophilic bionanocomposite coated on the PVDF membrane surface. Water/gasoline emulsions were tested for separation using a pressure process. The hydrophilic layer coating resulted in higher flux permeate and 99% total oil rejection of the modified membrane over the unmodified membrane.

## 6.3 REGENERATION STUDIES

Chemical procedures are commonly used to desorb pollutants from bionanocomposites. This approach has a number of benefits over the others; it is a less energy-intensive procedure that yields high adsorbate recovery without causing damage to the adsorbent. Solvent washing with well-chosen eluents or desorption agents is used to achieve chemical regeneration. The neutral ideal sorption pH of a bionanocomposite adsorbent needs an acidic desorption eluent, while bounded pollutants under acidic circumstances necessitate alkaline eluents. The presence of hydronium ions ($H_3O^+$)

in acidic eluents (the most widely used being HCl or $HNO_3$) lowers adsorbate attraction to adsorbent active groups and facilitates desorption via ion exchange processes. Because the adsorbate prefers Na+ ions over adsorbent active sites, basic eluents, such as NaOH, can minimize adsorbent–adsorbate interaction in alkaline medium; they can also give superior regeneration performance. Organic solvents such as acetonitrile, isopropanol, ethanol, acetone, and methanol were also studied as regeneration eluents. Due to their tiny and polar molecules, these organic solvents may easily enter the adsorbent interlayer region, interact with the adsorbate functional groups, and subsequently desorb the adsorbate molecules. Chelating compounds, such as EDTA-disodium ($Na_2EDTA$) and ethylene diamine tetraacetic acid (EDTA) were also employed to inhibit surface hydrolysis of the bionanocomposite matrix produced by acid eluents in bionanocomposite regeneration (especially for polysaccharides). These sodium salts, which include –COOH groups and N atoms, are prone to forming EDTA-metal complexes. Regeneration can also be done with sodium chloride (NaCl). Because of its ionic species ($Na^+$ and $Cl^-$), it can interact with the active sites of the adsorbent and release contaminants (Chkirida et al., 2021).

## 6.4 BENEFITS OF BIONANOCOMPOSITES

The biocompatibility of bionanocomposites favors their applications in medicine, food packaging, cosmetics, and wastewater treatment. Chitosan-based bionanocomposites do not show any chronic toxicity, when or where they are applied on animals. Biodegradable polymers get decomposed under aerobic/anaerobic conditions by action of microorganism/enzymes. This feature of bionanocomposites has attracted stakeholders to make it available for human usage, without causing any harm to the human body. Bionanocomposites like chitosan-based, cellulose-based bionanocomposites are capable of forming membranes and thin films to be used in water treatment and as food packaging. Tensile strength is a mandatory characteristic in the film-forming process because fragile films tend to have cracks thus causing contamination of food items. They must be capable of protecting the products from degradation.

The addition of nanoparticles to the biopolymer matrix increases the bionanocomposites' mechanical characteristics (rigidity and tensile strength). The mechanical characteristics of chitosan-based bionanocomposites can be improved by combining nanofillers with surface-functionalized nanofillers, according to researchers. Poly N-vinyl pyrrolidone, polyvinyl alcohol, polyethylene glycol, polyethylene oxide, collagen, cellulose, carboxymethyl cellulose, sulfonated cellulose fibers, and cellulose acetate are some examples of materials that might help improve mechanical properties. Antimicrobial activity of chitosan bionanocomposites has been shown against a wide range of microorganisms, including yeast, filamentous fungus, gram-negative and gram-positive bacteria which are very useful in water treatment. The activity of chitosan is dependent on the kind of chitosan used as well as other parameters (such as the quantity of free amino groups in the molecule). It is utilized to treat oxidative stress and some disorders because of its radical scavenging capacity and antioxidant qualities. Chitosan's biocidal effect on microorganisms has a lot of economic potential in the medical and agricultural fields. Chitosan-based bionanocomposites have antifungal qualities that aid in the management of fruit and agricultural diseases (Azmana et al., 2021).

## 6.5 CONCLUSION

Increased water demand and a scarcity of clean water necessitate the development of new technologies for removing contaminated products from wastewater. The most widespread issue affecting people all over the world is a lack of clean water. As a result, a significant amount of research has been dedicated to improving this global water pollution problem. Thus, the purpose of this chapter was to compile the most recent information available in the literature on the use of bionanocomposites as environmentally friendly materials, which play a significant role in removal of various contaminants commonly found in wastewater. Bionanocomposites have excellent mechanical properties, biodegradability, and biocompatibility, making them the ideal green nanocomposites for a variety of applications. Changes in the structures and properties of both nanomaterials and host matrices, as well as the pattern of nanomaterials within the host matrices (biopolymer) of the membranes/films, could be among the top priorities in the field of bionanocomposites in wastewater treatment.

## REFERENCES

Ahmad, M., Manzoor, K., Singh, S., and Ikram, S. (2017). Chitosan centered bionanocomposites for medical specialty and curative applications: A review. *Int. J. Pharm.* 529(1–2), 200–217.

Ahmad, N., Sultana, S., Kumar, G., Zuhaib, M., Sabir, S., and Khan, M.Z. (2019). Polyaniline based hybrid bionanocomposites with enhanced visible light photocatalytic activity and antifungal activity. *J. Environ. Chem. Eng.* 7(1), 102804.

Ahmad, R., and Mirza, A. (2015). Sequestration of heavy metal ions by methionine modified bentonite/Alginate (Meth-bent/Alg): A bionanocomposite. Groundw. *Sustain. Dev.* 1(1–2), 50–58.

Alay-e-Abbas, S.M., Mahmood, K., Ali, A., Arshad, M.I., Amin, N., and Hasan, M.S. (2020). Characterization techniques for bionanocomposites. In *Bionanocomposites* (pp. 105–144). Elsevier.

Ardeshiri, F., Akbari, A., Peyravi, M., and Jahanshahi, M. (2019). A hydrophilic-oleophobic chitosan/$SiO_2$ composite membrane to enhance oil fouling resistance in membrane distillation. *Korean J. Chem. Eng.* 36(2), 255–264.

Atef, M., Rezaei, M., and Behrooz, R. (2015). Characterization of physical, mechanical, and antibacterial properties of agar-cellulose bionanocomposite films incorporated with savory essential oil. *Food Hydrocoll.* 45, 150–157.

Azmana, M., Mahmood, S., Hilles, A.R., Rahman, A., Arifin, M.A.B., and Ahmed, S. (2021). A review on chitosan and chitosan-based bionanocomposites: Promising material for combatting global issues and its applications. *Int. J. Biol. Macromol.* 185, 832–848.

Bhatt, R., Ageetha, V., Rathod, S.B., and Padmaja, P. (2019). Self-assembled chitosan-zirconium phosphate nanostructures for adsorption of chromium and degradation of dyes. *Carbohydr. Polym.* 208, 441–450.

Bibi, S., Yasin, T., Hassan, S., Riaz, M., and Nawaz, M. (2015). Chitosan/CNTs green nanocomposite membrane: Synthesis, swelling and polyaromatic hydrocarbons removal. *Mater. Sci. Eng. C.* 46, 359–365.

Bonardd, S., Saldías, C., Ramírez, O., Radić, D., Recio, F.J., Urzúa, M., and Leiva, A. (2019). A novel environmentally friendly method in solid phase for in situ synthesis of chitosan-gold bionanocomposites with catalytic applications. *Carbohydr. Polym.* 207, 533–541.

Cherian, B.M., Leão, A.L., de Souza, S.F., Costa, L.M.M., de Olyveira, G.M., Kottaisamy, M., Nagarajan, E.R. and Thomas, S. (2011). Cellulose nanocomposites with nanofibers isolated from pineapple leaf fibers for medical applications. *Carbohydr. Polym.* 86(4), 1790–1798.

Chiew, C.S.C., Yeoh, H.K., Pasbakhsh, P., Krishnaiah, K., Poh, P.E., Tey, B.T., and Chan, E.S. (2016). Halloysite/alginate nanocomposite beads: Kinetics, equilibrium and mechanism for lead adsorption. *Appl. Clay Sci.* 119, 301–310.

Chkirida, S., Zari, N., and Bouhfid, R. (2021). Insight into the bionanocomposite applications on wastewater decontamination. *J. Water Process. Eng.* 43, 102198.

Dai, J., Tian, Q., Sun, Q., Wei, W., Zhuang, J., Liu, M., Cao, Z., Xie, W., and Fan, M. (2019). $TiO_2$-alginate composite aerogels as novel oil/water separation and wastewater remediation filters. *Compos. B Eng.* 160, 480–487.

Dar, O.A., Malik, M.A., Talukdar, M.I.A., and Hashmi, A.A. (2020). Bionanocomposites in water treatment. In *Bionanocomposites* (pp. 505–518). Elsevier.

Darder, M., Aranda, P., and Ruiz-Hitzky, E. (2007). Bionanocomposites: A new concept of ecological, bioinspired, and functional hybrid materials. *Adv. Mater.* 19(10), 1309–1319.

Debs, K.B., Cardona, D.S., da Silva, H.D., Nassar, N.N., Carrilho, E.N., Haddad, P.S., and Labuto, G. (2019). Oil spill cleanup employing magnetite nanoparticles and yeast-based magnetic bionanocomposite. *J. Environ. Manage.* 230, 405–412.

Dehaghi, S.M., Rahmanifar, B., Moradi, A.M., and Azar, P.A. (2014). Removal of permethrin pesticide from water by chitosan–zinc oxide nanoparticles composite as an adsorbent. *J. Saudi Chem. Soc.* 18(4), 348–355.

Dietrich, D.R., Webb, S., and Petry, T. (2005). *Hot Spot Pollutants: Pharmaceuticals in the Environment.* Elsevier.

Dinari, M., Shirani, M.A., Maleki, M.H., and Tabatabaeian, R. (2020). Green cross-linked bionanocomposite of magnetic layered double hydroxide/guar gum polymer as an efficient adsorbent of Cr (VI) from aqueous solution. *Carbohydr. Polym.* 236, 116070.

Dongre, R.S., Sadasivuni, K.K., Deshmukh, K., Mehta, A., Basu, S., Meshram, J.S., Al-Maadeed, M.A.A., and Karim, A. (2019). Natural polymer based composite membranes for water purification: A review. *Polym Plast Technol Eng.* 58(12), 1295–1310.

Dutta, T., Ghosh, N.N., Chattopadhyay, A.P., and Das, M. (2019). Chitosan encapsulated water-soluble silver bionanocomposite for size-dependent antibacterial activity. *Nano-Struct. Nano-Objects.* 20, 100393.

Faghihian, H., and Nejati-Yazdinejad, M. (2009). Sorption performance of cysteine-modified bentonite in heavy metals uptake. *J. Serbian Chem. Soc.* 74(7), 833–843.

Fan, J., Chen, D., Li, N., Xu, Q., Li, H., He, J., and Lu, J. (2018). Adsorption and biodegradation of dye in wastewater with $Fe_3O_4$@ MIL-100 (Fe) core–shell bio-nanocomposites. *Chemosphere.* 191, 315–323.

Fathi, N., Almasi, H., and Pirouzifard, M.K. (2019). Sesame protein isolate based bionanocomposite films incorporated with $TiO_2$ nanoparticles: Study on morphological, physical and photocatalytic properties. *Polym. Test.* 77, 105919.

Ghorai, S., Sarkar, A., Raoufi, M., Panda, A. B., Schönherr, H., and Pal, S. (2014). Enhanced removal of methylene blue and methyl violet dyes from aqueous solution using a nanocomposite of hydrolyzed polyacrylamide grafted xanthan gum and incorporated nanosilica. *ACS Appl. Mater. Interfaces.* 6(7), 4766–4777.

Ghourbanpour, J., Sabzi, M., and Shafagh, N. (2019). Effective dye adsorption behavior of poly (vinyl alcohol)/chitin nanofiber/Fe (III) complex. *Int. J. Biol. Macromol.* 137, 296–306.

Gomes de Souza Jr, F., Marins, J.A., Rodrigues, C.H., and Pinto, J.C. (2010). A magnetic composite for cleaning of oil spills on water. *Macromol. Mater. Eng.* 295(10), 942–948.

Habiba, U., Afifi, A.M., Salleh, A., and Ang, B.C. (2017). Chitosan/(polyvinyl alcohol)/zeolite electrospun composite nanofibrous membrane for adsorption of $Cr^{6+}$, $Fe^{3+}$ and $Ni^{2+}$. *J. Hazard. Mater.* 322, 182–194.

He, J., Bardelli, F., Gehin, A., Silvester, E., and Charlet, L. (2016). Novel chitosan goethite bionanocomposite beads for arsenic remediation. *Water Res.* 101, 1–9.

He, Y., Lin, H., Dong, Y., Li, B., Wang, L., Chu, S., Luo, M., and Liu, J. (2018). Zeolite supported Fe/Ni bimetallic nanoparticles for simultaneous removal of nitrate and phosphate: Synergistic effect and mechanism. *Chem. Eng. J.* 347, 669–681.

Huang, Z., Maness, P.C., Blake, D.M., Wolfrum, E.J., Smolinski, S.L., and Jacoby, W. A. (2000). Bactericidal mode of titanium dioxide photocatalysis. *J. Photochem. Photobiol. A: Chem.* 130(2–3), 163–170.

Iftekhar, S., Srivastava, V., and Sillanpää, M. (2017). Synthesis and application of LDH intercalated cellulose nanocomposite for separation of rare earth elements (REEs). *Chem. Eng. J.* 309, 130–139.

Javadian, H., Ruiz, M., Saleh, T. A., and Sastre, A.M. (2020). Ca-alginate/carboxymethyl chitosan/Ni0. 2Zn0.2$Fe_2$.6$O_4$ magnetic bionanocomposite: Synthesis, characterization and application for single adsorption of $Nd^{+3}$, $Tb^{+3}$, and $Dy^{+3}$ rare earth elements from aqueous media. *J. Mol. Liq.* 306, 112760.

Jiao, T., Zhao, H., Zhou, J., Zhang, Q., Luo, X., Hu, J., Peng, Q., and Yan, X. (2015). Self-assembly reduced graphene oxide nanosheet hydrogel fabrication by anchorage of chitosan/silver and its potential efficient application toward dye degradation for wastewater treatments. *ACS Sustain. Chem. Eng.* 3(12), 3130–3139.

Kazemi, J., and Javanbakht, V. (2020). Alginate beads impregnated with magnetic Chitosan@ Zeolite nanocomposite for cationic methylene blue dye removal from aqueous solution. *Int. J. Biol. Macromol.* 154, 1426–1437.

Khan, S.B., Ali, F., and Akhtar, K. (2019). Chitosan nanocomposite fibers supported copper nanoparticles based perceptive sensor and active catalyst for nitrophenol in real water. *Carbohydr. Polym.* 207, 650–662.

Kumar, J., Arland, S., and Gour, P. (2020). Mercury removal from simulated waste water by chitosan nano composite embedded with leaf extract of Brassica Gongylodes. *Mater. Today: Proc.* 26, 728–739.

Kyzas, G.Z., Bikiaris, D.N., Seredych, M., Bandosz, T.J., and Deliyanni, E.A. (2014). Removal of dorzolamide from biomedical wastewaters with adsorption onto graphite oxide/poly (acrylic acid) grafted chitosan nanocomposite. *Bioresour. Technol.* 152, 399–406.

Liu, J., Li, P., Chen, L., Feng, Y., He, W., and Lv, X. (2016). Modified superhydrophilic and underwater superoleophobic PVDF membrane with ultralow oil-adhesion for highly efficient oil/water emulsion separation. *Mater. Lett.* 185, 169–172.

Lü, T., Chen, Y., Qi, D., Cao, Z., Zhang, D., and Zhao, H. (2017). Treatment of emulsified oil wastewaters by using chitosan grafted magnetic nanoparticles. *J. Alloys Compd.* 696, 1205–1212.

Mahmoodi, N.M., Saffar-Dastgerdi, M.H., and Hayati, B. (2020). Environmentally friendly novel covalently immobilized enzyme bionanocomposite: From synthesis to the destruction of pollutant. *Compos. B Eng.* 184, 107666.

Makaremi, M., Lim, C.X., Pasbakhsh, P., Lee, S.M., Goh, K.L., Chang, H., and Chan, E.S. (2016). Electrospun functionalized polyacrylonitrile–chitosan Bi-layer membranes for water filtration applications. *RSC Adv.* 6(59), 53882–53893.

Mirza, A., and Ahmad, R. (2018). Novel recyclable (Xanthan gum/montmorillonite) bionanocomposite for the removal of Pb (II) from synthetic and industrial wastewater. *Environ. Technol. Innov.* 11, 241–252.

Mohammad, F., Arfin, T., and Al-Lohedan, H.A. (2017). Enhanced biological activity and biosorption performance of trimethyl chitosan-loaded cerium oxide particles. *J. Ind. Eng. Chem.* 45, 33–43.

Mostafa, M.K., Mahmoud, A.S., Saryel-Deen, R.A., and Peters, R.W. (2017). Application of entrapped nano zero valent iron into cellulose acetate membranes for domestic wastewater treatment. In Environmental Aspects, Applications and Implications of Nanomaterials and Nanotechnology 2017: Topical Conference at the 2017 AIChE Annual Meeting (pp. 27–34).

Narayanan, K.B., and Sakthivel, N. (2011). Heterogeneous catalytic reduction of anthropogenic pollutant, 4-nitrophenol by silver-bionanocomposite using *Cylindrocladium floridanum. Bioresour. Technol.* 102(22), 10737–10740.

Nehra, S., Dhillon, A., and Kumar, D. (2020). Freeze–dried synthesized bifunctional biopolymer nanocomposite for efficient fluoride removal and antibacterial activity. *Res. J. Environ. Sci.* 94, 52–63.

Nkinahamira, F., Alsbaiee, A., Zeng, Q., Li, Y., Zhang, Y., Feng, M., Yu, C.P. and Sun, Q. (2020). Selective and fast recovery of rare earth elements from industrial wastewater by porous β-cyclodextrin and magnetic β-cyclodextrin polymers. *Water Res.* 181, 115857.

Padilla-Ortega, E., Darder, M., Aranda, P., Gouveia, R.F., Leyva-Ramos, R., and Ruiz-Hitzky, E. (2016). Ultrasound assisted preparation of chitosan–vermiculite bionanocomposite foams for cadmium uptake. *Appl. Clay Sci.* 130, 40–49.

Pande, V.V., and Sanklecha, V.M. (2017). Bionanocomposite: A review. *Austin J. Nanomed. Nanotechnol.* 5(1),1045.

Pinem, J.A., Panjaitan, D.N.I., Siregar, M.R., Saputra, E., and Herman, S. (2019). Synthesis and characterization of chitosan-silica membranes for treating hotel wastewater treatment as affected by mass of poly ethylene glycol and poly vinyl alcohol. *J. Appl. Mater. Technol.* 1(1), 31–37.

Qiu, H., Liang, C., Yu, J., Zhang, Q., Song, M., and Chen, F. (2017). Preferable phosphate sequestration by nano-La (III)(hydr) oxides modified wheat straw with excellent properties in regeneration. *Chem. Eng. J.* 315, 345–354.

Rafati, L., Ehrampoush, M.H., Rafati, A.A., Mokhtari, M., and Mahvi, A.H. (2016). Modeling of adsorption kinetic and equilibrium isotherms of naproxen onto functionalized nano-clay composite adsorbent. *J. Mol. Liq.* 224, 832–841.

Ratner, B.D., Hoffman, A.S., Schoen, F.J., and Lemons, J.E. (2004). *Biomaterials Science: An Introduction to Materials in Medicine*. Elsevier.

Rhim, J.W., Park, H.M., and Ha, C.S. (2013). Bio-nanocomposites for food packaging applications. *Prog. Polym. Sci.* 38(10–11), 1629–1652.

Rosdi, N., Sokri, M. N.M., Rashid, N.M., Chik, M.C., and Musa, M.S. (2019). Chitosan/silica composite membrane: Adsorption of lead (ii) ion from aqueous solution. *J. Appl. Membr. Sci. Technol.* 23(1), 63–72.

Saifuddin, N., Nian, C.Y., Zhan, L.W., and Ning, K.X. (2011). Chitosan-silver nanoparticles composite as point-of-use drinking water filtration system for household to remove pesticides in water. *Asian J. Biochem.* 6(2), 142–159.

Selvakumar, D., Thenammai, A.N., Yogamalar, N.R., Hemamalini, R., and Jayavel, R. (2015). Enriched adhesion of talc/ZnO nanocomposites on cotton fabric assisted by aloe-vera for biomedical application. In AIP Conference Proceedings (Vol. 1665, No. 1, p. 050162). AIP Publishing LLC.

Soltani, R.D.C., Khataee, A.R., Safari, M., and Joo, S.W. (2013). Preparation of bio-silica/chitosan nanocomposite for adsorption of a textile dye in aqueous solutions. *Int. Biodeterior. Biodegrad.* 85, 383–391.

Song, W., Gao, B., Xu, X., Wang, F., Xue, N., Sun, S., Song, W., and Jia, R. (2016). Adsorption of nitrate from aqueous solution by magnetic amine-crosslinked biopolymer based corn stalk and its chemical regeneration property. *J. Hazard. Mater.* 304, 280–290.

Spoială, A., Ilie, C.I., Ficai, D., Ficai, A., and Andronescu, E. (2021). Chitosan-based nanocomposite polymeric membranes for water purification: A review. *Materials.* 14(9), 2091.

Tanahashi, M. (2010). Development of fabrication methods of filler/polymer nanocomposites: With focus on simple melt-compounding-based approach without surface modification of nanofillers. *Materials.* 3(3), 1593–1619.

Tang, X., Gan, L., Duan, Y., Sun, Y., Zhang, Y., and Zhang, Z. (2017). A novel Cd2+-imprinted chitosan-based composite membrane for $Cd^{2+}$ removal from aqueous solution. *Mater. Lett.* 198, 121–123.

Truong, T.T.C., Takaomi, K. and Bui, H.M. 2019. Chitosan/zeolite composite membranes for the eliminating of trace metal ions in the evacuation permeability process. *J. Serbian Chem. Soc.* 84(1), pp.83–97.

Urbina, L., Guaresti, O., Requies, J., Gabilondo, N., Eceiza, A., Corcuera, M.A., and Retegi, A. (2018). Design of reusable novel membranes based on bacterial cellulose and chitosan for the filtration of copper in wastewaters. *Carbohydr. Polym.* 193, 362–372.

Wang, L., Zhang, J., and Wang, A. (2008). Removal of methylene blue from aqueous solution using chitosan-g-poly (acrylic acid)/montmorillonite superadsorbent nanocomposite. *Colloids Surf. A Physicochem. Eng. Asp.* 322(1–3), 47–53.

Wang, X., Li, Y., Li, H., and Yang, C. (2016). Chitosan membrane adsorber for low concentration copper ion removal. *Carbohydr. Polym.* 146, 274–281.

Xiao, W., Xu, J., Liu, X., Hu, Q., and Huang, J. (2013). Antibacterial hybrid materials fabricated by nanocoating of microfibril bundles of cellulose substance with titania/chitosan/silver-nanoparticle composite films. *J. Mater. Chem. B.* 1(28), 3477–3485.

Yap, P.L., Hassan, K., Auyoong, Y.L., Mansouri, N., Farivar, F., Tran, D.N., and Losic, D. (2020). All-in-one bioinspired multifunctional graphene biopolymer foam for simultaneous removal of multiple water pollutants. *Adv. Mater. Interfaces.* 7(18), 2000664.

Yun, Y.H., Kim, E.S., Shim, W.G., and Yoon, S.D. (2018). Physical properties of mungbean starch/PVA bionanocomposites added nano-ZnS particles and its photocatalytic activity. *J. Ind. Eng. Chem.* 68, 57–68.

Zarei, S., Farhadian, N., Akbarzadeh, R., Pirsaheb, M., Asadi, A., and Safaei, Z. (2020). Fabrication of novel 2D Ag-$TiO_2$/$\gamma$-$Al_2O_3$/Chitosan nano-composite photocatalyst toward enhanced photocatalytic reduction of nitrate. *Int. J. Biol. Macromol.* 145, 926–935.

Zhuang, J., Dai, J., Ghaffar, S. H., Yu, Y., Tian, Q., and Fan, M. (2020). Development of highly efficient, renewable and durable alginate composite aerogels for oil/water separation. *Surf. Coat. Technol.* 388, 125551.

# 7 Critical Factors Affecting the Synthesis of Bionanomaterials and Biocomposites

*Rachita Sharma, Priya Singh, Ved Kumar Mishra, Naveen Dwivedi, and Nikita Singhal*

**CONTENTS**



DOI: 10.1201/9781003270959-7

## 7.1 INTRODUCTION

Nanotechnology is a rising field of biotechnology that deals with the study of particles at the nano size (1 to 100nm). Due to their small size and high surface-to-volume ratio, they exhibit various unique properties and, hence, have a wide range of applications in the medical sector as drug diagnostics and delivery tools, in information technology as transistors or nanochips, in the optical sector, in the food industry as packaging materials, and many more. There are two different approaches for the synthesis of nanoparticles, i.e., the top-down approach or bottom-up approach. There are several methods for the synthesis of nanomaterials: physical, chemical, and biological approaches. Initially, physical and chemical methods were the most used for the synthesis of nanomaterials. But due to their dependency on chemicals, release of toxic by-products, and harmful effects on the environment and humans, as well as high production cost and production of large amounts of waste, they are no longer the preferred methods. Instead, the use of biological methods for the synthesis of nanoparticles has been increasing over the past few years where we use biological approaches to synthesize the bionanoparticles or bionanocomposites. This method of synthesis is referred to as "green synthesis." In this process, we use biomaterials such as extracts of plants, viruses, algae, and bacteria to synthesize the nanomaterials. Nanomaterials synthesized through this green synthesis process are biologically safe, cost-effective, less toxic, and environmentally friendly with limited use of chemicals. And due to all these properties, bionanomaterials have a wide range of applications, such as in drug diagnosis and delivery systems, biosensor formation, nano-adsorbent, catalyst development, cancer treatment, etc.

Metal nanoparticles formed from biological materials using the green synthesis method have recently found new purpose in nanotechnology acting as reducing and stabilizing agents. These biomaterials use components of plant extracts, bacteria, viruses, or algae to create stable metal bionanoparticles. Synthesizing the bionanomaterials is a complicated process that requires the biological study of microbes and plants, as well as knowledge of charges and atomic confirmations of metals to fully understand the interactions between them.

Some basic steps in the green synthesis process are:

- Dissolving the solvent with biological materials.
- The reaction of liquid extract from biological materials with metal ions.
- Stabilizing and capping the synthesized nanomaterials and preventing them from accumulation and agglutination.

There are certain factors, like pH, temperature, volume of biomaterial used, type of process used, reaction time, pressure, and other environmental factors, that control or influence the synthesis of these nanoparticles. These factors influence the whole synthesis process in many ways, finally affecting their physical properties like size and shape. Therefore, these factors must be taken under consideration to maintain their different characteristics and properties. Bionanoparticles when treated with biopolymers like starch, chitosan, or cellulose form bionanocomposite composites.

Having the properties of bionanoparticles, they are also biodegradable (or biocompatible), nontoxic, and hence, they are slowly replacing the nonbiodegradable polymers used in the food industry as packaging materials. These bionanocomposites are of immense interest in biomedical technologies such as tissue engineering, bone replacement/repair, dental application, and controlled drug delivery.

## 7.2 SYNTHESIS OF NANOMATERIALS AND NANOCOMPOSITES

Nanoparticles are synthesized using two different approaches: the bottom-up and top-down approaches.

In the bottom-up approach, smaller elements are arranged into more complex, assembling atom-by-atom, molecule-by-molecule, cluster-by-cluster from the bottom;

whereas, the top-down approach uses traditional methods or microfabrication methods where the complex material is externally cut down and shaped into the desired form.

Different methods of synthesis of nanoparticles, physical, chemical, and biological, are outlined in the following sections (Figure 7.1).

### 7.2.1 Physical Method of Synthesis

Physical methods for the synthesis of nanoparticles include evaporation-condensation, laser ablation, and arc discharge methods.

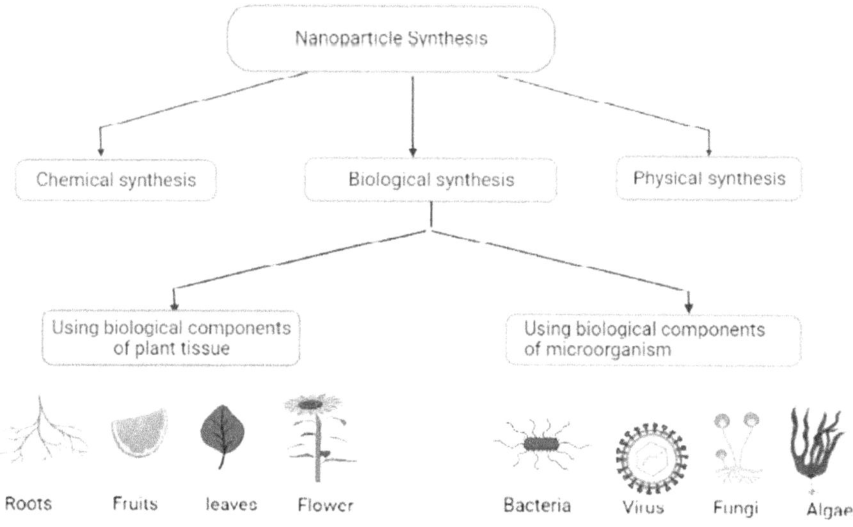


**FIGURE 7.1** Different methods for synthesis of nanomaterials.

The evaporation-condensation method starts with a gas medium for the production of nanoparticles. This method requires high temperatures which are used to achieve appropriate evaporation rates. The evaporation of a metallic source at high temperature in the presence of an inert gas is required for the production of nanoparticles.

In the arc discharge method, material is vaporized between cathode and anode electrodes by providing high voltage (50–100 V). Plasma is generated due to which metal atoms are evaporated and get condensed on a water-cooled substrate.

In the laser ablation method, a chamber filled with inert gas integrated with a high-power source is used at a high temperature of 6000°C. The high-power source releases Nd-YAG laser which is used to ablate the metal atom, resulting in evaporation and, ultimately, condensation on the water-cooled substrate.

### 7.2.2 Chemical Method of Synthesis

Chemical method for the synthesis of nanoparticles consists of colloidal synthesis done by microemulsion method or sol-gel method. Either way, the starting phase of the chemical method of synthesis of NPs is always liquid.

In microemulsion method, a solution of different ions is mixed under controlled temperature and pressure conditions to form an insoluble precipitate, namely micelle formation. If the dispersion medium is an aqueous solvent, then hydrophobic radicals of the micelle form the core while a hydrophilic group forms the surface layer of the micelle.

The sol-gel method involves "sol" formation in liquid medium and then the connection of different sols to form a network. Sols are the solid particles, and gel is the sol-containing liquid. The procedure consists of dissolving the precursors in the liquid resulting in sol formation. The sols are dehydrated with the medium gel where the sols are connected to each other, forming a network which can then be used for the synthesis of nanoparticles.

### 7.2.3 Biological Method of Synthesis

The biological method for the synthesis of nanomaterials, also known as "green synthesis" uses plant extracts, microorganisms (bacteria, fungi, algae, etc), enzymes, or agricultural waste. These syntheses are majorly done through the bottom-up approach, using materials such as stabilizing and capping agents. The basic process of green synthesis begins with dissolving the powdered biomaterial in the biodegradable solvent, then mixing that liquid biological material with the metal salts where the reaction takes place, producing nanomaterials. To prevent the accumulation or aggregation of synthesized nanoparticles, capping and stabilizing processes will be done at the end. These synthesized nanoparticles are highly stable and well characterized. During the reaction, certain factors affect the formation of a nano-scaled particle by changing their physical properties, like size and shape, and chemical properties, like charge. So, these factors are needed to be considered for desired bionanomaterial or bio nano-composite formation (Roy et al., 2019) (Figure 7.2).

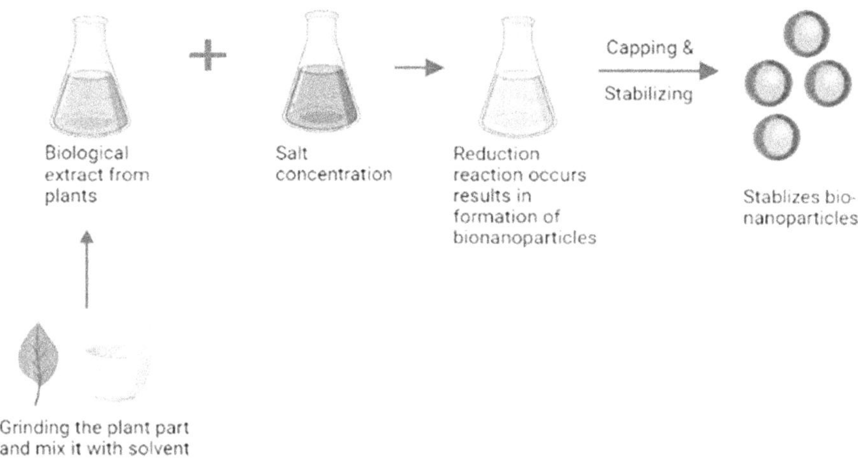


**FIGURE 7.2** Synthesis of nanoparticles from plant extract.

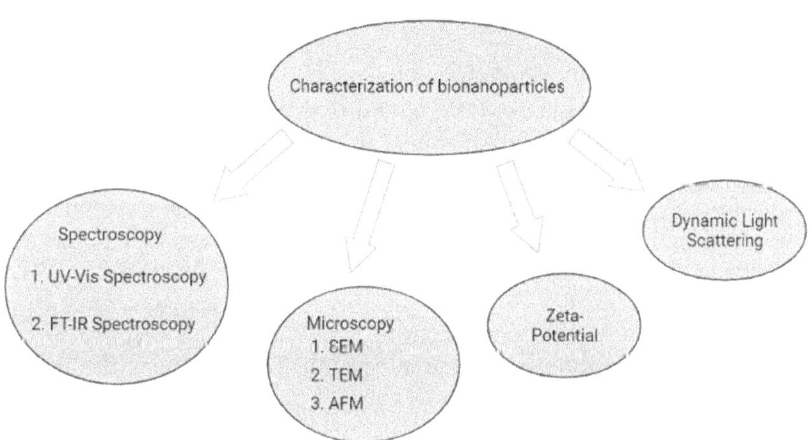


**FIGURE 7.3** Analytical techniques for the characterization of bionanoparticles.

## 7.3 CHARACTERIZATION OF SYNTHESIZED NANOPARTICLES

The size range of nanoparticles is 1–100 nm. That is at the atomic level, so we need some advanced analytical techniques to characterize them. The nano-sized particles exhibit different properties, compared to the bulk molecules, due to their high surface-to-volume ratio. Characterization of nanomaterial physical and chemical properties, like size, structure, solubility, aggregation, and stability, ultimately reveals their potential applications in fields like biomedical research, environmental sciences, food industries, etc. The characterization of bionanoparticles is done by different analytical techniques like scanning electron microscopy (SEM) and transmission electron microscopy (TEM) (Figure 7.3). These techniques are discussed below.

## 7.3.1 Spectroscopy

Spectroscopy is an analytical method used to determine the structure of atoms. Spectroscopy is also the study of the interaction of electromagnetic radiation with matter, resulting in the absorption or emission of wavelength or frequency of radiation. This interaction will help us to investigate the structure of matter on an atomic scale by using spectra. The absorption and emission of electromagnetic radiation will form a spectrum that gives the detailed study of the atomic configuration of matter. Absorption spectroscopy measures the absorption of wavelength or frequency of radiation when it interacts with the sample. When the electromagnetic radiation is absorbed, electrons will excite from a lower energy level to a higher energy level and form an "absorption spectrum." In emission spectroscopy, the electrons will be excited from a higher energy level to lower energy level when the matter interacts with electromagnetic radiations. This excitation will emit radiation of different frequencies or wavelengths that produce a line spectrum called an "emission spectrum." There are different types of spectroscopies used for the characterization of nanoparticles, including UV-Vis spectroscopy and FTIR spectroscopy.

### 7.3.1.1 Ultraviolet-Visible Spectroscopy

Ultraviolet-visible spectroscopy (UV-Vis) is based on the absorption of radiation in the range of ultraviolet and visible light. UV-Vis is a technique that is used to measure the absorbance or scattering of light (UV or visible) from the sample with the help of the absorption spectrum. In this technique, light is emitted from a source, whose wavelength will be selected by a wavelength selector, and passed into a liquid sample. When the light is passed through the sample, it will absorb a certain amount of energy to excite its electrons from the ground-state to the excitation state. The absorbance of a certain wavelength of light will help us find the difference in the wavelength of light compared to the control. The analysis of the amount of absorbance at different wavelengths will help us to draw a graph called "absorption-spectra," having absorbance on its y-axis and wavelength on its x-axis. Some metal nanoparticles, like gold or silver, show absorbance on the surface. Plasmon uses UV-Vis spectra for the characterization of the size and shape of nanoparticles. High absorption peaks obtained at certain wavelengths tell us the high size distribution, small size, particular shape, etc.

### 7.3.1.2 Fourier-Transform Infrared Spectroscopy

Fourier-transform infrared spectroscopy (FTIR) is a type of spectroscopy used to characterize the structure of nanoparticles at the molecular level and to study the effect of functional groups during the synthesis process of nanoparticles. In FTIR the infrared radiation is used to strike the sample, as some get absorbed while others get transferred. The absorption of infrared radiation will result in vibrational motion. Every functional group needs a different frequency to absorb radiation which will result in the formation of a spectrum (Faghihzadeh et al., 2016).

### 7.3.2 MICROSCOPY

Microscopy in an analytical technique which uses microscopes. Microscopes are the instruments used to visualize the objects or samples that cannot be seen through the naked eye by producing a high-resolution image. Microscopy is further classified into optical microscopy, electron microscopy, scanning probe microscopy, and x-ray microscopy based on the source used.

- **Optical microscopy.** In optical microscopy, the visible light travels through one or more lenses to form a magnified image.
- **Electron microscopy.** In electron microscopy, the beam of an electron with a small wavelength is used as the source of light. It is of two types: scanning electron microscopy (SEM) and transmission electron microscopy (TEM). Electron microscopy is the most preferred type of microscopy used to analyze nano-sized particles.
- **Scanning probe microscopy.** It includes atomic force microscopy (AFM) and scanning tunneling microscopy (STM). They both use a solid tip to physically scan the surface of the sample.

The types of microscopies used to determine the characteristics of nanoparticles are electron microscopy and scanning probe microscopy.

#### 7.3.2.1 Transmission Electron Microscopy

TEM increases the wavelength of the electron beam to increase the resolution of the formed image. That is, the longer the wavelength of the beam of electron used, the higher the resolution of an image is produced. It uses transmitted electrons that pass to the sample to get information of the internal structure of said sample.

TEM uses the transmitted electron beam to obtain a high-resolution two-dimensional (2-D) image. When the high voltage electron beam strikes the extra-fine thin sample, it will interact with the atoms and get transmitted depending on the atomic arrangement of materials. The transmitted electrons are used to project the image onto a screen. The darker areas represent the lower transmission regions of the sample, while the lighter areas represent the higher transmission regions. It is a widely used imaging technique to visualize the size, shape, distribution, and morphology of nanoparticles at the atomic level. To use them, the sample must be ultra-fine, so sample preparation is required.

#### 7.3.2.2 Scanning Electron Microscopy

When the electron beam (primary electrons) falls on the sample, some of the electrons are scattered or reflected back. The primary electrons are scanned across the sample to get images of particular areas. Detection of scattering electrons (secondary electrons) will give the structural information.

SEM uses the scattered electron from the surface of a sample to obtain a three-dimensional (3-D) image. When a beam of electron strikes the surface of the sample, the scattered electrons will generate signals that are recorded in detectors. The detector will visualize the image on the screen to study the characteristics of bionanoparticles. But their application is limited in some cases because they only give information about the presence, distribution, and number of particles produced but not about the exact size or shape.

#### 7.3.2.3 Atomic Force Microscopy

Atomic force microscopy (AFM) works on the principle of surface scanning. In AFM, a nano-sized sharp tip is used to scan the surface to develop a high-resolution surface image. The forces working between the sample and the tip are the force of attraction and the force of repulsion.

In atomic force microscopy, high-resolution 3-D images are formed by using a sharp tip cantilever scan over the surface of the material. The attraction and repulsion interaction between the tip and the surface cause deflection in the cantilever. This deflection will be obtained by the laser light that falls on the cantilever. An atomic-level image of the surface of the particles can be obtained by a photodiode. AFM is used to identify the size, shape, distribution, morphology, and physical properties, like the magnetic field, of nanoparticles.

### 7.3.3 Zeta Potential

Zeta potential is a net potential developed at the surface of the particle when it is in contact with the liquid medium. It gives the information of stability of the colloids. Colloids with higher values of zeta potential (whether negative or positive) are considered more stable. When the charges between the particles and liquid medium are opposite, they will attract each other, while the same charge between particle and medium will cause repulsion. Lower value of zeta potential results in agglomeration. This zeta potential is also used to determine the stability of metal oxide nanoparticles for better shelf-life and more reliable physical properties. The zeta sizer is an instrument used to calculate the zeta potential. It also helps to study the surface charge on metal oxide nanoparticles in suspension to determine the influence of concentration on their properties (Wang et al., 2013).

### 7.3.4 Dynamic Light Scattering

Dynamic light scattering (DLS) is also known as "photon correlation spectroscopy" and is used to determine the size of nanoparticles. The Brownian motion of the particles present in dispersed particles moving in different directions and colliding with each other. When light is passed to the sample, due to the motion of particles, it will scatter the light in all directions. The detection of scattered light at some angle after some time tells us the diffusion coefficient and size of the particle with the help of the Stokes–Einstein equation. Through DLS it is easy to determine the size distribution and the motion of nanoparticles in a solution.

## 7.4 FACTORS AFFECTING THE SYNTHESIS OF BIONANOMATERIALS

Bionanomaterial synthesis by the green method is affected by certain critical factors like pH, reaction time, temperature, volume of biomaterial, type of method used for synthesis, pressure, and other environmental factors. To study the impacts of these factors on the characteristics of bionanoparticles, several kinds of research have been going on in this field. Some of the factors affecting the synthesis of bionanoparticles are described below.

### 7.4.1 TEMPERATURE

Temperature is one of the most critical factors that needs to be considered while synthesizing nanoparticles. Physical and chemical synthesis of nanoparticles requires a high reaction temperature. On the other hand, green synthesis is a comparatively less energy-consuming process. In the green synthesis method, the optimum reaction temperature for obtaining the appropriate size of nanoparticles is observed to be less than 100°C. Liu et al. (2020) synthesized the silver nanoparticles from *Cinnamomum Camphor* leaf, and the influence of temperature (low temperature or high temperature) on the size of nanoparticles has been investigated. The reaction temperature can be controlled by the reaction kinetics of the biosynthesis process, which in turn determines the size. Another important factor that largely influence the size of nanoparticle is the silver ion precursor. Temperature affects the size of nanoparticle by two methods:

- At high temperatures, nucleation of nanoparticles starts; nanoparticles start degrading into smaller molecules.
- At low temperatures, growth of nanoparticles starts.

This means that the formation of smaller-sized nanoparticles at high temperatures is due to the increase in the nucleation rate constant ($k_1$); and formation of larger-sized nanoparticles at low temperatures is due to an increase in the growth rate constant ($k_2$). The nucleation effect is observed only when there are insufficient precursors of metal ions. Therefore, if precursors present in insufficient amounts, both growth rate constant and nucleation rate constant are taken into consideration. Hence, it is concluded that at low temperatures, the value of $k_2$ is larger than the value of $k_1$, meaning that the growth factor dominates over the nucleation factor, resulting in the growth of the nanoparticles (i.e., larger-sized nanoparticles are obtained). But at high temperature, there is a drastic increase in the value of $k_{1,}$ and it consumes a large amount of $Ag^+$ ions, resulting in the decrement in the number of silver ions for reaction. It can also be explained by comparing the values: if the value of $k_1$ is larger than the value of $k_2$, it means that now the dominating factor is the nucleation effect. This condition gives priority to the nucleation over growth, and hence, the particle size is smaller compared to those at the lower temperature. Nevertheless, what if we are providing a sufficient amount of $Ag^+$ precursor? In the presence of sufficient metal precursors, the size of the nanoparticles is observed to be increased with the increase of reaction temperature. This is because the growth rate constant ($k_2$) will increase with the increase in temperature under a sufficient amount of $Ag^+$ precursor. Only the growth rate constant ($k_2$) influences the size of bionanoparticles in this case. There is not any effect of nucleation on the bionanoparticles (i.e., the value of $K_1$ does not influence the whole reaction). So, metal ions largely control the size or shape of bionanomaterials. For green synthesis, the amount of metal precursor should be greater than that of biological materials for proper synthesis of nanoparticles. If the metal ions present in the reaction are sufficient, the size will increase with the increase in temperature. When there is an insufficient amount of metal precursor present in the reaction, the nucleation and growth kinetics affect the synthesis process, affect the size with increasing temperature. So, temperature plays a major role in determining the size of nanoparticles (Liu et al., 2020).

### 7.4.2 Metal Ion Concentration

The metal ion concentration used for the biosynthesis process has a major role in producing the desired size of bionanoparticles. It has been found in many experiments that when the concentration of the metal ions (i.e., silver nitrate) is increased, there is a variation in the size of the nanoparticles, indicated by the change in color. The concentration of metal ions in a reaction varies the size and shape of bionanoparticles. Different metal precursors are used in green synthesis according to their applications. The amount of metal precursor largely controls the size or shape of bionanomaterials. For green synthesis, the amount of metal precursor should be greater than that of the biological materials. In some research, it is also found that if there are an insufficient amount of metal ions, they will influence the size and shape of the formed bionanomaterials (Liu et al., 2020).

### 7.4.3 Reaction Time

Reaction time is another important factor influencing the process of bionanoparticle formation. Varying the reaction time can cause a change in the properties or characteristics of the nanoparticles. More interaction will result in narrow size distribution with a smaller average size of particle. But, after a very long period of time, nanoparticles start aggregating with each other, resulting in their size variation and hence affecting its properties, characteristics, and applications. There are other factors like storage conditions, environmental effect, or light intensity, which altogether over time will affect the size of nanomaterial.

The different particle colors obtained after a series of time durations indicate the formation of bionanoparticles. And this all can be determined with the help of UV-Vis spectrophotometric absorbance and $\lambda_{max}$. By studying the surface plasmon resonance (SPR) peak and the shift in wavelength with reaction time will result in producing the different sizes and shapes of nanoparticles (Ahmad et al., 2016). It has also been observed in many experiments that when the nanoparticles are kept for a long time, they will undergo aging. Aging will cause a change in physical and chemical properties like size, shape, surface reactivity, and many more. There are certain factors that influence the aging process like storage conditions, environment, stabilizing and capping agents efficiencies, and temperature (Nau, 2015).

Darroudi (2011) conducted bionanoparticle synthesis using gelatin to study the effect of time on bionanoparticles and the variation of size with time. Initially, they observed the broad range of size distribution of nanoparticles at the SPR peak as shown in Table 7.1. After a long time, they observed the narrow size distribution of particles, with an average size smaller than 20 nm and smaller particle diameter (Darroudi, 2011).

### 7.4.4 pH

pH is one of the important factors that affects the synthesis of nanoparticles. Research has shown that the pH of the solution media directly influences the size, shape, and texture of synthesized nanoparticles. Any change in pH directly relates to the electrical loads of biomolecules, which affects their ability to cap and stabilize and, hence, affects the growth of nanoparticles (Lade and Shanware, 2020).

Reaction rate also changes with changes in pH, which may cause agglomeration of nanoparticles (i.e., growth of nanoparticles into nanospheres or nanoclusters) and, hence, can affect their size. The effects of the pH of solution media on the nanoparticles can be studied by using UV-vis spectroscopy and TEM, where results show that the size of synthesized nanoparticles in acidic or basic medium varies; the amount formed also varies.

The UV-vis spectra of silver nanoparticles shows different SPR peaks at different wavelengths under different pH conditions. Table 7.2 shows that the SPR peaks for acidic pH appear at relative short wavelengths ~330 nm, whereas the SPR peaks for basic pH conditions tend toward relatively longer wavelengths ~420 nm. This shift in SPR peaks is called red shift. This red shift toward longer wavelength under basic conditions indicates the relatively large size of synthesized Ag nanoparticles,

**TABLE 7.1**
**SPR Peaks at Different Wavelengths at Different Reaction Times**

| Different Reaction Times (Hours) | SPR Peaks at Different Wavelengths (nm) |
|---|---|
| 1 | ~375 |
| 2 | ~380 |
| 3 | ~385 |
| 4 | ~390 |
| 5 | ~395 |
| 6 | ~400 |

**TABLE 7.2**
**SPR Peaks at Different Wavelengths at Different pH Values**

| pH Values | SPR Peaks at Different Wavelengths (nm) | Size of Silver Nanoparticles |
|---|---|---|
| 4 | ~330 | Small |
| 5 | ~330 | Small |
| 6 | ~330 | Small |
| 7 | ~350 | High |
| 8 | ~400 | High |
| 9 | ~410 | High |
| 10 | ~420 | High |

as compared to nanoparticles synthesized under neutral or acidic conditions. It also shows that, with the increase in pH, the intensity of SPR peaks for acidic conditions decreases whereas the intensity of SPR peaks for basic conditions increases. This effect of pH on nanoparticle synthesis can be explained on the fact based on dissociation, agglomeration, isolation, interfacial free energy, and net charge of the complexing agent.

In acidic conditions, the particles are small because of the balancing of the repulsive force with that of high-driving force for silver nanoparticle dissolution. Whereas, in basic conditions, NaOH dissociates in water, resulting in the production of negatively charged ions (i.e., $OH^-$ ions). These negative ions enhance the reduction of $Ag^+$ ions to AgNPs. At high ion density, Ag atoms adsorb on the nanoparticle surface and start forming bonds with adjacent neighboring atoms, leading to the increase of particle size shown in Table 7.2.

### 7.4.5 Pressure

It is generally believed that pressure has negligible effects, but it has a significant effect on the size of synthesized nanoparticles. It is one of the controlled variables during nanoparticle synthesis. It is applied during the green synthesis process to obtain nanoparticles of a specific size and shape. Pressure applied to the reaction media also affects the magnetic properties of the synthesized nanoparticles.

At high pressure, the size of nanoparticles increases. The reason for the increment in the size of the nanoparticle can be explained by the change in the Gibbs free energy value during crystallization. Pressure influences the high surface-to-volume ratio of nanoparticles and also some other physical properties, such as surface tension and supersaturation (Yazdani and Edrissi, 2010). Drabik et al. (2011) studied the effect of pressure on the Haberland style cluster source. He found that with pressure increasing from 50 to 100 Pa, the size of nanoparticles also increases, but when the pressure increases from 100 to 150 Pa, the size of nanoparticles decreases.

Gunnarsson et al (2016) studied the effect of pressure on titanium oxide nanoparticles via sputtering a hollow cathode in argon (Ar) inert gas. Here, at low pressure

(109 Pa), the nanoparticles appear to be almost of spherical shape at low magnification. As the pressure increases from 109 Pa to 163 Pa, the nanoparticle size increases and their shape also resembles a single crystal cubic structure. Now, upon further increasing the pressure to 189 Pa, the size of nanoparticles increases and retains a cubic shape; but here, nanoparticles do not have a clear-cut structure with sharp edges but shows a fractured appearance. At the highest pressure of 263 Pa, many morphologies of nanoparticle can coexist.

### 7.4.6 Size and Shape

Nanoparticles are the simplest structures in the nanometer range (1 to 100 nm). Most of the physical properties, such as elasticity, conductivity, optical and catalytic properties, of nanoparticles depend on the dimension of the nanoparticle, likely the number of atoms present in nanomaterial and, hence, dimension plays a major role in determining the properties of nanoparticles (Cao, 2004; Giri et al., 2011). With the decrement in size, the surface area increases, and so now nanoparticles will have more surface area for the reaction. Hence, reactivity increases with the decrease in size. The surface-to-volume ratio increases with a decrease in size and increase in surface ratio. Besides different sizes of nanoparticles, there can be different kinds of structures (shapes) of nanoparticles with different bonds and bond strengths. Properties, such as magnetic behavior, redox potentials, and different colors of nanoparticles can be influenced by the change in the shape and size of nanoparticles (Singh et al., 2017). The shape and size of nanoparticles are also affected by many factors, such as temperature, pressure, pH, and many more. Of these factors, temperature affects strongly. Raza et al. (2016) studied the antibacterial activity of silver nanoparticles of different shapes and sizes. Here, sterile paper disks impregnated with different sized silver nanoparticles were placed on nutrient agar medium, on which bacteria (*P. aeruginosa* and *E. coli*) were spread. After incubation at 37°C for 24 h, growth was detected. He observed that antibacterial efficiency of silver nanoparticles against *E. coli* is low compared to *P. aeruginosa* and, hence, shows more susceptibility than *E. coli*. He also investigated the effect of size on the bactericidal property. The smallest sized NPs exhibited maximum bactericidal activity against both the strains, namely *E. coli* and *P. aeruginosa*; whereas larger sized NPs showed very low bactericidal activity against the strains. This is due to the fact that small-sized AgNPs release more $Ag^{+}$ ions, thereby killing and destroying both bacterial strains by affecting the bacterial DNA replication functions, deactivating the production of some specific enzymes, and by affecting the synthesis of high-energy ATP molecules compared to large-sized nanoparticles (Raza et al., 2016).

### 7.4.7 Environmental Factors

The nature of nanoparticles also depends on their surrounding environment. Environment affects the synthesis of nanoparticles in many ways, such as different light conditions (artificial light or natural sunlight), the impact of sound, and others. Light at different wavelength activates certain enzymes, which shows different results due to different reactions occurring at different conditions, such as room temperature. Some studies have shown that antimicrobial activity is not linked to photoactivity. These studies have also indicated that nanoparticles synthesized in dark conditions are supposed to be more stable than the nanoparticles synthesized in UV light (Lade and Shanware, 2020). Other environmental factors affect the synthesis of nanoparticles by effecting the size of nanoparticles, where a single nanoparticle may change into a core-shell nanoparticle after reacting with other particles by oxidation-reduction or a corrosion process or simply can be coated with other substances that will make them larger (Baek et al., 2014). Nanoparticles can also adsorb the moisture from the surrounding environment. The multilayer physical adsorption mechanism works for humidity adsorption based on the adsorption isotherm fitting method. The water molecules will bond or adsorb on the surface of nanoparticles by means of hydrogen bonding. This adsorption can be reached up to many layers, sometimes as many as 23, and hence will influence the size of synthesized nanoparticles. The adsorption also influences the porosity of nanoparticles by decreasing the tensile strength of a synthesized nanoparticle surface. Nowadays, researches are also investigating the effect of sound on the synthesis of nanoparticles (Lade and Shanware, 2020).

Traiwatcharanon et al. (2017) demonstrated the influence of light radiation on synthesis of Ag NPs. He studied the UV-vis spectra of AgNPs which shows different absorbance peaks for light radiations of different intensity (red, blue, green, white). Blue light radiation shows higher absorbance peaks compared to other radiations. Furthermore, the absorbance peak of blue light radiation also shows a small red shift in wavelength compared to the absorbance peaks of other light radiations. This shows that with shorter wavelength of light radiation, there will be increase in the synthesis of AgNPs. This fact can be explained by the principle of photocatalytic reactions, which states that at high energy radiation (shorter wavelength), photons produce more energetic electrons on the excitation of surface plasmons. Due to which, $Ag^{+}$ ions will be more readily reduced to $Ag^{0}$ particles, resulting in the increase of AgNP synthesis.

### 7.4.8 Biogenic Materials

Different types of biological components like proteins, metabolites, lipids, and polysaccharides obtained from biomaterials like plants or microorganisms are utilized for the green synthesis process. These biological materials are used as reducing agents or stabilizing agents for bionanoparticles. For plants, different types of plant species, plant parts, like leaves, flowers, fruits, produce different compositions of biological metabolites resulting in a variety of properties in bionanomaterials. For microorganisms, different proteins or enzymes produced inside or outside of the cells gives a variety of characteristics in nanoparticles produced (Zhang et al.,2020).

The biological method of nanoparticle synthesis mostly uses a bottom-up approach. In this, the intracellular or extracellular components of biomaterials will interact with the metal precursors and act as a reducing agent and give stability to the formed metal bionanoparticles. However, the overall mechanism at work during the biosynthesis process is still unknown. Different biological materials use different biological component that works differently to produce bionanoparticles. Not only the type of biological materials used but also the amount and size to be used influence the whole biosynthesis process.

In microorganisms, the extracellular and intracellular products of cells are responsible for the biosynthesis process. In many research works, the NADH-dependent reductase enzymes are found to be an efficient reducing agent and capping agent used in the green synthesis process (Bhargava et al.,2016). In the case of bacteria, other factors, like temperature, pH, and ionic concentrations, also play an important role in maintaining the physical and chemical properties of bionanomaterials produced by bacteria. In the case of plants, various types of biomolecules present and metabolites produced in plants are utilized for the green synthesis process. To obtain uniform size or shape of nanoparticles we use the same plant species and plant part because they give similar compositions of biological metabolites, resulting in similarity of properties with limited variation. Certain types of microalgae are also found to be used as a biomaterial in the green synthesis process. The components of their cell walls, like alginate and laminarin, give them the ability to synthesize bionanoparticles. Also, amino acids present in algae cells act as capping or stabilizing agents (Choudhary et al., 2020).

## 7.5 CONCLUSION

Green synthesis, i.e., biological synthesis of nanoparticles, depends on certain factors for their synthesis. From these factors, temperature plays a major role in nanoparticle synthesis at different steps. Some other critical factors, such as pH, size and shape, reaction time, pressure, surface-to-volume ratio, may affect the synthesis of nanoparticles/nanomaterials. These factors affect their synthesis by affecting their physical and chemical properties, some of which are melting point, pore size, absorbing capacity, magnetic property, etc. The nanoparticles synthesize at different rates, express different properties, and hence find numerous applications around us.

## REFERENCES

Ahmad, T., Irfan, M., Bustam, M. A., & Bhattacharjee, S. (2016). Effect of reaction time on green synthesis of gold nanoparticles by using aqueous extract of Elaiseguineensis (oil palm leaves). *Procedia Eng.*, 148, 467–472.

Bhargava, A., Jain, N., Khan, M. A., Pareek, V., Dilip, R. V., & Panwar, J. (2016). Utilizing metal tolerance potential of soil fungus for efficient synthesis of gold nanoparticles with superior catalytic activity for degradation of rhodamine B. *J. Environ. Mngmt.*, 183, 22–32.

Cao, G. (2004). *Nanostructures & Nanomaterials: Synthesis, Properties & Applications.* Imperial College Press.

Chaudhary, R., Nawaz, K., Khan, A. K., Hano, C., Abbasi, B. H., & Anjum, S. (2020). An overview of the algae-mediated biosynthesis of nanoparticles and their biomedical applications. *Biomolecules*, 10(11), 1498.

Darroudi, M. (2011). Time-dependent effect in green synthesis of silver nanoparticles publisher and licensee Dove Medical Press Ltd. *Int. J. Nanomed.*, 2011(6), 677–681.

Drábik, M., Choukourov, A., Artemenko, A., Kousal, J., Polonskyi, O., Solař, P., Kylián, O., Matousek, J., Pesicka, J., Matolínová, I., Slavínská, D., & Biederman, H. (2011). Morphology of titanium nanocluster films prepared by gas aggregation cluster source. *Plasma Process. Polym.*, 8(7), 640–650.

Faghihzadeh, F., Anaya, N. M., Schifman, L. A., & Oyanedel-Craver, V. (2016). Fourier transform infrared spectroscopy to assess molecular-level changes in microorganisms exposed to nanoparticles. *Nanotech. Environ. Engg.*, 1(1), 1.

Giri, N., Natarajan, R. K., Gunasekaran, S., & Shreemathi, S. (2011). 13C NMR and FTIR spectroscopic study of blend behavior of PVP and nano silver particles. *Archives of Applied Sci. Research*, 3(5), 624–630.

Gunnarsson, R., Pilch, I., Boyd, R. D., Brenning, N., & Helmersson, U. (2016). The influence of pressure and gas flow on size and morphology of titanium oxide nanoparticles synthesized by hollow cathode sputtering. *J. Appl. Phys.*, 120(4), 044308.

Lade, B. D., & Shanware, A. S. (2020). Phyto nanofabrication: Methodology and factors affecting biosynthesis of nanoparticles. In *Smart Nanosystems for Biomedicine, Optoelectronics and Catalysis*. Intech Open.

Liu, H., Zhang, H., Wang, J., & Wei, J (2020). Effect of temperature on the size of biosynthesized silver nanoparticle: Deep insight into microscopic kinetics analysis. *Arabian J. Chem.*, 13(1), 1011–1019.

Nau-Izak, E. (2015). Impact of storage conditions and storage time on silver nanoparticles' physicochemical properties and implications for their biological effects. *RSC Adv.*, 2015, 84172–84185.

Patra, J. K., & Baek, K. H. (2014). Green nanotechnology: Factors affecting synthesis and characterization techniques. *J. Nanotech.*, 2014, 12 pages.

Raza, M. A., Kanwal, Z., Rauf, A., Sabri, A. N., Riaz, S., & Naseem, S. (2016). Size-and shape-dependent antibacterial studies of silver nanoparticles synthesized by wet chemical routes. *Nanomater*, 6(4), 74.

Roy, A., Bulut, O., Some, S., Mandal, A. K., & Yilmaz, M. D. (2019). Green synthesis of silver nanoparticles: biomolecule-nanoparticle organizations targeting antimicrobial activity. *RSC Adv.*, 9(5), 2673–2702.

Singh, A. K., Srivastava, O. N., & Singh, K. (2017). Shape and size-dependent magnetic properties of Fe 3 O 4 nanoparticles synthesized using piperidine. *Nanoscale Res. Lett.*, 12(1), 1–7.

Traiwatcharanon, P., Timsorn, K., & Wongchoosuk, C. (2017). Flexible room-temperature resistive humidity sensor based on silver nanoparticles. *Mater. Res. Express*, 4(8), 085038.

Wang, N., Hsu, C., Zhu, L., Tseng, S., & Hsu, J. P. (2013). Influence of metal oxide nanoparticles concentration on their zeta potential. *J. Colloid Interface Sci.*, 407, 22–28.

Yazdani, F., & Edrissi, M. (2010). Effect of pressure on the size of magnetite nanoparticles in the coprecipitation synthesis. *Mater. Sci. Engg: B*, 171(1–3), 86–89.

Zhang, D., Ma, X. L., Gu, Y., Huang, H., & Zhang, G. W. (2020). Green synthesis of metallic nanoparticles and their potential applications to treat cancer. *Front. Chem.*, 8.

# 8 Air and Water Pollution Monitoring and Control Through Bionanomaterial-Based Sensors

*Monika Singh, Doli, Amit Yadav, Sumit Kaushik, Namrata Gupta, Gyanendra Singh, Piyush Gupta*

**CONTENTS**



DOI: 10.1201/9781003270959-8

## 8.1 INTRODUCTION

Environmental pollution, particularly developing pollutants, toxic heavy elements, and other dangerous agents, is a major concern around the world. Given the worldwide nature of the problem, it is critical to create and develop strategic measuring approaches that are more effective and precise in detecting a wider range of contaminants. The development of accurate equipment can also aid in real-time and in-process monitoring of environmental pollution creation and release from many industrial sectors. Furthermore, real-time monitoring can eliminate the use of various harsh chemicals and reagents, with the added benefit of on-site pollutant composition assessment prior to discharge into the environment. Electrochemical biosensors have received a lot of attention as a result of recent scientific breakthroughs in order to overcome this challenge. Electrochemical biosensors can be a great analytical tool for monitoring law implementation activities. The current trends in the application of electrochemical biosensors as novel methods to detect diverse pollutant types, including dangerous heavy metals, were discussed in this article. The importance of screen-printed electrodes, nanowire sensors, and paper-based biosensors in pollution detection methods was highlighted. The work is completed with final remarks and future perspectives near the finish. In conclusion, electrochemical biosensors and related fields such as bioelectronics and (bio)-nanotechnology appear to be developing fields that will have a significant impact on the development of novel bio-sensing technologies in the future (Hernandez-Vargas et al., 2018).

As we speak, pollution can be categorized into a variety of categories. Each type of pollution, whether in the air, water, or land, has a different impact on the ecosystem. Although recent advances in science and technology have made it easier to detect the complicated impacts of pollution on our environment, it is still not possible to recycle or completely remove the harmful effects of this increasing toxicity on our ecosystem. However, it should be mentioned that pollution's detrimental impacts can be decreased not only through garbage recycling, but also through increased public awareness. Such knowledge may be promoted through multimedia and social networking, and the impacts of pollution can be avoided by avoiding the dumping of toxic trash in our environment (Ibanez et al., 2007). It should be emphasized that air pollution is one of the most difficult and dangerous types of pollution. Air pollution has a major toxicological and carcinogenic influence on living beings in this period of growing urbanization. The majority of air pollution is caused by emissions from industrial processes and automobiles. Because of its capacity to occupy enormous spaces and disperse uniformly throughout the environment, air pollution is difficult to regulate. Respiratory, skin, cardiovascular, and neuropsychiatric illnesses are all caused by air pollution (Gaddi and Capello, 2018). If fresh air is necessary for a living creature to live a healthy and happy life, freshwater plays an equally crucial role in any living being's life cycle. Fresh water is a vital resource that is easily contaminated in today's period of urbanization, to the point that it is termed polluted (Kumar et al., 2021).

## 8.2 TYPES OF POLLUTION

Increased population and profit-making lead to rapid change in our ecosystem, which is deemed important to be increased through conservation and sustainable use in order to strengthen all elements of human life.

### 8.2.1 Air Pollution

The compounds that accumulate in the atmosphere contribute to air pollution, which leads to an increase in unacceptable concentrations, which can be harmful to human health or have other unintended consequences on living matter and other materials. Major causes of pollution include electricity and heat-generating enterprises, solid waste burning, and industrial activities, particularly transportation (Adetunde and Glover, 2010). Carbon monoxide, hydrocarbons, nitrogen oxides, particulates, sulphur dioxide, and photochemical oxidants are the most common pollutants. Nitrogen oxides coming out of industries, power plants, and vehicles and may cause problems in respiratory systems: lungs, asthma, and bronchitis. Carbon monoxide released from fossil fuels emission is responsible for severe headaches, irritation to mucous membranes, unconsciousness, and may even cause death. Carbon dioxide coming out of suspended vehicular emission or burning of fossil fuels is responsible for vision problems; lung irritation reduces development of red blood cells (RBC), sometimes pulmonary malfunctioning, severe headache, and heart strain. Sulphur oxide is also one of the harmful gases which we get from industries, vehicles, and power plants, again, a cause for irritation of the eyes and throat, allergies, cough, respiratory, and eye problems. Hydrocarbons are generated from burning fossil fuels and lead to kidney problems, irritation in eyes, nose and throat, asthma, hypertension, and carcinogenic effects on the lungs. Lastly, chlorofluorocarbons released from refrigerators is the major cause of ozone layer depletion and global warming.

### 8.2.2 Water Pollution

Water quality is deteriorating day by day as a result of physical, chemical, and biological changes negatively affecting water and living organisms. Pesticides, heavy metals, and nondegradable, bioaccumulative, chemical compounds dissolved or suspended solid pollutants, which are the most insidious and persistent toxic pollutants of water, can be categorized as conventional and nonconventional.

*Conventional/Classical Pollutants.* These are typically associated with undeviating input via human waste. Because treatment systems have not kept up with demand, sewage problems have arisen as a result of increasing urbanization and population growth. Untreated and partially treated sewage from municipal wastewater systems and septic tanks in undeveloped areas contributes significant amounts of nutrients, suspended solids, dissolved solids, oil, metals (arsenic, mercury, chromium, lead, iron, and manganese), and biodegradable organic carbon to the water environment. High quantities of suspended particles can clog rivers and navigational channels, requiring dredging on a regular basis. Water with too many dissolved solids is unfit for drinking and agricultural irrigation (Kumar et al., 2005; Mian et al., 2010).

*Nonconventional Pollutants.* These are highly toxic untested chemicals that are discharged into water pollutants, ranging from biologically inert materials like clay and iron residues from construction and demolition wastes to the most toxic and dangerous materials like halogenated hydrocarbons (DDT, kepone, mirex, and polychlorinated biphenyls [PCBs]). Because of their widespread nature and chemical durability, chronic low-level pollutants are proving to be the most difficult to repair and abate (Bhatnagar and Minocha, 2006; Murad, 2010).

### 8.2.3 Soil Pollution

The environment is detrimental to humans and other living species when soil is contaminated with varying quantities of heavy metals. Soil contamination is caused by pesticide usage in agriculture, excessive industrial activity, poor waste management, and inefficient waste disposal, all of which are difficulties that require more remediation (Chibuike and Obiora, 2014). The mobility and biological impact of xenobiotics, which are carcinogenic and may accumulate in the environment, causing hazardous consequences to humans largely through food, inhalation, or drinking water, is affected by the presence of these toxins in soil (Giller et al., 1998).

## 8.3 IMPACT OF POLLUTION ON EARTH'S ECOSYSTEM

Population growth and economic growth accelerate changes in our ecosystem, which needs conservation and sustainable usage to strengthen all elements of human existence. Over time, humans have transformed natural ecosystems in ways that benefit human health and economic growth. But their contributions have not benefitted everybody. In addition, these dependencies make individuals more conflicted toward resource usage, which negatively impacts global climate change. But the natural ecosystem should be preserved for the future.

### 8.3.1 Effects of Air Pollution on Ecosystem and Human Health

Inhaling particulate matter (PM) has been linked to an increased risk of death in persons with diabetes, chronic lung illness, and inflammatory diseases, according to clinical research. Fine suspended PM, nitrous oxide ($N_2O$), sulphur dioxide ($SO_2$), volatile organic compounds (VOCs), and ozone ($O_3$) are the most common and dangerous pollutants in the urban environment, prompting studies and policies such as cap-and-trade, carbon credits, pollution credits, and others to reduce greenhouse gas (GHG) emissions (Vaseashta, 2008). Various researches on the relationship between air pollution and species variety clearly reveal that environmental contaminants have a negative impact on the extinction of flora and fauna (Camargo and Alonso, 2006). Animal reproductivity may be affected by poisoned air (Veras et al., 2010). Climate change, as a result of greenhouse gas emissions, as well as acid rain and temperature inversion, are some of the main ecological effects of air pollution (Schneider, 1989). Particle pollutants are the most common kind of air pollutants. The majority of these are linked to respiratory and cardiovascular disorders, as well as mortality (Sadeghi et al., 2015; Sahu et al., 2014). Their size ranges from 2.5 to 10 μm. Furthermore, numerous scientific studies have shown that fine particle pollutants might lead to early death in those with cardiac or pulmonary disorders, such as nonfatal heart attacks, cardiac dysrhythmias, and deteriorated lung function. The most prevalent clinical symptom of pulmonary disease caused by air pollution is coughing and restriction of activity owing to lung difficulties (Bentayeb et al., 2013; Guo et al., 2014).

$O_3$ is the most abundant element in the atmosphere. Ground-level ozone is produced by the chemical reaction of nitrogen and VOCs, as well as by anthropogenic activities, and is associated with an increased risk of lung disorders, including asthma (Gorai et al., 2014). These are considered to trigger lipid peroxidation of cellular membranes and macromolecules by increasing the quantity of free radicals. They also impact DNA, resulting in cellular dysfunction (McCarthy et al., 2013). Carbon monoxide is a dangerous gas that is produced when coal, wood, and other fossil fuels are burned incompletely. It has a far stronger affinity for hemoglobin than for oxygen, roughly 250 times stronger. Carbon monoxide poisoning causes severe headaches, nausea, and loss of consciousness, among other symptoms (Akyol et al., 2014). Sulfur dioxide ($SO_2$) is a highly reactive gas that is mostly released as a result of fossil fuel consumption, volcanic activity, and industrial processes. Skin and pulmonary disorders are more common in patients with lung diseases, babies, and persons who are exposed to $SO_2$. Due to its water solubility, $SO_2$ is responsible for acid rain and acidification of soils as a sensory irritant that can cause bronchospasms and mucus discharge in people.

### 8.3.2 Effects of Water Pollution on the Ecosystem and Human Health

If heavy metals, such as fluoride, arsenic, lead, cadmium, mercury, and others, are present in excessive concentrations in water, it is hazardous to human health. Dental caries and mottling of teeth are caused by concentrations of fluoride below 0.5 mg/L, while exposure to greater levels above 0.5 mg/L for 5–6 years can produce fluorosis, a disease that affects human health (Murad and Krishnamurthy, 2004). Toxic chemicals, like arsenic and cadmium, are extremely harmful to human health, causing cancer of the lungs and skin lesions. Long-term exposure to these carcinogens can cause bladder and lung cancer. Pipes, fittings, solder, and domestic plumbing systems all contribute to lead pollution in drinking water. It should be emphasized that lead exposure has negative effects on the blood, central nervous system, and kidneys of humans, which can be fatal to children and pregnant women's health (Murad and Krishnamurthy, 2008).

## 8.4 NANOSENSORS

Flow injection analysis, liquid chromatography, mass spectroscopy, and gas chromatography are the most used traditional pollutant detection techniques. These strategies are undeniably vital and effectively suit the aim. However, because of the time-consuming analysis, high cost, specific operation procedures, and time-consuming sample preparation, a practical need to research alternate approaches has long been felt (Begum et al., 2015; Peng et al., 2016). For qualitative and quantitative analysis of analytes – such as vapors, pesticides, explosives, heavy metals, and bacteria – methods based on fluorescence, infrared spectroscopy, chromogenic detection, electrochemical detection, and surface plasmon resonance (in the form of nanosensors), have been proposed. These alternative approaches are not only portable, but they also enable analyte detection with appropriate sensitivity, selectivity, fast response time, and ease of use (Manoli and Samara, 1999; Patra et al., 2010).

A nanosensor's primary function is to gather data/signals at the atomic level in order to achieve high detection sensitivity. A nanosensor's performance parameters, such as detection sensitivity, detection selectivity, reaction time, detection limit and range, portability, affordability, and recyclability, can all be studied and evaluated. High sensitivity, dynamic range, selectivity, and stability, as well as low detection limits, strong linearity, tiny hysteresis, fast response time, and a long life cycle, are all characteristics of an ideal sensor. In the literature, various types of nanosensors have been presented with the goal of meeting some, if not all, of the desired properties. Nanosensors for diverse analytes for both research and industrial purposes have been successfully developed in recent decades. In most cases, a sensor does not need to meet all of the perfect features at the same time because unique applications require device specifications that depend on the needs of the application. A sensor device used to monitor the concentration of a component in an industrial process, for example, does not always need a detection limit of parts per billion, but a quick response time is preferable. When it comes to environmental monitoring applications, where pollutant concentrations change slowly, the detection limit can be substantially higher, but a reaction time of a few minutes is frequently sufficient. Gas/vapor sensing, heavy metal monitoring, pesticide analysis, bacterium identification, and explosive detection are just a few of the highly desirable domains where portable nanosensors are needed. Different scientific societies, such as biologists, chemists, engineers, and government organizations, have expressed a need for nanosensors in these applications (Kaur et al., 2019). Although the use of nanosensors has the potential to revolutionize the way environmental pollutant analysis is conducted, the International Food Policy Research Institute (IFPRI, 2011) has correctly recommended the following policy: "conduct risk analysis so decision-makers understand the cost effectiveness of using certain nanotechnology applications to improve food and water safety compared to other technologies" (Rich and Perry, 2011).

## 8.5 BIONANOMATERIAL-BASED SENSORS

Biosensors combined with new molecular biology, microfluidics, and nanomaterial technologies have applications in agricultural production, food processing, clinical care, and environmental monitoring for rapid, specific, sensitive, low-cost detection of pesticides, antibiotics, pathogens, toxins, proteins, microbes, plants, animals, foods, soil, air, and water in the field, on-line, and or real-time. As a result, ultra-sensitive biosensors are ideal analytical tools for pollution monitoring, allowing us to work toward enacting legislation to protect our biosphere. Future emerging biosensor development trends in the context of bioelectronics, nanotechnology, miniaturization, and especially biotechnology appear to be all growing areas that will have a significant impact on the development of new biosensing platforms to solve our future clinical diagnostics and severe pollution problems affecting not only human health but all living entities (Singh and Choi, 2010).

Many technologies for detecting and monitoring environmental samples have been developed and used over the years, including sample analysis using techniques such as flow injection analysis, atomic absorption spectroscopy (AAS), X-ray fluorescence (XRF), high performance liquid chromatography (HPLC), gas chromatography (GC), inductively coupled plasma mass spectrometry (ICP-MS), and others, but they all have drawbacks, such as expert handling, time-consuming sample preparation, and high sensitivity. As a result, monitoring, assessing, and detecting any changes in environmental systems necessitates the use of specific equipment that can respond to certain chemical, biological, and physical stimuli and produce an output that can be recorded and evaluated. Sensors are a term used to describe certain types of equipment. Currently, technological and scientific growth has led to the development of a variety of sensors, but significant progress in the field of sensors has been noticed with advances in nanotechnology, which works with materials and systems on the nano-scale (10 to 100 nm or $10^{-9}$ m). The term "nano" refers to anything incredibly little, or one billionth of the stated unit, i.e., $10^{-9}$. One nanometer is about equal to 5 silicon atoms or 10 hydrogen atoms brought into alignment (Kumar et al., 2021). Nanosensors are described as sensors with at least one dimension less than 100 nm and the ability to gather information at the nano-scale and transform it to analyzable data, according to most definitions. Nanosensors do not have to be tiny devices; they can also be huge devices that use the special properties of nanomaterials (NMs) to perform nano-scale measurements. Nanomaterials have distinct physical and chemical properties, including optical, mechanical, and electrical properties, due to their nano-scale dimensions. They have a high ratio of surface area to volume, giving rise to distinct physical and chemical properties, including optical, mechanical, and electrical properties that differ from bulk materials. Nanosensors make use of the unique properties of nano-scaled materials, such as their ability to interact with the surrounding environment on a nano-scale level. Furthermore, because they have similar dimensions to the analytes of interest in many circumstances, they can provide information for nano-scale matrices that is significantly more accurate and sensitive (Grieshaber et al., 2008) than that acquired from conventional sensors. Hence, the main advantages of nanosensors over conventional sensors are:

- Opportunities to design miniaturized and portable sensors.
- High selectivity and sensitivity, low analyte detection limit.
- Ease of surface functionalization to design target specific sensors.
- Fast response time.
- Applications in real-time sensing.
- Tuneable size-dependent properties making them useful for efficient detection of a large variety of samples. In addition, recognizing many biological systems' innate detecting capacities, the merging of nanotechnology with biotechnology has resulted in the development of nanobiosensors with improved sensitivity and selectivity (Dubey and Mailapalli, 2016).

The collective response of a nanomaterial–medium system that is attributable to reduced dimensions, viz. size (area and distribution), surface structure (groups, functionalization), physical (electronic, optical, photo activation, luminescence), and chemical (crystallinity, purity, solubility) is vital to developing a scientific model that predicts its response as a sensor, bio-adverse response pathways for toxicity, adsorption pathways for materials for storage, and interaction with light for optical response. It is an exceedingly complex task due to a large matrix of parameters consisting of nanomaterials (metals, metal oxides, semiconductors, macromolecules, and self-assembled), the surrounding environment, and influencing mechanisms of interactions (Vaseashta, 2008).

### 8.5.1 Classification of Nanosensors

Nanomaterials have introduced an interesting approach to the development of low-cost, portable fluorescent devices (Tao et al, 2020). Nanomaterials, such as metallic nanoparticles (NP), multi-walled carbon nanotubes (MWCNT), graphene (G), graphene oxide (GO), quantum dots (QD), carbon dots (CD), nanosheets, and metal-organic frames (MOFs), can boost biosensor performance dramatically. This is owing to features such as a larger surface area-to-volume ratio and exceptional physicochemical qualities. (Tao et al., 2020). Nanomaterials have unique optoelectronic features that should enable the creation of sophisticated systems for the simultaneous detection of many analytes at a low cost (Zhang et al., 2014). The main components of a basic sensor are: (1) sensor material and (2) transducer.

Nanomaterials have special properties. In the presence of the target analytes, the sensor materials are responsive and sensitive to changes in environmental stimuli, such as thermal (temperature, heat flow, entropy, specific heat, etc), chemical (concentration, pH, composition, oxidation/reduction potential, rate of reaction, etc), magnetic (magnetic moment, flux density, permeability, field intensity, etc), and optical (absorbance, transmittance, fluorescence, etc) (Middlehoek and Noorlag, 1981). These variations are converted to mostly electrical signals by the transducer to give an output that can be analyzed.

Nanosensors are classified based on the signal transmission techniques they employ. Despite the fact that signal transduction methods vary, nanosensors can be classed using the following primary transduction mechanisms: nanosensors include optical nanosensors, electrochemical nanosensors, mechanical/acoustic nanosensors (Khalil et al., 2019), and magnetic nanosensors.

### 8.5.2 Optical Nanosensors

Optical nanosensors are based on nanomaterials' distinctive optical behavior when light signals interact with them. The approaches utilized to detect optical phenomena determine the sensitivity of optical nanosensors (Qu et al., 2012). These nanosensors respond to changes in optical signal transduction and can be classified further depending on optical properties including absorption and emission, as well as surface characteristics. Electrochemical biosensors use screen printed electrodes (SPEs), nanowire (Hernandez-Vargas, et al., 2018), and aptamers, DNAzymes, and aptazymes for environmental pollution monitoring (Palchetti and Mascini, 2008). Typical examples of commercially available sensing devices for environmental monitoring: (QRAE 3; Water quality; KANE101 Indoor Air Quality Analyzer; Radioactivity Meter PCE-RAM 10; QRAE 3- Water quality; MATTS Monitors e Air monitoring system; Air Particle Counter PCE-PCO 1; Electromagnetic Radiation Detector PCE-EMF 823; Orion Star A326 pH/Dissolved Oxygen Portable Meter.)

## 8.6 ADVANTAGES OF NANOSENSORS

### 8.6.1 Response Time

The response time of a detector relates to how quickly it reacts to a signal change. This is especially relevant for nanosensors used in applications such as air quality and heavy metal monitoring. For example, target gases (e.g., dangerous gases $NO_2$, $SO_2$, and $CO_2$) must be identified promptly in gas sensing applications in order to avert accidents or raise public/government awareness. For gas detection, high-performance liquid chromatography, spectroscopy, enzyme electrodes, and other classic methods are used. On the other hand, these instrumental procedures do not provide a speedy analysis. Excess nitrogen dioxide ($NO_2$) gas emissions have become a serious air quality issue in our environment. This pollutant can cause major health problems, such as lung damage.

Even modest levels of $NO_2$ (e.g., 50 ppb) can be hazardous to human health. For these reasons, it is vital to diagnose $NO_2$ as soon as feasible. Nanostructured metal oxide semiconductor-based sensors have been created for the detection of $NO_2$ with a fast reaction time. Urasinska-Wojcik et al., 2017 developed a nanosensor that detects $NO_2$ at the parts-per-billion level using a tungsten oxide ($WO_3$) transducer. Changes in DC resistance or conductance could be measured using a low-power micro-hotplate platform based on micro-electromechanical systems (MEMS). The electric resistance of tungsten oxide in air, nitrogen dioxide, and nitrogen was used to test the sensor response (S). $NO_2$ levels ranging from 10 to 250 ppb were recorded in air with a relative humidity of 50%. When the sensor's results were compared to those of commercial gas sensors (SGX Sensortech's MiCS-2714), it was discovered that the resistive gas sensor not only had a high sensitivity toward $NO_2$ (3.4% /ppb vs 0.2%

/ppb for commercial metal oxide-based devices), but it could also operate reliably even at lower oxygen levels (down to 0.5%). Furthermore, $NO_2$ had a substantially longer detection time, with response times of 100 and 200 seconds for trials done in air and low oxygen, respectively. Another significant air contaminant that can cause sick building syndrome and other health problems is formaldehyde (HCHO). The tin oxide ($SnO_2$) microsphere-based sensor detected 100 ppm of HCHO gas in 17 seconds and recovered in 25 seconds, which was significantly faster than most other methods previously described. The nanosensors' rapid reaction and recovery periods are assumed to be influenced by the NMs' favorable form (high surface-to-volume ratio, high reactivity). Because of the porous structure of the transducer material used, the nanosensors also have strong permeability, allowing for fast adsorption and diffusion of analytes. In most cases, nanosensors have been found to have a significantly faster response time than conventional techniques, and they are capable of giving quick test results (a few seconds to minutes).

### 8.6.2 Detection Limit

Rapid economic development at the expense of the environment causes pollution. Even in little amounts, a few pollutants might cause serious health problems. For the analysis of such pollutants, highly sensitive technologies are required. Heavy metal contamination is one such example. Traditional heavy metal analysis methods (e.g., several types of atomic absorption spectroscopy, optical emission spectroscopy, mass spectroscopy, X-ray fluorescence, and ion chromatography) are obviously available, but they are complex and expensive. The necessity for widely available and cost-effective solutions for the monitoring and detection of heavy metals in today's environment prompted research on nanosensors. To be a viable alternative to present technologies, any conceivable heavy metal nanosensor must have a high sensitivity (low detection limits). The ability to detect heavy and dangerous metal ions such as Pb(II), Cd(II), Hg(II), and As(III) is critical for determining their toxicity and environmental buildup. Because of its simplicity, low cost, and mobility, heavy metal sensors have historically been based on electrochemical stripping analysis background technology. Modified screen-printed electrodes (SPEs) have recently emerged as interesting alternatives for heavy metal nanosensor development. For example, gold nanoparticles with a graphene quantum dot (GQD) function have been employed for electrochemical detection of heavy metal ions. GQDs with cysteamine-capped gold nanoparticles have been functionalized for detecting $Hg^{2+}$ with a detection limit of 0.02 nM. (Ting et al., 2015). In addition, carbon dots have been utilized to detect mercury ions optically. PEG200 was used to increase the fluorescence response of the carbon dots before they were triggered. They were pale yellow in hue and fluoresced with a 565 nm emission wavelength. Gonçalves et al. (2010) found that this optical detection system could detect $Hg^2$ ions at a concentration as low as 2.69 mM. The simultaneous and sensitive electrochemical detection of $Pb^2$ and $Cd^2$ ions has been reported using a graphene-based composite (graphene oxide [GO], a multi-walled carbon nanotube [MWCNT]). Because of the good synergy between GO and

MWCNTs, the abovementioned sensor platform allowed for a faster and enhanced electron transfer rate, allowing for very sensitive detection of $Pb^{2+}$ (0.2 mg/L) and $Cd^{2+}$ (0.1 mg/L) ions using anodic stripping voltammetry (Huang et al., 2014). A cadmium-based MOF material was used to provide highly sensitive optical detection of Hg2 in pure water. Within 15 seconds of analysis time, this radiometric fluorescence sensor was reported to detect 0.55 mM of $Hg^{2+}$ (Wu et al., 2015).

### 8.6.3 Portability

Environmental monitoring requirements have altered in recent years, and more regular analyzes are frequently required for quality assessment and control. Traditional instruments are excellent for providing research-grade analysis, but in the real world, field-deployable technologies and devices are more in demand. A routine, user-friendly, and cost-effective analysis of pollutants necessitates the availability of portable instruments that can be utilized by untrained, non-laboratory staff. On-field analysis of gases, water, and air quality parameters can be done with technologies that have a small footprint, need little energy, are environmentally tough, and are inexpensive. The transition of research-level nanosensing technology to product-level availability has occurred in recent years. Although more work remains to be done to bring the various nanosensors to market, the current trends are quite promising. People in the research community are also looking for ways to combine nanosensing technology with tiny and portable platforms. The performance of the tiny platforms has often matched the desired criteria of a practical-level sensing application.

### 8.6.4 Sensitivity/Resolution

Nanosensors have the distinct benefit of displaying high detection sensitivity and resolution. The magnitude/degree of detected signal per analyte concentration unit is used to calculate the sensitivity of a nanosensor. The slope of the corresponding calibration graph can be used to calculate its numerical value. The detection limit is sometimes mistaken with sensitivity/resolution. The use of NMs has allowed for high levels of signal resolution in many applications, and such sensors are expected to have a superior signal-to-noise ratio. Nanostructures have remarkable electric characteristics due to their ultra-high surface-to-volume ratios. Such sensors can detect any surface-adsorbed species, such as gas molecules, with great sensitivity. Chemical nanosensors made of carbon nanotubes, functionalized SiNWs, and metal nanowires are particularly promising since they can detect very low pollutant concentrations with high sensitivity.

### 8.6.5 Selectivity

Sensor selectivity refers to its capacity to measure a the concentration of a substance in the presence of other interfering chemicals. A successful nanosensor should also be able to detect the desired analyte from complicated medium selectively. For example, a single gas from a mixture of gases or a single heavy metal from a sample containing other metal ions. Several nanosensors with the key attribute of selectivity have been reported in this regard.

## 8.7 LIMITATIONS OF NANOSENSORS

### 8.7.1 Price/Cost

By applying surface modification tools and procedures, NMs can be paired with a wide range of tiny organic ligands and big bio-macromolecules, resulting in highly selective pollution sensing systems. Highly sensitive pollution detection devices are achievable thanks to the interplay of NM characteristics with specific sensors. Fabrication of NMs of various sizes and forms has become possible thanks to recent manufacturing improvements. These breakthroughs will serve as the foundation for additional innovation to produce unique features tailored to specific applications. When compared to other technologies, the cost-effectiveness of using some nanotechnology applications is relatively low-cost. NMs and field application technologies, like other technologies, are required for environmental sensing applications. Because the synthesis procedures and characterization tools utilized in the development and deployment of NMs are rather expensive, there is always the risk that NM-based sensing systems will not be cost-effective. For pathogen detection and heavy metal ion/anion analysis, the majority of the industry and companies still choose classic cellulose/fiber/carbon-based platforms to create and market their point-of-care devices. Despite significant advancements in the field of gas nanosensors, the majority of gas sensors on the market are still built on metal oxide substrates. The development of the next generation of heavy metal sensors with cost-effective production and a user-friendly format with an in-field readable result has sparked a lot of interest. In terms of device cost, very sensitive and selective colorimetric detection is definitely a good option right now for achieving these next-generational goals. These data suggest that nanosensors using NMs like carbon nanotubes, graphene, molybdenum disulfide, CdTe quantum dots, and SiNWs can be expensive, and that attempts to reduce their cost are still needed.

Nanosensors can be especially sensitive to diverse species present in a test solution/media due to the inclusion of highly surface-active NMs. As a result, utilizing particular tactics to reduce cross-interference or false-positive signals becomes beneficial. Using the correct sample preparation methods is probably one of the best ways to get accurate signals from nanosensors. Complex analytes, such as pathogens, germs, heavy metals, and food poisons, for example, generally necessitate effective analyte isolation and purification prior to analysis. Prior sample preparation and preconcentration activities can increase the cost of the analysis while also making the process less convenient. To achieve sample purification and preconcentration, nanosensors are now being coupled with microfluidic channels. However, one can debate the importance of durable gadget packing in such situations.

### 8.7.2 Self-Standardization and Validation

Almost all traditional pollution analysis methods operate on the basis of calibration test runs and self-standardization. The majority of available or reported nanosensors do not include this characteristic on a large scale. Calibrating nanosensors with only one or two standard test samples is always going to raise questions about the final test result's dependability. In fact, the majority of nanosensors (or the sensing portion of a nanosensor) are designed to be disposable. In typical analytical applications, the inter assay precision of several of these sensor electrodes/chips/test kits is frequently overlooked. As goods based on nanotechnology are produced and pushed into broader use, standardization will play an increasingly important role in ensuring a smooth transition from the laboratory to the marketplace. The research and market for nanosensors have grown dramatically in recent decades. Nonetheless, in most nanosensor research reports, validation with real samples under real settings is a crucial missing factor. The majority of the time, researchers demonstrate the operation of their nanosensors in synthetic solutions/media in the lab. Additional data is needed to verify the functionality of nanosensors in complicated test environments, such as the outdoors (for gas sensors), river and supply water (for heavy metals, bacteria, pesticides sensors), and food (for toxins, heavy metals, and pathogen samples) (Kaur et al., 2019).

### 8.7.3 Potential Concerns of Nano-Toxicity

The use of NMs is always accompanied by concerns about unknown toxicity profiles. The implications of making NMs available to the general public for usage, as well as the discharge/disposal of NMs after use must be thoroughly investigated. For instance, (1) accurate characterization of NMs in biological matrices for in-depth understanding of their toxicity in biological systems, (2) study of NM interaction with the environment, (3) exposure assessment, and (4) determination of permissible limits in terms of concentration and amount are all major issues that need to be addressed in relation to the application of NMs. Models that forecast NM exposure limits and scenarios with specifics of transport, transformation, and toxicity evaluation of each individual NM, similar to pesticide drift assessment models, are needed. Detailed studies on various elements of NMs, such as the level of safe limits of nanoparticles in diverse contexts, their fate in the environment, and absorption on cellular membranes and cell walls, must be done and made available to the general public and interested enterprises (Khot et al., 2012). Nanosensors are frequently designed as disposable electrodes, kits, or papers. It's also crucial to create processes for the proper disposal of such sensors once they've been used, which have been thoroughly tested and certified. It is not possible to dispose of the sensor parts in municipal garbage. As a result, providers must take steps to educate the public on the safe use and disposal of nanosensors at the time or before they are made commercially available.

Modern biosensing technologies include multiple pieces of information, such as the use of nanomaterials for bio-conjugation and signal enrichment, as well as sensing techniques spanning from optical to electrochemical to electromechanical. Nucleic acid-based biosensors (NA) adds to the excitement because of its physicochemical properties. Aptamers/NAs are simple to make and functionalize with functional groups that help them stick to the electrode surface. The redox chemical groups of the electrochemical biosensor enable the transformation of molecular interactions, such as DNA hybridization, or contaminants into electrical signals correlating a target analyte(s). Due to their rapid analysis, low cost, and manufacture, these biosensors have a wide range of applications in environmental monitoring, including direct detection of toxins with low detection limits and excellent precision for as low as single-molecule detection. However, there are still some limitations that NA-based electrochemical biosensors must overcome, such as the development of a methodical approach for NA or aptamer selection for biosensors, molecule immobilization on the conductive surface, semi-conductor substance quality, and signal-to-noise ratio perfection. Furthermore, multiplex and complicated sample conditions should be used in electrochemical NA biosensors. The next step should be to develop robust, regenerative, and long-lasting sensing elements for long-term use. Combining nanomaterials, nanocomposites, and nanopolymers could result in hybrid devices that can be used in a variety of biosensors. Although a few NA-based electrochemical biosensors have made the leap from research lab benchtop to portable point-of-care devices, the sensitivity and selectivity of the majority of these biosensors are inadequate. At this time, the electrochemical NA biosensor application is the only target-oriented technique, and the changed transducer cannot be reused. It is, however, parallelizable and, hence, has enormous promise for monitoring several targets at the same time. Overall, the use of NA biosensors in electrochemical biosensing is still in the early stages of research. As a result, various issues must be resolved before electrochemical aptasensors may be used to detect several analytes in complicated samples. Several homologue/analogue compounds may have nonspecific interactions with DNA aptamers, such as attaching to the sugar-phosphate group spine of DNA and so competitively hinder the exact conjugation of the target analytes. Aside from these factors, nontarget NA in biological fluids may hybridize with the NA probe, reducing the biosensor's ability to maintain its suitable binding site. As a result, more attention should be placed on aptamer or NA targeting in the real sample. Because the electrochemical biosensor is susceptible to salt or ion composition, the solvent composition may also have an impact on the biosensor's performance.

## 8.8 APPLICATIONS IN POLLUTION MONITORING, DETECTION, AND REMEDIATION

For atmospheric pollution research, satellite picture data has typically been underutilized. Earth radiances are impacted by temperature, emissivity of the ground surface, and the air column above, as well as its surroundings, as recorded by satellite sensors in various bands. By making observations over vast spatial domains with 3D models, satellite image data can aid in the detection, tracking, and understanding of pollutant sources and transport. Aerosol optical thickness can also be used to examine pollution's optical effects on atmosphere. The Global Ozone Monitoring Experiment (GOME) on ERS-2 and the SCIAMACHY on ENVISAT data can be analyzed using the Differential Optical Absorption Spectroscopy (DOAS) method to retrieve NO2 columns in the 425–450 nm areas with high precision. Data from ground-based stations, such as the US Environmental Protection Agency (EPA), can be modeled and matched with city-specific data from the Texas Commission on Environmental Quality (TCEQ), Airparif (France), the National Institute for Public Health and Environment (RIVM, Netherlands), the British Atmospheric Data Centre (BADC), and the Natural Environmental Research Council (NERC, UK), among others. There is evidence, based on observations, that these aerosols can change cloud properties. Pollution increases scattering albedo, which reduces back scatter fraction. The hygroscopicity of aerosols is determined by their photochemistry. In both ambient air and smog systems, infrared absorption spectroscopy effectively identifies trace contaminants. These contaminants were analyzed using satellite photos and data from the Advanced Spaceborne Thermal Emission and Reflection Radiometer (ASTER). It was made up of high spatial and spectral resolution images from the TIR and SWIR bands, and the data were compared to EPA air quality and spectroscopic study data on atmospheric pollution (Vaseashta et al., 2007).

The ability of a variety of nanomaterials to detect and mitigate contaminants in air, water, and soil has been investigated. We have looked into a variety of ways that nanomaterials can be used to successfully reduce/isolate atmospheric pollution, including using carbon nanotubes filled in a polymer composite matrix to create a static discharge to remove PM from incoming air, chemical protective clothing, and breathing filtration masks. We investigated the elimination of indoor scents under weak UV illumination using immobilized $TiO_2$ films made by a sol-gel technique. In the presence of $TiO_2$ photocatalysts, weak UV light of 1 W/cm2 was sufficient to degrade such compounds at 10 ppm. In one experiment, 150 mL of an *E. coli* slurry containing 30,000 cells was placed on a $TiO_2$-coated glass plate with a weak UV (1 mW/cm2) illumination and showed antibacterial activity. We're concentrating on reducing water contaminants like arsenic (As), which can be present in the natural environment as well as through human activity. Long-term exposure to low levels of As has been related to cancer, as well as cardiovascular, pulmonary, immunological, neurological, and endocrine impacts, according to several studies, necessitating research into efficient ways for eliminating As from drinking water. We used a technique based on sorption of iron(III) oxides, such as amorphous hydrous ferric oxide (FeOOH), weakly crystalline hydrous ferric oxide (ferrihydride), and goethite, to remove both As(V) and As(III) from aqueous solutions. According to preliminary findings using synthetic akaganeite, the sorbents' maximal capacity is roughly 75 mg of As per gm of akaganeite. Also, with a maximal capacity of approx. 40 mg of As per gm of sorbent, rather good results were obtained using synthetic magnetite/maghemite. The use of nanostructures based on iron oxides/oxyhydroxides in zeolites, as well as nano-scale zero-valent iron (nZVI) technology, paves a way for high-capacity arsenic removal.

Nanomaterial properties drive research into lightweight materials and structures that can withstand harsh operating conditions, such as high temperatures, dynamic loading, thermal and mechanical stresses, improve blast and shock resistance and improve ballistic performance, without sacrificing structural integrity. In the transportation and aviation industries, lightweight components will cut fuel consumption and emissions. Carbon fibers, protective coatings, honeycomb, and bonded structures are among the materials under investigation. We generated nanofibers utilizing electrospinning and high-performance polymers impregnated with carbon nanotubes (CNTs) (Vaseashta, 2008).

Satellites are used for a variety of purposes, including communication, navigation, climatology, surveillance, and environmental monitoring. Nanotechnology and micro/nano electromechanical systems (MEMS/NEMS) have the potential to transform the field of environmental pollution monitoring satellite design. We can extract pollution and optical data for any city using image processing of satellite data. To detect pollution, we conducted experiments with nano-scale metal-oxide sensors. The sensors are linked to a web server and a personal digital assistant (PDA) with GPS and wireless communication. We used data from this prototype device to overlay satellite data in order to cross-reference pollutant coordinates. We expanded our research to look into how nano-scale materials might pollute the air and water. The application of such transdisciplinary and revolutionary technologies has sparked a wide-ranging public discussion, ranging from support for more study to a moratorium in the absence of scientific risk assessment. This part offers a quick overview of ethical and legal issues, fate and transportation, and *in-silico* risk assessment.

Because of the strong link between the environment and human well-being, environmental monitoring is one of the most emphasized sectors in the ecosphere (Justino et al, 2017). The discharge of harmful chemicals into the environment, such as pesticides, insecticides, heavy metals, and pharmaceutical xenobiotics, is a global problem. Numerous chromatographic methods, such as gas chromatography, high-performance liquid chromatography, and mass spectrometry, are used in traditional analytical approaches for determining environmental pollutants. However, these procedures necessitate expensive chemicals, time-consuming sample preparation, sophisticated equipment (Justino et al, 2017; Lang et al. 2016; Hassani et al., 2020), and specialized human resources and, thus, are not suitable for monitoring at the site of application. In this arena, less expensive, rapid, *in-situ*, easy to handle, portable, and real-time analytical approaches, such as biosensors, are urgently needed to monitor such toxins in order to avoid further escalation of ecological problems (Justino et al, 2017). In the next sections, we'll look at some of the most current uses of electrochemical NA biosensors in various fields. The qualitative and quantitative pollution detection is relevant, emphasizing the benefits and fundamental methodologies for capturing and identifying environmental pollutants, such as steroids, mycotoxins, heavy metals, pesticides, and antibiotics and environmental pathogen detection based on DNA hybridization techniques (Hashem et al., 2021).

### 8.8.1 Qualitative and Quantitative Analysis of Pollutants

Environmental pollution analysis, both qualitative and quantitative, is a critical topic in identifying and dealing with environmental dangers. Aptamer or NA-based electrochemical biosensors have gotten a lot of attention in this situation because of their ability to tackle a lot of problems and issues in environmental pollution (Hayat and Marty, 2014) and other systems. Toxins, steroids, insecticides, heavy metals, and antibiotics are examples of target analytes that can be found in diverse forms in ecological samples (Hayat and Marty, 2014). Using aptamers and NAs as sensing components, electrochemical NA biosensors have been proven to recognize and monitor a variety of environmental pollutants (Justino et al, 2017). The advantages of these NA biosensors over photosensitive, piezoelectric, or thermal processes are numerous. Fast, sensitive, and selective electrochemical transduction is disposable, vigorous, well-suited to new microfabrication technologies, simple to miniaturize, high throughput, and unaffected by sample turbidity. Normally, electrochemical reactions provide an electrical response without the use of an expensive signal transduction device (Hayat and Marty, 2014). As a result, we concentrate on electrochemical DNA biosensors, also known as aptasensors. In the realm of NA-based electrochemical biosensors, two fundamental strategies have emerged. The first strategy uses the hybridization of NAs. A single-stranded oligonucleotide can look for complementary strands in a sample and hybridize to generate a double strand when coupled to a transducer surface. The second group of DNA biosensors, known as aptasensors, is effective for monitoring environmental materials that lack NA. The interaction of tiny pollutants with NA conjugated to a transducer is detected by biosensors. These biosensors could be used as a general indicator of contamination because they can provide quick and accessible information on the availability of such substances (Guo, 2014). Small molecule binding to an aptasensor has recently been electrochemically recognized on a variety of systems for environmental monitoring. Among them, label-free electrochemical impedance spectroscopy (EIS) has appeared to be a promising method for identifying biomolecule activated substrates and a sensitive method for monitoring aptamer-target conjugation on the transducer surface. More importantly, EIS is nondestructive, making it ideal for aptamer-based tiny contaminant detection (Hayat, 2013 and Hayat et al., 2013).

### 8.8.2 Detection of Heavy Metals

Pollution including heavy metals has the potential to harm human health and the environment. Heavy metals, such as mercury and arsenic, for example, produce significant toxicity in the neurons and endocrine system, as well as heart problems, skin damage, and cancer (Hayat and Marty, 2014). Arsenic is a toxic carcinogen that is widely dispersed throughout the world. Humans can be exposed to arsenic by both direct and indirect ingestion, such as drinking arsenic-contaminated water or eating foods grown on arsenic-contaminated land. While arsenic can have serious and long-term health consequences, chronic ingestion or long-term arsenic exposure can lead to malignancies, skin lesions, arsenicosis, and cardiovascular disease. Kim et al. described a single-stranded DNA aptasensor for the detection of arsenic as a viable device to be used in a remarkable procedure (Kim et al, 2009). A small amount of lead (Pb) in the body can induce neurological, cardiovascular, reproductive, and developmental problems (Zhang et al., 2016). The lead ion ($Pb^{2+}$) was determined precisely and sensitively using an electrochemical DNAzyme biosensor based on DNAzyme catalysis. DNA-Gold bio-bar codes to maintain signal increment when $Pb^{2+}$ binds to DNAzyme. A thiol-Gold interface was used to create a unique DNAzyme for $Pb^{2+}$ on a Gold electrode surface. The DNAzyme hybridizes with a substrate strand with an overhang on the other side. The DNA-gold bio bar code hybridizes with the overhang. $Ru(NH_3)_6{}^{3+}$is a redox chemical that attaches to the negatively charged phosphate moiety of DNA via an electrostatic interaction and performs activities. After conjugation of $Pb^{2+}$ with the DNAzyme, the DNAzyme catalyses the hydrolytic destruction of the substrate. The substrate strand, the DNA bio-bar code, and the $Ru(NH_3)_6$ are all eliminated from the gold electrode surface as a result. The removal of $Ru(NH_3)_6{}^{3+}$ from the surface reduced the electrochemical signal of $Ru(NH_3)_6{}^{3+}$. The $Pb^{2+}$ can be measured numerically using a differential pulse voltammetry (DPV) pulse of $Ru(NH_3)_6$, with a linear detection range of 5 nM to 0.1 M. The use of DNA-gold bio-bar codes boosts detection sensitivity by 5 times, allowing recognition of $Pb^{2+}$ concentrations as low as 1 nM. The DPV signal of this DNAzyme biosensor was insignificant for other metal ions, implying $Pb^{2+}$ detection selectivity (Shen et al, 2008). Heavy metals are non-biodegradable and last a long time in the environment. That is why detecting heavy metal pollution, such as copper, is so important. As a result, there is a pressing need to develop precise ion biosensors for determining harmful heavy metals quickly. On the basis of AuNPs, a very sensitive and selective electrochemical aptasensor for the measurement of $Cu^{2+}$ ions was mentioned. AuNPs provided a large surface area for fabricating a large number of aptamers as well as useful electrochemical signal transduction. As a $Cu^{2+}$ aptasensor, it demonstrated great sensitivity, a lower detection limit, and a large detection range. With a detection limit of 0.1 pM, the electrical signal amplified proportionately with the concentration of $Cu^{2+}$ from 0.1 nM to 10μM. The detection of the $Cu^{2+}$ ion was unaffected by the presence of other divalent cations, indicating that the sensor has a high selectivity.

### 8.8.3 Detection of Mycotoxins

Mycotoxins, a category of naturally occurring chemicals created by mold living on crops, are the most common food contaminants. Even though they are natural, they have certain negative consequences on human health, such as renal damage, stomach problems, and immunological suppression. One of the most common food-polluting mycotoxins, ochratoxin A (OTA), is potentially carcinogenic to humans (Hayat and Marty, 2014; Han et al., 2016). The Wang group used an AuNPs-conjugated reduced graphene oxide (AuNPs-rGO) as a signal magnification material to recognize OTA. The reporter DNA was immobilized using AuNPs-rGO, which worked as a carrier for reporter DNA, and the capture DNA was linked to the gold electrode. This was an impedimetric aptasensor that could detect OTA concentrations as low as 0.74 pM (Jiang et al, 2014).

### 8.8.4 Detection of Steroid Compounds

Since the beginning of global industrialization, steroidal estrogens have been a critical and rising environmental hazard. Estrone, estradiol, and estriol are steroid compounds that represent a serious hazard to soil, water resources, plants, and humans. Several studies have shown that increased levels of natural and artificial estrogens feminize male fish. The consequences include shrinking testes, influencing reproductive aptness, lowering sperm count, inducing vitellogenin (VTG) production in males, and altering other reproductive characteristics (Adeel et al., 2017). Kim YS et al. (2007) and Kim et al. (2009) described one of the earliest electrochemical aptasensors for the detection of 17β-estradiol. They discovered that 17β-estradiol has a linear detection range of 0.01 to 1 nM, with a detection limit of 0.1 nM. Another label-free competitive electrochemical aptasensor, based on competition inside the BTX-beads and BTX-horseradish peroxidase (HRP) bioconjugate, was reported for the detection of brevetoxins (BTXs). In this aptasensor, the detection limit for BTXs was found to be 106 pg/mL. Elshafey et al. developed an aptamer incorporated in an EIS aptasensor to determine the smallest powerful neurotoxin, anatoxin-a (ATX). The detection limit of this aptasensor was 0.5 nM, with a linear detection range of 1 to 100 nM (Eissa et al., 2015; Elshafey et al., 2015). Electrochemical aptasensors paired with nanostructured substances have recently shown a significant potential for detecting tiny combinations for environmental monitoring. Fan et al. demonstrated a highly sensitive photoelectrochemical (PEC) aptamer platform based on TiO 2 nanotubes and CdSe QDs for the detection of 17β-estradiol (E2).

### 8.8.5 Detection of Pesticides

To keep pests and insects at bay, pesticides and insecticides are utilized. The impacts on nontarget creatures, on the other hand, are a source of worry. Pesticides are chemical compositions that are used to kill fungi and animals. Due to spraying or spreading throughout entire agronomic areas, more than 98% of applied pesticides and 95% of sprayed herbicides spread to an endpoint other than their targets (Miller, 2004). Atrazine is one of the most widely used herbicides for controlling the growth of weeds. Similarly, the insecticide acetamiprid is used in crops to protect them from pest assault. These can cause toxicity in people and wildlife after they seep into the ecosystem. Wang et al. created an oligonucleotide aptamer that can detect up to four very lethal organophosphorus insecticides, including isochroous, proxenos's, omethoate and phorate (Shen et al., 2008). Based on EIS, Fan et al. developed another aptasensor for sensitive and specific identification of acetamiprid (Fan et al., 2013).

### 8.8.6 Detection of Antibiotics

Antibiotics are utilized in animal farming, as well as in animal feed as a preventative and therapeutic medication. Only a small amount of the antibiotics consumed by fauna are metabolized, allowing a large portion to accumulate in tissues or be excreted and discharged into the environment. Antibiotic resistance may develop in the ecosystem, with the risk of transmission to humans via the food chain

(Hayat and Marty, 2014; Chang et al., 2008; Boxall, 2010). Metha et al., 2011 have devised and developed a DNA-based electrochemical aptasensor for chloramphenicol detection. They used the SAM method to immobilize the aptamers on the gold electrode surface, resulting in an aptasensor that was highly sensitive and selective for the detection of chloramphenicol (Mehta et al, 2011). Tetracyclines are a type of broad-spectrum antibiotic that protects us from germs by preventing them from completing their life cycle. This antibiotic has been proven hepatoxic to pregnant women. Prussian blue (PB) was used as an indicator in a label-free electrochemical aptasensor for tetracycline identification. Streptomycin (STR) is an aminoglycoside antibiotic used to treat gram-negative bacterial infections in people and animals. Human health is severely harmed by STR residue (Ghanbari and Roushani, 2018). Based on a signal augmentation approach, a highly sensitive and selective electrochemical aptasensor for identifying STR was developed. Porous carbon nanorods (PCNRs) produced from porous carbon nanospheres and multifunctional graphene nanocomposites (GR- $Fe_3$ $O_4$ –AuNPs) were used as sensing materials in the aptasensor.

### 8.8.7 Pollutant Detection Based on DNA Hybridization

When DNA/RNA is the analyte, nucleic acid hybridization biosensors can be utilized to direct contaminant(s). The term genosensor refers to this type of NA biosensor. A genosensor is made by gluing a short synthetic oligonucleotide sequence (probe) to a transducer surface, which generates a signal. The biorecognition element is the immobilized probe, which detects the target DNA or RNA. Probe design and manufacture are the first steps in developing a genosensor, followed by hybridization response creation and identification (Palchetti and Mascini, 2008). The biological recognition event in these types of biosensors is DNA base-pairing between complementary stands. The incorporation of a single-stranded DNA probe onto supporting materials, such as nanoparticles or nanocomposites, improves the sequence-specific identification of DNA hybridization by these sensors (Hlongwane et al., 2019).

## 8.9 FUTURE TRENDS AND POTENTIALS

The sensors used in environmental monitoring follow a similar pattern. Because most monitoring is done on-site, the market for portable devices and accompanying sensors has grown rapidly. Biomedical applications (cholesterol monitoring, blood gas analysis, pregnancy test strips, screening of infectious diseases, and glucose monitoring) as well as industrial process control, food toxicity monitoring, environmental monitoring, and agriculture are expected to increase demand for biosensors in the future. Sensors for monitoring applications may also see an increase in use. The biosensor sector has become an appealing endeavor in terms of profit margins as a result of increased market acceptance. The global biosensors market was worth USD 22.4 billion in 2020, and it is predicted to increase at a 7.9% compound annual growth rate (CAGR) from 2021 to 2028 (Biosensors, 2021–28). Many devices are available or will be available soon for the analysis of a wide range of analytes, including gases, disease biomarkers, water contaminants (such as heavy metal ions, bacteria, organic compounds, drug residues, and pesticides); remote sensing of airborne bacteria; pathogen identification and quantification; analysis of toxic substance levels before and after bioremediation processes; and measurements of vital body fluid constituents (such as mycotoxins).

Biosensors with increased functionality and performance are emerging as a result of the introduction of novel manufacturing techniques derived from nanotechnology. This is owing to novel nanofabrication techniques, as well as the integration of new nanomaterials and chemical processing techniques for biosensor surface processing. Plasmonic, fluorescence, electrochemical, pH sensing, and nanomechanical biosensors are all benefiting from new nanomaterials such as gold nanoparticles, quantum dots, carbon nanotubes, and graphene.

## 8.10 CONCLUSION

The authors attempted to incorporate all of the fundamental aspects and recent progress about the various types of bio-inspired micro and nanomaterials, including cellulose, hemicellulose, lignin, and their composites, as well as a brief history of development, processing methods, structural properties, and notable scientific applications, in this review article. We've concentrated our efforts on cellulose, hemicellulose, and lignin, as well as their newly produced composites for mechanical, thermal, electrical, and biodegradation applications. Not only are these materials abundant on our planet, but they are also biocompatible and suited for biological purposes (owing to the natural intrinsic structures, which make them first choice for the transformative device fabrications). Furthermore, the antioxidant activity of lignin as a nanofiller in various polymer matrices, as well as its impact on various cell types cultivated on electro-spun nanofibers, were investigated, suggesting that these materials have the potential to be exploited in healthcare applications. In addition, the structural characteristics of all of the aforementioned bio-materials, as well as recent advancements, are thoroughly discussed in this article. We concentrated on lignin/cellulose-based, low cost, environmentally friendly hydrogels in the last section of this review for a variety of applications such as tissue engineering, medication delivery, and healthcare systems (Mishra et al., 2018).

This chapter has highlighted several advantages associated with the advent of nanosensors for pollutant analysis and quantification. Nanosensors are the latest detection technology driving convenience, practice, and portability of pollutant analysis in many sectors, including air and water quality measurements, food analysis, strategic purposes, and quality control assessment. Apart from their portability and on-site deployment capabilities, nanosensors have a number of other advantages (such as high sensitivity and selectivity) that combine to make them one of the most popular detection technologies in the future.

On a compact device, it is now possible to establish many sensing zones. Researchers have created chips and electrodes with multiplexing capabilities, allowing for the simultaneous measurement of numerous analytes. Multiplexed nanosensor technology can detect VOCs and other harmful substances. NMs have also been used to adapt dipstick and lateral-flow platforms into nanosensors for environmental analytes. This technology has a lot of potential, especially in food and water quality studies, where pollutants like viruses, heavy metals, and pesticides must be tracked at multiple phases of procurement, processing, packaging, and consumption. Because nanosensor technology is still a relatively new path, some pressing difficulties must be addressed. Product price, standardization, validation, and potential environmental toxicity are all important considerations. Nonetheless, both researchers and business are working hard to find solutions to these problems.

## REFERENCES

Adeel, M., Song, X., Wang, Y., Francis, D., and Yang, Y. (2017). Environmental impact of estrogens on human, animal and plant life: a critical review. *Environ. Int.* 99, 107–119.

Adetunde, L.A., and Glover, R.L.K. (2010) Bacteriological quality of borehole water used by students' of University for Development Studies, Navrongo campus in upper-east region of Ghana. *Curr. Res. J. Biol. Sci.*, 2(6), 361–364.

Akyol, S., Erdogan, S., Idiz, N., Celik, S., Kaya, M., and Ucar, F. (2014). The role of reactive oxygen species and oxidative stress in carbon monoxide toxicity: an in-depth analysis. *Redox. Rep.* 19,180–189.

Begum, S.S., Sushmaa, B., and Vijayaraja, S. (2015). Bioanalytical techniquesean overview. *Pharma. Tutor.* 3 (9), 14–24.

Bentayeb, M., Simoni, M., Norback, D., Baldacci, S., Maio, S., and Viegi, G. (2013). Indoor air pollution and respiratory health in the elderly. *J. Environ. Sci. Health. Part A*, 48(14), 1783–1789.

Bhatnagar, A., and Minocha, A.K. (2006). Conventional and non-conventional adsorbents for removal of pollutants from water: a review. *Indian J. Chem. Technol.* 13, 203–217.

Boxall, A.B. (2010). Veterinary medicines and the environment. In: *Comparative and Veterinary Pharmacology.* Springer, 291–314.

Camargo, J.A., and Alonso, A. (2006). Ecological and toxicological effects of inorganic nitrogen pollution in aquatic ecosystems: a global assessment. *Environ. Int.* 32, 831–849.

Chang, Z., Fan, H., Zhao, K., Chen, M., He P., and Fanga, Y. (2008). Electrochemical DNA biosensors based on palladium nanoparticles combined with carbon nanotubes. *Electroanalysis.* 20 (2), 131–136.

Chibuike, G. U., and Obiora, S. C. (2014). Heavy metal polluted soils: effect on plants and bioremediation methods. *Appl. Environ. Soil Sci.*, 2014,1–12.

Dubey, A., and Mailapalli, D.R. (2016). Nanofertilisers, nanopesticides, nanosensors of pest and nanotoxicity in agriculture. In: Lichtfouse E. (eds) *Sustainable Agriculture Reviews.* Springer, 19, 307–320.

Eissa, S., Siaj, M., and Zourob, M. (2015). Aptamer-based competitive electrochemical biosensor for brevetoxin-2. *Biosens. Bioelectron.* 69, 148–154.

Elshafey, R., Siaj, M., and Zourob, M. (2015). DNA aptamers selection and characterization for development of label-free impedimetric aptasensor for neurotoxin anatoxin-a. *Biosens. Bioelectron.* 68, 295–302.

Fan, L., Zhao, G., Shi, H., Liu, M., and Li, Z, (2013). A highly selective electrochemical impedance spectroscopy-based apta sensor for sensitive detection of acetamiprid. *Biosens. Bioelectron.* 43, 12–18.

Gaddi, A.V., and Capello, F. (2018) Pollution and air pollution–related diseases: an overview. In: Capello F, Gaddi A (eds) *Clinical Handbook of Air Pollution-related Diseases.* Springer, 1–5.

Ghanbari, K., and Roushani, M. (2018). A novel electrochemical aptasensor for highly sensitive and quantitative detection of the streptomycin antibiotic. *Bioelectrochemistry.* 120, 43–48.

Giller, K.E., Witter, E., and Mcgrath, S.P. (1998). Toxicity of heavy metals to microorganizms and microbial processes in agricultural soils. *Soil. Biol. Biochem.* 30(10–11),1389–1414.

Gonçalves, H., Jorge, P.A., Fernandes, J., and da Silva, J.C.E. (2010). Hg (II) sensing based on functionalized carbon dots obtained by direct laser ablation. *Sensors and Actuators B: Chemical.* 2, 702–707.

Gorai, A.K., Tuluri, F., and Tchounwou, P.B. (2014). A GIS based approach for assessing the association between air pollution and asthma in New York state, USA. *Int. J. Environ. Res. Public. Health.* 11, 4845–4869.

Grieshaber, D., MacKenzie, R., Vörös, J., and Reimhult, E. (2008). Electrochemical biosensors: sensor principles and architectures. *Sensors*, 8, 1400–1458.

Guo, Z. (2014). *Development of electrochemical biosensors for environmental pollutant and food safety monitoring* (Doctoral dissertation, Université Claude Bernard-Lyon I).

Han, K., Liu, T., Wang, Y., and Miao, P. (2016). Electrochemical aptasensors for detection of small molecules, macro- molecules, and cells. *Rev. Anal. Chem.* 35 (4), 201–211.

Hashem, A., Hossain, M. A. M., Marlinda, A. R., Mamun, M. Al, Simarani, K., and Johan, M. R. (2021). Nanomaterials based electrochemical nucleic acid biosensors for environmental monitoring: A review. *Applied Surface Science Advances.* 4, 100064.

Hassani, S., Rezaei Akmal, M., Salek Maghsoudi, A., Rahmani, S., Vakhshiteh, F., Norouzi, P. and Abdollahi, M. (2020). High-performance voltammetric aptasensing platform for ultrasensitive detection of bisphenol A as an environmental pollutant. *Front. Bioeng. Biotechnol.* 8, 1055.

Hayat, A. (2013). Electrochemical grafting of long spacer arms of hexamethyldiamine on a screen printed carbon electrode surface: application in target induced ochra- toxin A electrochemical aptasensor. *Analyst.* 138 (10), 2951–2957.

Hayat, A., and Marty, J.L., (2014). Aptamer based electrochemical sensors for emerging environ- mental pollutants. *Front. Chem.* 2, 41.

Hayat, A., Andreescu, S., and Marty, J.L. (2013). Design of PEG-aptamer two piece macromolecules as convenient and integrated sensing platform: application to the label free detection of small size molecules, *Biosens. Bioelectron.* 45, 168–173.

Hernandez-Vargas, G., Sosa-Hernández, J. E., Saldarriaga-Hernandez, S., Villalba-Rodríguez, A. M., Parra-Saldivar, R., and Iqbal, H. M. N. (2018). Electrochemical biosensors: a solution to pollution detection with reference to environmental contaminants. *Biosensors.* 8(2). https://doi.org/10.3390/BIOS8020029

Hlongwane, G.N., Dodoo-Arhin, D., Wamwangi, D., Daramola, M. O., Moothi, K., Iyuke, S. E. (2019). DNA hybridisation sensors for product authentication and tracing: state of the art and challenges, *S. Afr. J. Chem. Eng.* 27, 16–34.

Huang, H., Chen, T., Liu, X.,and Ma, H. (2014). Ultrasensitive and simultaneous detection of heavy metal ions based on three-dimensional graphene-carbon nanotubes hybrid electrode materials. *Analytica Chimica Acta.* 852, 45–54.

Ibanez, J.G., Hernandez-Esparza, M., Doria-Serrano, C., Fregoso-Infante, A., and Singh, M.M. (2007). Physicochemical and physical treatment of pollutants and wastes. In *Environmental chemistry: Fundamentals.* Springer, 237–245.

IFPRI. (2011) *2011 Annual Report.* https://ebrary.ifpri.org/utils/getfile/collection/p15738coll2/id/126935/filename/127146.pdf

Jiang, L., Qian, J., Yang, X., Yan, Y., Liu, Q., Wang, K., and Wang, K. (2014). Amplified impedimetric aptasensor based on gold nanoparticles covalently bound graphene sheet for the picomolar detection of ochratoxin A, *Anal. Chim. Acta.* 806, 128–135.

Justino, C., Duarte, A., and Rocha-Santos, T. (2017). Recent progress in biosensors for environmental monitoring: a review. *Sensors.* 17 (12), 2918.

Kaur, R., Sharma, S. K., and Tripathy, S. K. (2019). Advantages and limitations of environmental nanosensors. In: *Advances in Nanosensors for Biological and Environmental Analysis.* Elsevier, 119–132.

Khalil, M., Berawi, M., Heryanto, R. and Rizalie, A. (2019). Waste to energy technology: the potential of sustainable biogas production from animal waste in Indonesia. *Renew. Sust. Energ. Rev.* 105, 323–331.

Khot, L. R., Sankaran, S., Maja, J. M., Ehsani, R., and Schuster, E. W. (2012). Applications of nanomaterials in agricultural production and crop protection: a review. *Crop Protection.* 35, 64–70.

Kim, M., Um, H.J., Bang, S., Lee, S.H., Oh, S.J., Han, J.H., Kim, K.W., Min, J., Kim, Y.H. (2009) Arsenic removal from Vietnamese groundwater using the ar- senic-binding DNA aptamer. *Environ. Sci. Technol.* 43 (24), 9335–9340.

Kim, Y.S., Jung, H.S., Matsuura, T., Lee, H.Y., Kawai, T., and Gu, M.B. (2007). Electrochemical detection of 17 b-estradiol using DNA aptamer immobilized gold electrode chip. *Biosens. Bioelectron.* 22(11), 2525–2531.

Kumar, R., Singh, R.D., and Sharma, K.D. (2005). Water resources of India. *Curr. Sci.* 85(5), 794–811.

Kumar, R., Kumar, R., and Kaur, G. (2021). New frontiers of nanomaterials in environmental science. In: *New Frontiers of Nanomaterials in Environmental Science.* Springer Nature. https://doi.org/10.1007/978-981-15-9239-3

Lang Q., Han L., Hou C., Wang, F., and Liu, A. (2016). A sensitive acetylcholinesterase biosensor based on gold nanorods modified electrode for detection of organophosphate pesticide. *Talanta.* 156, 34–41.

Li, Y., Chen, Y., Yu, H., Tian, L., and Wang, Z. (2017). Portable and smart devices for monitoring heavy metal ions integrated with nanomaterials. *Trends Anal. Chem.* 98, 190–200.

Manoli, E., and Samara, C. (1999). Polycyclic aromatic hydro- carbons in natural waters: sources, occurrence and analysis. *Trends Anal. Chem.* 18 (6), 417–428.

McCarthy, J.T., Pelle, E., Dong, K., Brahmbhatt, K., Yarosh, D., and Pernodet, N. (2013). Effects of ozone in normal human epidermal keratinocytes. *Exp. Dermatol.* 22,360–361.

Mehta, J., Dorst, B.V., Martin, E.R., Herrebout, W., Scippo, M.L., Blust, R., and Robbens, J. (2011). In vitro selection and characterization of DNA aptamers recognizing chloramphenicol, *J. Biotechnol.* 155 (4), 361–369.

Mian, I.A., Begum, S., Riaz, M., Ridealgh, M., McClean, C.J., and Cresser, M.S. (2010). Spatial and temporal trends in nitrate concentrations in the river Derwent, North Yorkshire, and its need for NVZ status. *Sci. Total Environ.* 408, 702–771.

Middelhoek, S., and Noorlag, D.J.W. (1981). Signal conversion in solid-state transducers. *Sens. Actuators.* 2, 211–228.

Miller, G.T. (2004). *Sustaining the Earth: An Integrated Approach.* Books/Cole, ISBN 0534496725, 9780534496722

Mishra, R.K., Ha, S.K., Verma, K., and Tiwari, S.K. (2018). Recent progress in selected bionanomaterials and their engineering applications: an overview. *J. Sci. Adv. Mater. Dev.* 3(3), 263–288.

Murad, A., and Krishnamurthy, R.V. (2008). Factors controlling stable oxygen, hydrogen and carbon isotope ratio in regional groundwater of eastern United Arab Emirates (UAE). *Hydrol. Process.* 22, 1922–1931.

Murad, A.A. (2010). An overview of conventional and non-conventional water resources in arid region: assessment and constrains of the United Arab Emirates (UAE). *J. Wat. Resour. Protect.* 2, 181–190.

Murad, A.A., and Krishnamurthy, R.V. (2004). Factors controlling groundwater quality in eastern united Arab emirates: a chemical and isotopic approach. *J. Hydrol.* 286, 227–235.

Palchetti, I., and Mascini, M. (2008). Nucleic acid biosensors for environmental pollution monitoring. *Analyst.* 133(7), 846–854.

Patra, K.C., Pareta, S.K., Harwansh, R.K., and Kumar, K.J. (2010). Traditional approaches towards standardization of herbal medicines: a review. *J. Pharmaceut. Sci. Technol.* 2(11), 372–379.

Peng B.J., Tang, F., Zhou, R., Xie, X., Li, S., Xie, F., Yu, P., and Mu, L. (2016). New techniques of on-line biological sample processing and their application in the field of biopharmaceutical analysis. *Acta. Pharmaceut. Sini. B.* 6(6), 540–551.

Qu, S., Wang, X., Lu, Q., Liu, X., and Wang, L. (2012). A biocompatible fluorescent ink based on water-soluble luminescent carbon nanodots. *Angew. Chem. Int. Ed. Engl.* 51(49), 12215–12218.

Rich, K.M., and Perry, B.D. (2011). The economic and poverty impacts of animal diseases in developing countries: new roles, new demands for economics and epidemiology. *Prev. Vet. Med.* 101(3–4), 133–47.

Sadeghi, M., Ahmadi, A., Baradaran, A., Masoudipoor, N., and Frouzandeh, S. (2015). Modeling of the relationship between the environmental air pollution, clinical risk factors, and hospital mortality due to myocardial infarction in Isfahan, Iran. *J. Res. Med. Sci.* 20, 757–762.

Sahu, D., Kannan, G.M., and Vijayaraghavan, R. (2014). Carbon black particle exhibits size dependent toxicity in human monocytes. *Int. J. Inflam.* 2014, 1–10. https://doi.org/10.1155/2014/827019

Schneider, S.H. (1989). The greenhouse effect: science and policy. *Science.* 243, 771–781.

Shen, L., Chen, Z., Li, Y., He, S., Xie, S., Xu, X., Liang, Z., Meng, X., Li, Q., Zhu, Z., Li, M., Le, X.C., and Shao, Y. (2008). Electrochemical DNAzyme sensor for lead based on amplification of DNA-Au bio-bar codes. *Anal. Chem.* 80(16), 6323–6328.

Singh, R.P., and Choi, J.W. (2010). Bio-nanomaterials for versatile bio-molecules detection technology. *Adv. Mater. Letts.* 1(1), 83–84. https://doi.org/10.5185/amlett.2010.4109

Tao, H., Fan, Q., Ma, T., Liu, S., Gysling, H., Texter, J., and Sun, Z. (2020). Two-dimensional materials for energy conversion and storage. *Prog. Mater. Sci.* 111, 100637.

Ting, S.L., Ee, S.J., Ananthanarayanan, A., Leong, K.C., and Chen, P. (2015). Graphene quantum dots functionalized gold nanoparticles for sensitive electrochemical detection of heavy metal ions. *Electro. chimica. Acta.* 172, 7–11.

Urasinska-Wojcik, B., Vincent, T.A., Chowdhury, M.F., and Gardner, J.W. (2017). Ultrasensitive $WO_3$ gas sensors for $NO_2$ detection in air and low oxygen environment. *Sens. Actuators, B.* 239, 1051–1059.

Vaseashta, A. (2008). Nano-scale materials, devices, and systems for chem-bio sensors, photonics, and energy generation and storage. In: *Functionalized Nano-scale Materials, Devices and Systems.* Springer, 3–27 (ISBN: 978-1-4020-8901-5).

Vaseashta, A., Vaclavikova, M., Vaseashta, S., Gallios, G., Roy, P., and Pummakarnchana, O. (2007). *Sci. Technol. Adv. Mat.* 8, 47–59.

Veras, M.M., Caldini, E.G., Dolhnikoff, M., and Saldiva, P.H. (2010). Air pollution and effects on reproductive-system functions globally with particular emphasis on the Brazilian population. *J. Toxicol. Environ. Health.* 13, 1–15.

Wu, P., Liu, Y., Liu, Y., Wang, J., Li, Y., Liu, W., and Wang, J. (2015). Cadmium-based metal–organic framework as a highly selective and sensitive ratiometric luminescent sensor for mercury (II). *Inorg. Chem.* 54(23), 11046–11048.

Zhang, C., Lai, C., Zeng, G., Huang, D., Tang, L., Yang, C., Zhou, Y., Qin, L., and Cheng, M. (2016). Nano porous Au-based chrono coulometric apta sensor for amplified detection of $Pb^{2+}$ using DNAzyme modified with Au nanoparticles. *Biosens. Bioelectron.* 81, 61–67.

Zhang, W., Asiri, A.M., Liu, D., Du, D., and Lin, Y. (2014). Nanomaterial-based biosensors for environmental and biological monitoring of organophosphorus pesticides and nerve agents. *Trends Anal. Chem.* 54, 1–10.

# 9 Bionanotechnology in Resource Recovery from Wastewater and Agro Waste into Valuable Products

*Shruti Awasthi and Preethi Rajesh*

## CONTENTS



DOI: 10.1201/9781003270959-9

## 9.1 INTRODUCTION

Water is very important in the day-to-day activities of life, but the supply of fresh drinking water is dwindling worldwide due to contamination and salination. The important bases of water pollution are toxic industrial waste, agricultural runoff, personal care products, inorganic compounds, pesticides, insecticides, fungicides, anions, etc, which are untreated and dumped into ponds, rivers, seas, and oceans. Immediate attention is required for the development, resource recovery, attribute utilization, and preservation of resources (Pontius, 1990) (Figure 9.1).

In response to the effects of globalization and industrialization, numerous new technologies have been developed for the treatment of wastewater with the aims of lowering contaminant production and mitigating water pollution, However, these techniques have many limitations; for instance, wastewater treatment processes are

**FIGURE 9.1** Water is polluted by industrial dumping.

**FIGURE 9.2** Wastewater treatment technology.

complicated, maintenance and operational costs are high, the rate of resource recovery is very low, and the process produces a lot of toxic compounds (Pontius, 1990).

### 9.1.1 Adsorption Method

In comparison to various other techniques, the absorption method research has proven to be the best alternative for wastewater treatment because of its convenience, operational costs, and simple design. The adsorption process is a good treatment for various kinds of organic and inorganic pollutants because of this reason this method is widely used for water remediation. In this method, first carbon is stimulated without any doubt it is thought to be a worldwide good adsorbent it has the capability to remove toxic compounds from wastewater (Figure 9.2).

### 9.1.2 Biosorption

Biosorption is another economically and environmentally friendly solution to wastewater treatment. It uses biological resources like animal tissue, plant parts, fungi, bacteria, algae, and yeast, which act like biosorbents. They have the capability to absorb metal ions present in water as well as wastewater, thereby reducing the pollutants in them.

This technique is preferred mostly by industry for detoxification and recovery of heavy metals since it requires low capital. Low manufacturing and operational costs, the ability to remove a variety of metals, and no waste generation make biosorption an excellent wastewater treatment option (Davis et al., 2003).

### 9.1.3 Waste Materials from Industrial or Agricultural Operations

In the present scenario because of globalization, both agricultural and industrial waste materials may act like powerful substitutes as biosorbents (Mahvi et al., 2005). Waste from agriculture is renewable due to rare availability of chemical proportions, and its abundance in nature make waste material a good option for organic and inorganic treatment. Agricultural waste material can be used either in its natural form or after modification by either by physical or chemical methods; this change seems to be promising for the removal of toxic material (Tarley and Arruda, 2004).

Research studies have proved that the adsorption capability of these adsorbents may be enhanced if they are treated with chemical agents: e.g., raw lignocellulosic biosorbents were modified by various physical and chemical methods to increase their absorption capability by binding metal ions with functional groups such as carboxyl, amino, or a phenolic group of lignocellulosic biosorbents. Recently, researchers have been modifying industrial and agricultural waste materials using physical and chemical agents to increase the capacity of adsorbents (Tarley and Arruda, 2004).

### 9.1.4 Role of Bionanotechnology in Wastewater Treatment

Nanomaterials are defined as materials having nanoscale materials of which a single unit small-sized between 1 and 100 nm. One dimension is smaller than 100 nm, for example, thin films or surface coatings (Xiaolei Qu et al., 2013). Two-dimensional nanomaterials include nanowires and nanotubes. Nanomaterials in three dimensions are particles, for example, precipitates, colloids, and quantum dots. The two main factors in which nanomaterials differ are surface area and quantum effects.

#### 9.1.4.1 Nano-adsorbents

Nano-adsorbents can be introduced into existing water treatment methods in slurry reactors or absorbers. Whenever nano-adsorbents are used in a nano-powder form on slurry reactors their efficiency is highly increased. Nano-adsorbents can be used as pellets, beads, and porous granules which would be loaded with nano-adsorbents either in a fixed or fluid form. These nanoparticles are used for the removal of Pb(II), Cu(II), and As(III), which are metal contaminants found in wastewater.

#### 9.1.4.2 Nanocatalysts

In the present scenario, nanocatalysts are used in chemical processes which are beneficial for human beings. These catalysts are extensively used in wastewater treatment which has the capability to increase the catalytic activity at the surface because of higher surface area which is dependent on the property of shape. This property will enhance the reactivity and contaminant degradation. Catalytic nanoparticles are made of semiconductor materials, zero valance metal, and bimetallic nanoparticles for environmental contaminant degradation.

### 9.1.5 Membrane and Membrane Process

Membranes that can provide a physical barrier and have the capability to remove pollutants from water are extensively used for water treatment. The membrane performance largely depends on the membrane material used. The incorporation of functional nanomaterials into membranes offers a great opportunity to improve membrane permeability, fouling resistance, and mechanical and thermal stability (Xiaolei Qu et al., 2013). The electrospinning method is used to generate a nonwoven web of micro- or nanofiber mats which have complex pore structures. In this technique, high voltage electricity is applied to the liquid solution and a collector, which lets the solution extrude from a nozzle forming a jet. The jet-formed fibers are deposited on the collector during drying. Nanofibers are commercially used for air filtration; they can be used as pretreatment prior to ultrafiltration or reverse osmosis.

### 9.1.6 Disinfection and Microbial Control

Nanoparticles like silver, zinc oxide, titanium oxide, carbon nanotubes, and fullerenes have antimicrobial properties and don't have strong oxidation; therefore, they have little ability to form disinfection by-products. So, they can improve water quality by performing disinfection, membrane biofouling control, and biofilm control on other relevant surfaces (Figure 9.3).

## 9.2 ROLE OF BIONANOTECHNOLOGY IN RESOURCE RECOVERY

Natural resources are being depleted day by day, which is changing the current social condition of the manufacturing cycle and changing the focus from wastewater treatment toward the development of resource recovery methods. Bionanotechnological techniques give us the opportunity to concentrate on the development of newer techniques so that resources can be transformed from waste/wastewater into products that are good for society, reducing the number of pollutants, which is essential for the development of the bio-based economy. Upcoming new methods like nanotechnologies will make possible resource recovery from wastewater treatment more efficient, e.g., bioenergy, biohydrogen, biogas, etc.

Currently, the production system is totally based on raw materials which are extracted and transformed into products; this is becoming a long-term major issue

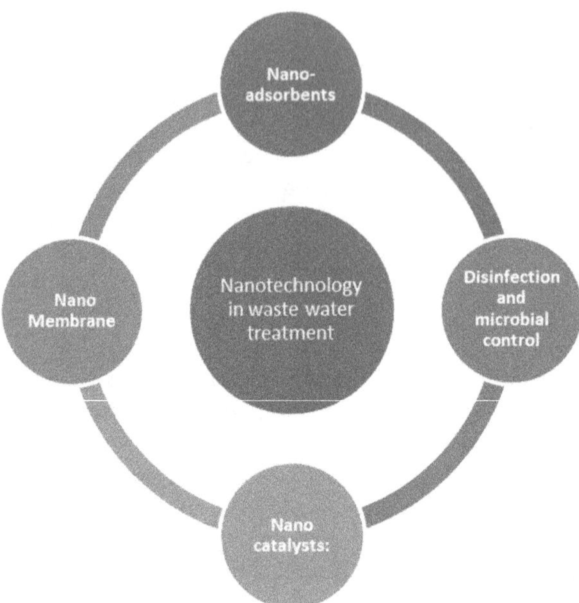


**FIGURE 9.3** Nanotechnology in wastewater treatment.

because resources are depleting (Lovins, 2008). The main concern is that we are using nonrenewable resources, so in the near future it is possible resources like fossil fuels, agricultural nutrients like nitrogen, phosphorus, and scarce metals which are used in electronics, will completely run out. This would mean that manufacturing would shut down. unit and authorize all researchers of the world to join together and search for the new resources for the treatment and recovery of the waste so that they can go for the bio-based economy. These strong drivers will change the current production systems, and the next two decades would be key to enabling a sustainable technological society. Economists of the world predict that if the manufacturing sector were to become auto-regenerative and the waste generated totally converted into raw materials, like organic, inorganic, and metal, then our global economy would be bio-based, which is the need of the hour for sustainable development. Traditionally, it is termed as the cradle-to-cradle concept; it is substituted by the triple-R model, that is, reduce, reuse, and recycle.

Treatment of wastewater is a key platform to bring the changes in technological development that are driving the changes of the production system. Near off 50–100% of lost waste resources are in wastewater. Therefore, a major concern for the economy and environment expertise is to recover and regain all the lost substances. The EU has converted substantial resources into bioproduct for which a specific research and innovation program was recently undertaken by the Bio-based Industries Joint Undertaking, which was funded by the European Commission under the Horizon 2020 framework.

The USA has played a major role in pushing the resources into the bio-economical products under the blueprint of the global bio-based economy (House, 2012).

President of Barack Obama claimed in 2011, "The world is shifting to an innovation economy and nobody does innovation better than America." This statement clearly signifies that the USA has a clear mindset for leading the evolution of the global economy toward a bio-based society. This change is possible if we develop new technologies where, after the treatment of water/wastewater, the maximum amount of resources, like metals, organic, and inorganic contaminants, would be recovered and reused as valuable product. This would, indeed, be a remarkable first step toward making the economy bio-based and making our goal of sustainable development stronger (Batstone et al., 2015).

### 9.2.1 Domestic Wastewater as Nutrient and Energy Recovery

The activated sludge process has completed 100 years anniversary in the year 2014 and has seen significant development in human health, environment, and standard of living when it was allowed by activated sludge (Jenkins and Wanner, 2014). Domestic wastewater cannot completely fulfill fertilizer requirements, as there is a consequential diversion to both domestic animal manufacture and the environment. Worldwide domestic wastewater consists of approximately 20% of nitrogen and phosphorous (Batstone et al., 2015), because of urban concentration it can be recovered. The condition is more enervated for energy (Batstone et al., 2015) (Figure 9.4).

Rapid economic growth, agriculture practices. globalization, industrialization, urbanization, and the population which is increasing at a faster pace resulting in the depletion of natural resources like water, minerals, and organic compounds subsequently resulting in the discharge of an enormous quantity of waste to mother nature. Global outlook by the rule of resource dissipation in the present scenario domestic wastewater is the important resource which needs to be fully used up because wastewater alone cannot fulfill the demand of energy requirements of the society.

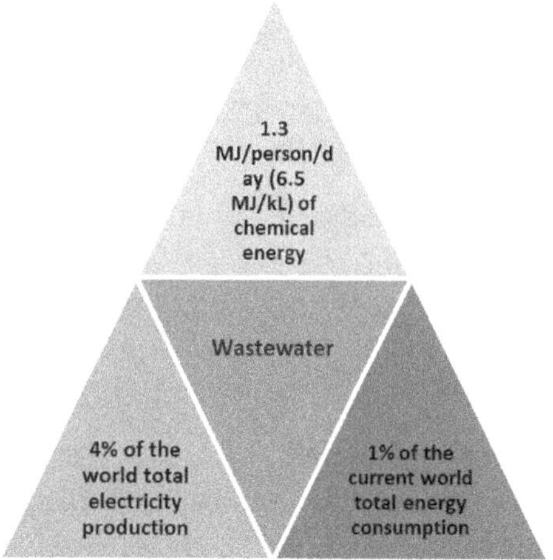


**FIGURE 9.4** Domestic wastewater.

Finally, traditionally domestic wastewater is taken as a benchmark for the development of new technology for the treatment of water and wastewater due to the enormous number of financial resources available in comparison to the agri-industrial waste recovery. The latest technology could be used up in resource recovery in both the industrial and agricultural cycles.

Many more new processes have been discovered or researched by using the latest technologies, like passive systems to reduce the amount of energy used for treating wastewater e.g., wetlands, lagoons, mainline anaerobic (Wett et al., 2013).

### 9.2.2 Partition-Release-Recover

The streams were categorized (Verstraete et al., 2009) into two that is major which is more concentrated and minor which is more diluted. There were many new technologies identified, like filtration-based treatment (also known as gravity-microfiltration-reverse osmosis), by the anaerobic digestion of solids and concentrate, recovery of the nutrients from digested materials (e.g., the process of electro-dialytic can recover nitrogen) (Mondor et al., 2008), and precipitation of phosphate metal. Verstraete found many substitute methods, like inoculating the biological sample with higher microorganism concentrations which will grow at a faster pace, like heterotrophic activated sludge organisms.

Partition-release-recover is the upgraded method, where biological agents are used to selectively remove carbon and nutrients from the liquid phase. This process is combined and is scalable, which has the capability to treat wastewater with no energy input and can fully recover nitrogen, phosphorous, organics or microbial and value-added products from the wastewater (Batstone et al., 2015).

### 9.2.3 Important Discharge Areas of Wastewater

#### 9.2.3.1 Discharge Area for Water

This area is the main hydraulic load known as the "partition" stage from where the nutrients are partially discharged. Then they are scattered via recyclable water, with a quantified discharge of nitrogen and phosphorous, depending on recycling requirements and taking local regulations and the latest techniques into consideration.

#### 9.2.3.2 Discharge Area for Biogas

This sink stream is the main discharge area for surplus chemical energy which is known as the "release" stage from where maximum energy would be produced. This is a comparatively low-value energy stream, and its main goal is to recover organics so that they can be converted to high-value products.

#### 9.2.3.3 Discharge Area for Biosolids

This discharge area to a great extent is composed of inert organic compounds, non-retrievable nutrients, and surplus metals which are released from the release stage as a by-product. In the surplus sludge manufacturing process, most of the properties are lost. Biosolids are best applied as organic fertilizers (Tonetti et al., 2016).

#### 9.2.3.4 Discharge Area for Fertilizer Stream

This discharge area to a great extent is composed of nitrogen, phosphorous, and potassium, which are released from the recovery stage as a by-product. Again, due to the low-value goods chemicals, we can attain our goal of generation of valuable products and recovery of organic product which would have a better value for this product (Figure 9.5).

### 9.2.4 The Partition Stage

The partition stage consists of key differentiating features they have a number of different agents that can be used for many purposes. These are outlined in the following sections.

#### 9.2.4.1 A-Stage Treatment

Wastewater gives energy and electron equivalents to heterotrophic bacteria, which are needed for the growth and development of heterotrophic bacteria. This is termed as "A-stage" treatment and has been in use for almost 20 years (Jetten et al., 1997).

#### 9.2.4.2 Domestic Treatment Option in the Laboratory

- The energy for the growth and development of phototrophic anaerobic bacteria is acquired from light, whereas carbon electrons and nutrients are derived from either water or wastewater. This phenomenon is called the "domestic treatment option in the workshop" (Hulsen et al., 2014).
- In algae and oxygenic photosynthetic bacteria, energy is derived from light for the growth and development of bacteria, whereas electrons are drawn via molecular water, nutrients, and carbon dioxide. Extraordinarily for the case of heterotrophic treatment, this will greatly involve the association of aerobic bacteria, which will nitrify and oxidize carbon to carbon dioxide (Cai et al., 2013).
- Phototrophic is a budding stage in nature and algae is still under fatal development this has limited application. These have many rudimentary limitations, especially, energy input and carbon utilization efficiency. All three generate valuable products in the form of biomass which are good for the sustainable development of the environment that will direct toward resource recovery.

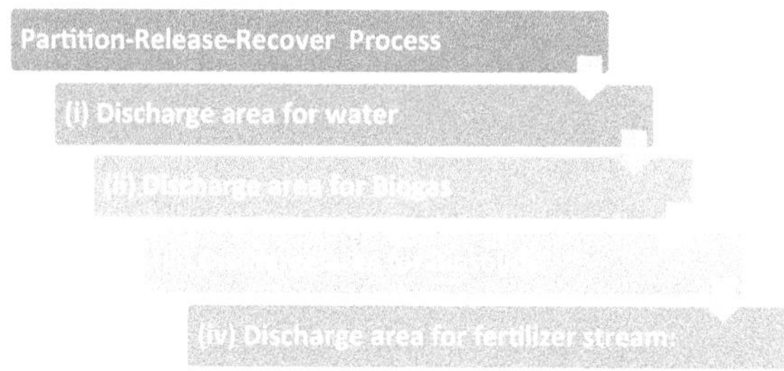


**FIGURE 9.5** Partition-Release-Recover process.

#### 9.2.4.3 Biofactory of Wastewater

A new concept for conventional activated sludge is "re-engineering," especially by recognizing new valuable by-products they are having a higher value than the raw material from wastewater. Activated sludge in a natural manner increases the concentration of organics in a stream of sludge. The goal of partition-release-recover is to maximize this. However, a parallel aim is to enhance the formation of conventional activated sludge, which will increase the transition of this into valuable products.

Conventional activated sludge which will increase the production of valuable products can be traditionally broken down into various chemicals. The synthesis of these chemical allows the purest form of organic chemicals and complex organic materials to be diverted into manufacturing, agriculture, and consumer uses.

The industry of commodity chemicals consists of organic acids, alcohols, carbon dioxide, nutrients, and metals, which are in totally purified forms. Organic compounds are diverted into two routes, fermentation and fermented product extractions (Kleerebezem and Van Loosdrecht, 2007), recovery of the organic compounds so that their concentration is increased, later converting into syngas for successive transformations (Batstone and Virdis, 2014). Both of these organic forms are better applied to the streams of sludge.

From the perspective of sustainable development, it's very important to manufacture a greater number of composite products, which can be done by exploiting activated sludge to produce polyhydroxyalkanoates (PHAs) (Kleerebezem and Van Loosdrecht, 2007). PHA composite, a generation of long chain of microbial exopolysaccharides, considering even alginates (Sam and Dulekgurgen, 2016) especially by using aerobic granular sludge which needs to be activated, and then direct recovery of fibers (e.g., cellulose in water and wastewater). Production of this utilizes only a fragment of the resource-carbon available in the wastewater, but definitely, this should be part of a large resource recovery strategy.

### 9.2.5 Recovery of Metals

Heavy metal contamination in water is garnering more attention recently, as it causes cancer, organ damage, etc. The origin of this contamination is mainly related to human activities like mining, burning fossil fuels, electroplating, leather tanning, cement production, metallurgical operations, and other manufacturing activities like plastics, fertilizers, anticorrosive agents, batteries, drugs, dyes, photovoltaic devices, pigments, pesticides, and insecticides. Minerals are also found in domestic wastewater treatment plants, the sources of which are industrial effluents, domestic waste, soil leachates from ponds that are highly polluted, landfills, and mines. The metals are specifically recovered from domestic wastewater and domestic sewage sludge where more concentrations of these metals, like Na, Fe, Al, Ti, Zn, Cu, Sn, Mn, Cr, Mo, Ag, Ni, U, and V have been recognized. We know that the composition depends on the geomorphology and human activities. Globalization has increased environmental pollution, and human concern increases day by day. The concern around the heavy concentration of metal contaminants, in particular, has resulted in efforts to recover these resources so that they can be reused. The recovery of metals by using biological techniques, like biomining bioremediation, have been researched for decades.

#### 9.2.5.1 Recovery of Nutrients and Metals

Metals and nutrients can be recovered through bio-electrochemical systems (BES). We can now recover phosphorous, whereas removal and recovery of nitrogen can be done either by electrodialysis or by ammonium bicarbonate. BES also proffers the possibility of recovering metals from wastewater.

## 9.3 TRANSFORMING THE ENERGY SECTOR TO SUSTAIN GROWTH

The global demand for energy is increasing rapidly, due to huge increases in population and the emerging economies in developing and underdeveloped countries. requires energy as the primary input to economic efficiency and sustainability. The usage of energy has witnessed an enormous shift after the industrial revolution from manual and biomass to coal (solid), oil (liquid), and natural gas (gas). In the late 20th century, the world thrived on fossil fuel as the primary source of energy (Figure 9.6). The transportation sector is one of the largest consumers of petroleum products. Consumption of fossil fuels accounts for most of the $CO_2$ emissions adversely affecting climate and ecological balance. The growing threat posed by this conventional energy has necessitated the transition of fossil fuel-based energy systems into one which is greener, cheaper, and economically efficient. Technological developments have enabled the trapping of new forms of renewable energy resources like wind, solar, hydroelectric, and even nuclear reactions. Advances in biotechnology and its immense application, emphasize the potential of biofuel as an alternative source of

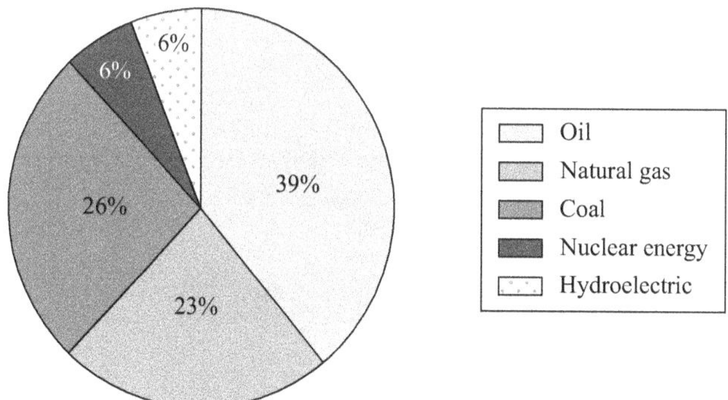


**FIGURE 9.6** Energy sources.

energy in addition to the marginal increments in the wind and solar energy. Nuclear energy can serve as a very cost-effective, zero-emission, and reliable source of energy once concerns about its safety are settled. Studies also suggest that the use of hydrogen can generate electricity in a fuel cell to power generators, vehicles, and other portable devices.

The projected electricity generation worldwide from 2018 to 2050, by energy source, shows a gradual increase in renewable energy including biomass-based energy and hydroelectric power systems (Figure 9.7). A study by the International Atomic Energy Agency projected that electricity generated by most energy sources worldwide, with the exception of liquids, will increase in the coming years. Electricity from renewable sources is expected to experience the largest growth of up to 21.66 trillion kilowatt-hours in 2050, from almost 7 trillion kilowatt-hours in 2018. This projection sees increased consumption from all fuel sources except for coal, where demand seems to have plateaued. Conventional sources like coal are increasingly being replaced by natural gas, renewables, and nuclear power. Agricultural industrial waste is an important source of biomass-based energy production.

### 9.3.1 Agriculture as an Industry

Agro-based industries are industries that use plant and animal-based agricultural output as their raw material. Also, they add value to agricultural output by processing and producing marketable and usable products. Some examples of agro-based industries in India include textiles, sugar, vegetable oil, tea, coffee, and leather goods.

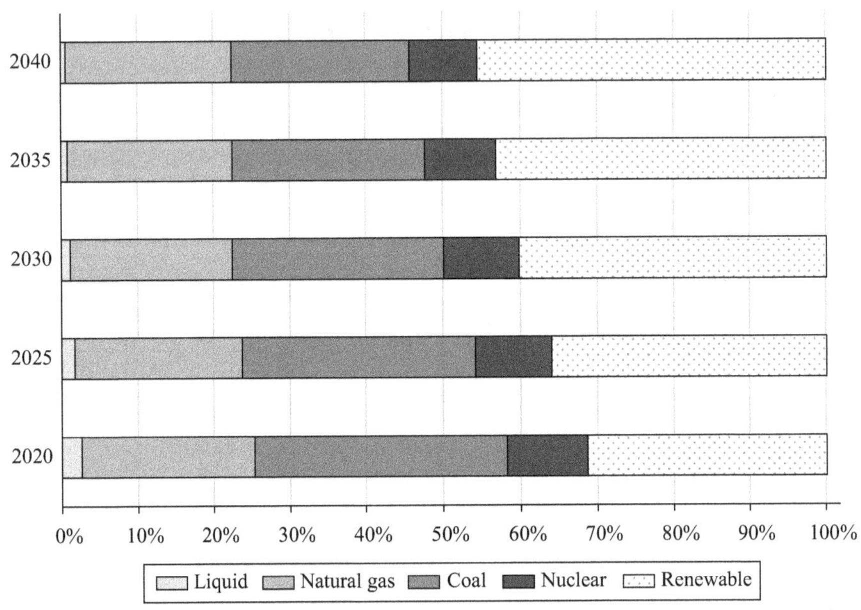


**FIGURE 9.7** Electricity generation worldwide from 2020 to 2040.

#### 9.3.1.1 Importance of Agro-Based Industries in India

All agro-based industries are important because they

1. Help in increasing industrial production.
2. Provide employment to landless agricultural labor and tribal population from rural and backward areas.
3. Ensure the development and stability of rural economies through diversification and reduced dependence on agriculture.
4. Ensure the alleviation of poverty by providing steady sources of income and livelihood.
5. Earn much required foreign exchange for the country.
6. Improve the standard of living in rural areas.
7. Help in reducing the extreme inequalities in the distribution of income and wealth.
8. Are easy to establish.
9. Support balanced growth between agriculture and industry.
10. Help in avoiding wastage of perishable agricultural products.

### 9.3.2 Scenario and Scope of Agro-Based Industries in India

The scope of agro-based industries in India is pretty high because the country is predominantly dependent on agriculture. According to the statistical data for the year 2020, the agriculture sector in India contributes about 18% to India's gross domestic product (GDP). Also, approximately 42% of the Indian population is employed in the agricultural sector alone. Although the share of the population employed in the agriculture sector has been declining year after year for various reasons it still remains the largest employment sector. The agro-based industry is regarded as the "sunrise" sector of the Indian economy because of its huge potential for growth, likely socioeconomic impact (specifically on employment and income generation), and for the ability to generally keep itself recession-proof. Also, approximately 70% of the population is dependent on agriculture and agro-based industries. According to the economic survey conducted by the Central Statistical Office (2014–15), agro-based industries consistently grew in India during the period from 2009–10 to 2013–14. Some estimates also suggest that in developed economies, approximately 14% of the total workforce engages in the agro-processing sector directly or indirectly, whereas, in India, only about 3% of the workforce finds employment in this sector.

## 9.4 TYPES OF AGRO-BASED INDUSTRIES IN INDIA

Agro-based industries in India can be broadly classified into the following types.

### 9.4.1 Agro-Produce Processing Units

These units are not involved in manufacturing and mainly deal with the preservation of perishable products and utilization of by-products for other uses. Rice and dal processing mills are examples of these kinds of units.

### 9.4.2 Agro-Produce Manufacturing Units

These units engage in the manufacturing of new products where the finished goods are entirely different from the raw materials used. Sugar factories, solvent extraction units, and textile mills are some of the examples of these kinds of units.

### 9.4.3 Agro-Inputs Manufacturing Units

These units are engaged in the manufacturing of products, either for the mechanization of agriculture or for increasing agricultural productivity. Some examples of these units include agricultural implements, seed, fertilizer, and pesticide manufacturing units.

### 9.4.4 Agro Service Centers

Agro service centers are workshops and service centers that are engaged in the repair and service of pump sets, diesel engines, tractors, and other types of farm equipment.

## 9.5 AGRO-INDUSTRIAL WASTES

Agricultural wastes (AW) can be defined as the waste generated by agricultural activities including growing and processing of raw agricultural products like fruits and vegetables, dairy products, crops, meat, and poultry. AW can be in the form of solid, liquid, or slurries depending on the nature of the agricultural activities. About 30% of agricultural and food industry residues and wastes constitute a significant proportion of worldwide agricultural productivity (Sarmah, 2009).

### 9.5.1 Biodegradable and Nonbiodegradable Agro-Industrial Waste

#### 9.5.1.1 Biodegradable Agro-Industrial Waste

Those industrial wastes which can be decomposed into non-poisonous matter by the action of certain microorganisms are biodegradable wastes. They are even comparable to house wastes. These kinds of waste are generated from food processing industries, dairy, textile mills, slaughterhouses, etc. Some examples are paper, leather, wool, animal bones, wheat, etc. They are nontoxic in nature, and they do not require special treatment either. Their treatment processes include combustion, composting, gasification, bio-methanation, etc.

#### 9.5.1.2 Nonbiodegradable Agro-Industrial Waste

Those industrial wastes which cannot be decomposed into non-poisonous substances are nonbiodegradable wastes. Examples are plastics, fly ash, synthetic fibers, gypsum, silver foil, glass objects, radioactive wastes, etc. They are generated by iron and steel plants, fertilizer industries, chemical, drug, and dye industries. It is estimated that about 10–15% of total industrial wastes are non-biodegradable and hazardous, and the rate of increase in this category of waste is only increasing every year. These wastes cannot easily be broken down and made less harmful.

Hence, they pollute the environment and cause a threat to living organisms. They accumulate in the environment and enter the bodies of animals and plants, causing diseases. However, with the advancement in technology, several disposals and reuse methods have been developed. One method involves waste from one industry being treated and utilized in another industry. For example, the cement industry uses slag and fly ash generated as waste by steel industries. Landfill and incineration are other methods that are being resorted to for hazardous wastes.

The term AW relates to all leftovers and residuals of agriculture production which do not have direct economic value for the farmer and are meant for disposal. Special processes and know-how are needed to convert these wastes into valuable products to be used in the next agricultural production. In most cases (and mainly in field crops and vegetables) the economic breakpoint is in doubt considering the costs of removal, transport, and processing of these wastes. Although the quantity of wastes produced by the agricultural sector is significantly low compared to wastes generated by other industries, the pollution potential of agricultural wastes is high in the long term. For instance, the spreading of manures and slurries on land can cause nutrient and organic pollution of soils and waters. Given animal excreta also contains a plethora of organic chemicals and pathogens, the risk for surface and groundwater contamination can be high (Sarmah, 2009) (Table 9.1).

**TABLE 9.1**
**Characterization of AW Depending on the Agricultural Activity**

| Agricultural activity | Types of waste | Method of disposal |
|---|---|---|
| Harvest and crop yield | Straw, stover | Land application, burning, |
| Vegetables and fruit purifying | Biological sludges, trimmings, peels, leaves, stems, soil, seeds, and pits | Landfilling, animal feed, land application, burning |
| Purification of sugar | Biological sludges, pulp, lime mud | Landfilling, burning, composting, animal feed |
| Production of animal | Blood, bones, feather, litter, manures, liquid effluents | Land application, fertiliser |
| Dairy product processing | Biological sludges | Landfilling, land spreading |
| Leather tanning | Fleshing, hair, raw and tanned trimmings, lime and chrome sludge, grease | By-product recovery, landfilling, land spreading |
| Rice manufacturing | Bran, straw, hull | Feeds, mulch/soil conditioner, packaging material for glass and ceramics |
| Coconut manufacturing | Stover, cobs, husk, leaves, coco meal | Feeds, vinegar, activated carbon, coir products |

*Source:* Loehr (1978).

## 9.6 MANAGEMENT OF AGRO WASTE INTO VALUABLE PRODUCTS

AW treatment technologies aim to recover useful by-products and energy, minimize environmental impacts, and produce "cleaner" wastes for safe disposal.

### 9.6.1 Technologies for the Recycling of Agricultural Wastes

Population growth, an increase in quality of life, and changes in behavioral patterns have caused an increase in the quantity of waste produced by modern society. Possible treatment methods include:

- Landfilling.
- Waste incineration or combustion to produce energy.
- Recycling.

The most common waste treatment method used today throughout the world is landfilling. The advantage of this method is that up until the last decade, its cost was low. However recently with an increase in environmental awareness, the desire to move waste disposal sites far away from population centers, the problems of finding new sites, and the consideration of external costs, landfilling costs have increased significantly. The second most common waste treatment alternative is waste incineration to produce energy. Although it is an effective method, the costs of establishing and operating incineration plants are very high. In addition, air quality laws and regulations are limiting the incineration of waste due to the serious concern of air pollution. An increase in the quantity of municipal, industrial, and agricultural waste simultaneously with a reduction in the volumetric capacity of incineration sites and an increase in incineration costs are motivating the recycling of organic components in waste.

### 9.6.2 Biofuel Production

A biofuel is any liquid fuel derived from biological material such as trees, agricultural wastes, crops, or grass. Biofuel can be produced from any carbon source that can be replenished rapidly, such as plants.

### 9.6.3 Biogas

This is the gaseous form of biofuel. It burns just like natural gas and, for this reason, is slowly but steadily taking its place. Biogas is mainly composed of methane gas produced from the process of anaerobic breakdown of biomass. Most agricultural firms use biogas, and the fuel is currently being packaged in gas cylinders for household use.

### 9.6.4 Biodiesel

Biodiesel has become a key source as substitution fuel and is making its place as a key future renewable energy source. As an alternative fuel for diesel engines, it is becoming increasingly important due to diminishing petroleum reserves and the environmental consequences of exhaust gases from petroleum-fuelled engines. To minimize the biofuel cost, cooking is used as feedstock. The used cooking oils are used as raw material, the adaption of continuous transesterification process and recovery of high-quality glycerol from biodiesel by-product (glycerol) are primary options to be considered to lower the cost of biodiesel. There are four primary ways to make biodiesel, direct use and blending, micro-emulsions, thermal cracking (pyrolysis), and transesterification. The utilization of liquid fuels such as biodiesel produced from the used cooking oil transesterification process represents one of the most promising options for the substitution of conventional fossil fuels. However, as biodiesel is produced from vegetable oils and animal fats, there are concerns that biodiesel feedstock may compete with food supply in the long term. Currently, the higher greenhouse gas emissions from fossil fuels have persuaded policy makers, investors, and researchers to think more about the substitution of fossil fuels to save the planet.

#### 9.6.4.1 Bioethanol

Ethanol from lignocellulosic biomass seems to be a promising substitute for traditional fossil fuels, which can be used to run dedicated engines or gasoline-based engines, or in fuel blends. This biofuel is also liquid in nature and is produced from the biomass of both plants and animals, but mostly plants. It is made through the process of fermentation of high carbon content biomass, mainly sugars, and cellulose. Currently, there are several routes to ferment lignocellulosic materials to ethanol using ethanologenic microorganisms. This includes mainly simultaneous saccharification and fermentation (SSF) and separate hydrolysis and fermentation (SHF).

### 9.6.5 Preparation of Handmade Paper from Jute Waste

The development of a technology of making handmade paper from jute fiber especially jute residue will open up a new area where a substantial quantity of jute waste can be used for making handmade paper of good commercial value. A new avenue has opened for the utilization of jute waste, which would otherwise be burned or thrown away by farmers, creating a disposal problem (Figure 9.8).

The handmade paper/paper board is made from low-grade jute fiber and can be suitably blended with other lignocellulosic fibers by adopting an inexpensive pulping process with minimum use of cooking chemicals like caustic soda, sodium carbonate, lime, etc. Most of the properties are the same as normal handmade paper and even better in some cases. There are diversified uses of it such as in files, folders, greetings card, shopping bags, visiting cards, posters, writing grade paper, paper boards, file covers, greeting cards, etc.

### 9.6.6 Lac Mud as Organic Manure

Lac mud is the waste product of lac processing industries which is obtained to a tune of about 2.5% to 4.5% on a dry and wet weight basis, respectively, of the raw material

**FIGURE 9.8** Preparation of handmade paper from jute waste.

(stick lac). The lac mud produced is mostly dumped due to lack of a proper method of disposal, creating pollution hazards. Application of decomposed enriched lac mud in vegetables has recorded 22.0%, 22.5%, and 18.3% higher yields of brinjal, tomato, and spinach, respectively, and over 100% N through inorganic source (farmers practice). This technology ensures the saving of 48% of N and P fertilizers and 65% of K fertilizer in brinjal and tomato. In spinach, similar savings in N and P fertilizers along with 36.6% savings of K fertilizers was recorded. Application of fortified lac mud in floriculture has recorded 31.7% and 38.5% higher flower yields of rose and chrysanthemum, respectively, over the conventional method of manuring. This technology ensures the savings of 5kg and 700 g of manure per plant in rose and chrysanthemum, respectively. It may also improve the soil fertility status as well as decrease dependency on other manures. The technology developed ensures quality flower and vegetable production, savings of inorganic fertilizers, improvement in soil fertility status, and moreover, it may give another diversified dimension to the lac industry which may be helpful in sustaining the lac production system. Technology developed has been tested and demonstrated under farmer's field conditions (Figure 9.9).

### 9.6.7 Protein Isolates/Concentrates from Deoiled Cakes/Meals using a Novel Process

The process of producing protein isolates/concentrates from oilseed cakes/meals (e.g., soy meal, groundnut cake) without the addition of strong or diluted acid is a novel one. The protein yield from this process is about 35–36% of the total weight of soymeal and 25% of the total weight of groundnut cake used, whereas, in the existing process, a maximum 30% protein yield from soymeal was obtained. The developed method comprises novel bacterial strains isolated from a food sample for producing protein from deoiled meal/flour (Figure 9.10).

**FIGURE 9.9** Lac mud as organic manure.

**FIGURE 9.10** Soy protein supplements.

The obtained supernatant after precipitation of protein from a particular batch may be used for precipitation of another batch and so on. This process is acid-free and yields higher protein about 5% over the conventional process. Superior protein quality in terms of solubility, wettability, water absorption capacity, degree of hydrolysis, and digestibility. The plant protein is used in protein supplements, texturized vegetable proteins, imitation dairy products, seafood products, beverage industry, infant food formulations, bakery products, etc.

### 9.6.8 Microbial Protein Using Corn Cob

Microbial protein is the dried cells of microorganisms that can be used as a dietary protein supplement. This is a nonconventional and alternative source of plant and animal protein produced by bacteria, fungi, yeast, and algae. The advantages include its high protein content and short growth times, leading to rapid biomass production, which can be continuous and is independent of the environmental conditions.

The process could be made economical by using lignocellulosic waste, such as corn cob, generated from agricultural and industrial activities, which serve as an inexpensive carbon source for microbial protein production. This fulfills the dual purpose of effective handling of agro-residues and its adequate bioconversion for the development of microbial protein. This protein is suitable for human consumption and thus can be used as an ingredient or a substitute for fortification.

### 9.6.9 Utilization of Corn Cob Powder for Kulhad Making

Traditionally mud cups *(*kulhad*)* are prepared using red mud. Continuous depletion of source has affected the availability of this mud and consequently, there is a need to develop alternatives to fully or partially replace red soil in Kulhad making. Corn cobs, otherwise a waste product, are utilized for making mud cups to convert this waste into a value-added product (Figure 9.11).

The corn cob powder is mixed in different ratios with mud and cups are prepared by baking the molded cups in a direct fire. Physical properties, thermal properties, and Newton's cooling constant of corn cob powder-blended mud cups are measured for assessing the feasibility of partial replacement of mud by corn cob powder as an agro biomass utilization process.

### 9.6.10 Biochar

Biochar is a carbon-rich material produced by incomplete combustion of biological materials in the absence of oxygen or with a limited amount of oxygen. Biochar is produced from agricultural waste and weed by using the pyrolysis method. Agricultural biomass can be converted into biochar within two hours with a conversion efficiency

**FIGURE 9.11** Corn cob powder for kulhad making.

**FIGURE 9.12** Biochar.

of 25–35% depending on the type of biomass by using a modified portable metallic kiln. Biochar stores carbon in the soil for hundreds to thousands of years and thus, the level of greenhouse gases, like $CO_2$ and $CH_4$, can be reduced significantly from the atmosphere. Biochar can be produced from biomass of *Ageratum conyzoides, Lantana camera, Gynura sp., Setaria sp., Avenafatua,* Maize stalk, and pine needle (Figure 9.12).

It improves soil fertility and water retention, aeration, and soil tilth. Application of biochar improves soil pH and soil nitrogen and potassium availability.

### 9.6.11 Production of Enzymes Cellulases at Industrial Scale by Microbial Fermentation Utilizing Groundnut Shell as Substrate

Lignocellulosic food industry waste is easily available and the cheapest form of carbohydrates. Agro-industrial waste has tremendous nutritious value and facilitates microbial growth. Fungal and bacterial species can be used for the production of enzymes with commercial application. Agro-industrial wastes, such as rice bran, corn cob, and sugar cane bagasse, are widely used for the production of enzymes utilizing fermentation strategies.

Groundnut shell, one of the richest sources of cellulose (up to 70%), is difficult to decompose. In India, about 2 MT of groundnut shells available annually at nominal cost, are used in poultry feeds, for the production of briquettes, etc. At present, about 10% of the domestic need for cellulase is met from local production, and 90% is imported and therefore a part of the available groundnut shell can be utilized for the production of cellulase at an industrial scale by microbial fermentation. The cellulase enzyme is used in industries engaged in bio-polishing fabrics/garments, bio-stone washing, animal feed, textile, food industry, enzymatic saccharification of agro waste, etc. (Figure 9.13)

### 9.6.12 Bio Compost Technology

Bio compost is a carrier-based formulation of a consortium of four lignocellulolytic fungi namely *Phanerochaete chrysosporium, Trichoderma viride, Aspergillus*

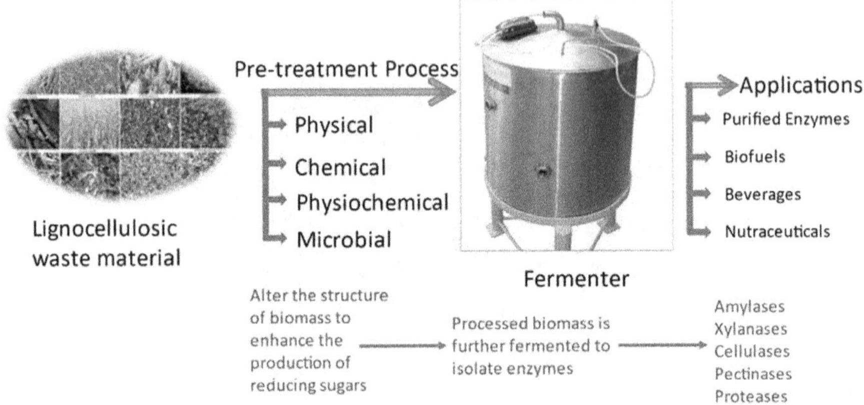


**FIGURE 9.13** Bioconversion of agro-industrial wastes to industrially important enzymes.

*awamori*, and *Pleurotus florida*. The consortium can be used to convert diverse agricultural residues of almost all field crops like wheat, rice, mustard, maize, and soybean into mature compost in 65–70 days by pit or windrow methods. The dimension of the compost pit is approximately 3 x 2 x 1 meters. The windrows are kept in a trapezoidal shape with a bottom width of 2 m and top width of 1.5 m and a height of 1 to 1.2 m. The windrow method of composting involves layering the material in the form of piles and subsequent mixing together with the application of culture. Each windrow is prepared using straw mix (substrates for composting) + Cow dung + good quality soil + old compost in the ratio of 8: 1: 0.5: 0.5. To enhance the process of composting, bio compost is added at a rate of 5000 ml per ton of the material. All the constituents are mixed thoroughly. In case P-enriched compost is to be prepared, rock phosphate at 1% is also added as a source of insoluble P. At all times, 65–70% moisture is maintained in the biomass. The piles containing raw material should be turned after every 15 days. The mature compost needs to be sieved so as to have uniform size particles and to get rid of any undecomposed material.

### 9.6.13 Tamarind Seed Husk Reduces Enteric Methane Emissions

Tamarind (*Tamarindus indica*, common name Imli) is grown in more than 50 countries. India alone annually produces more than 98 thousand metric tons of tamarind. Tamarind seed husk, an agricultural waste, constitutes about 35% of the decorticated roasted seed. The tamarind seed husk contains 13–15% tannins and is very effective in the modulation of rumen fermentation. The use of tamarind seed husk at a 5% level in the diet of cattle can reduce enteric methane emissions by 17–20%. Tamarin Plus, a product developed using tamarind seed husk can be fed to the growing (>4 months old), lactating, and adult animals *ad libitum*. Tamarind seed husk is an inexpensive material and costs only 3–4 rupees per kilogram and, therefore, can be used as an ingredient of total mixed ration along with straw and concentrate items at the above-specified level. Annually, $38 x 10^8$ Giga calories of energy is wasted from livestock due to methane emissions in India. About $26.28x10^6$ Giga calorie energy can be saved annually with a 20% confirmed methane reduction by the adoption of the product. The saved energy would lead to additional production of 10–12 MMT from the livestock in the country. The patent for this technological process was bagged by the ICAR-National Institute of Animal Nutrition and Physiology, Bengaluru (Figure 9.14).

During the process of oil extraction from oil palm fresh fruit bunches (FFB), oil palm bunch refuse and oil palm mesocarp waste is available in sterilized form. These waste materials have proved to be ideal substrates for growing edible mushrooms, like Paddy straw mushroom (on FFB), Oyster mushrooms, summer white milky mushroom, and summer white button mushroom (on mesocarp waste) either directly or after developing compost. Farmers near the factory zone can take up cultivation of mushrooms as a small-scale industry within the premises of oil palm factories or in fields. It can provide regular income to rural population, farmers, womenfolk involved in mushroom production thus improving their economic status. It also helps in the eco-friendly disposal of oil palm factory wastes by utilizing for edible mushroom production and bioconversion of waste material into edible food rich in high-quality protein (Figure 9.15).

**FIGURE 9.14** Tamarind seed husk reduces enteric methane emission and oil palm factory waste for mushroom production.

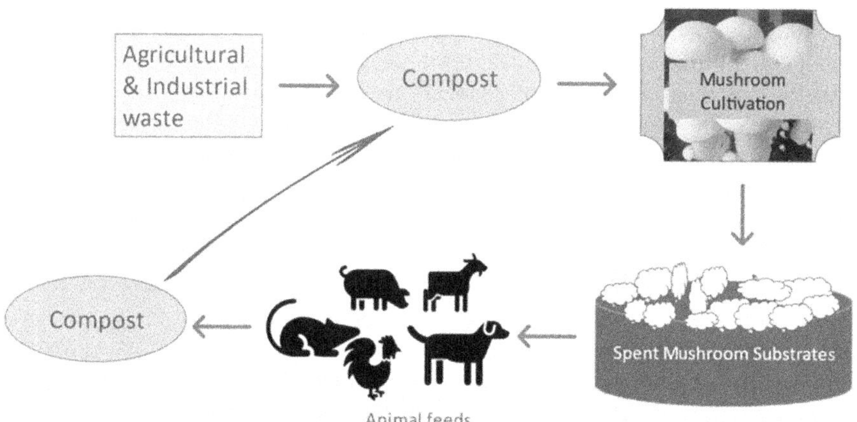


**FIGURE 9.15** Mushroom: A Potential Tool for Food Industry Waste.

## 9.7 FUTURE OF AGRO INDUSTRY

India is very diverse in its climate across the nation and in its languages, dialects, and culture. The climate varies from tropical to temperate in the south and alpine in the north which allows diversity in agricultural product output. The Indian contribution to the world food trade has immense potential in the food processing industry as it is the largest producer of jute, pulses, and milk. India stands second in the production of cotton, vegetables, fruits, groundnut, rice, and wheat. It is also the chief producer of spices, plantation crops, and livestock.

The food processing industry is one of the largest in India accounting for 12.8% of total GDP. For FY2022, food grain production is projected to increase by 2% with 307.31 million tonnes of food grains against 303.34 million tonnes in FY2021. An increase of 10.5 million metric tonnes was observed in the production of horticultural crops with a record 331.05 million metric tonnes (MMT) in 2020–21. Milk production has registered a 10% increase in FY21 from 198 MT in FY20 to 208 MT. Similar growth is seen in sugar cane production and in the generation of livestock. The food market in India is estimated to grow to Rs. 3,451,352.5 crore by 2025 from Rs. 1,931,288.7 crore in FY20. The food processing industry employs about 1.93 million people and supports more than 37–50 million farmers. The Department for Promotion of Industry and Internal Trade (DPIIT) has opened the Indian food processing industry to attract 100% Foreign Direct Investment (FDI) which further augments the growth in the agro-industrial sector.

## 9.8 CONCLUSION

Rapid economic growth, agriculture practices, globalization, industrialization, urbanization, and the increased population are factors that threaten natural resources like water, minerals, and organic compounds, and subsequently result in the discharge of an enormous quantity of waste into Mother Nature. Various methods engaged in the treatment of wastewater, purification of water, and resource recovery require engineering expertise and good infrastructure. But currently there is no global outlook that covers waste and resource management; so it is the need of the hour to develop novel methodologies for the treatment of wastewater.

In the present scenario, bionanotechnology could be one of the promising alternatives, as we can now synthesize nanomaterials which could be used for the treatment of wastewater. These nanostructure materials are rich in incomparable properties, like high sensitivity, high adsorption capability, and high surface-to-volume ratios, and have the capability to overcome all the problems associated with the old traditional methods.

However, there is one major problem associated with nanotechnology, that is, the risk which is involved. As we know, there is a possibility that these tiny, nanoscale particles could be transmitted to humans or aquatic animals, causing toxicity. Further, the extent of toxicity depends on surrounding conditions such as pH, concentration, and time of contact. Take socks, for example: AgNPs are embedded in the fibers of socks (to remove foot odor), which would be released during repeated washings with some number of negative consequences. The silver nanoparticles have the capability to not only destroy harmful bacterial but also destroy beneficial bacteria, which are important for breaking down organic compounds in wastewater.

Scientists and engineers need to work together to explore the field of nanotechnology to the core, so that they can explore each and every corner of this new technology for sustainable development. The use of nanotechnology seems to be very interesting for wastewater management and recovery of resources which could pose a great future in this regard.

India is expected to achieve the ambitious goal of doubling farm income by 2022. The agriculture sector in India is expected to generate better momentum in the next few years due to increased investment in agricultural infrastructures, such as irrigation facilities, warehousing, and cold storage. Furthermore, the growing use of genetically modified crops will likely improve the yield for Indian farmers. India is expected to be self-sufficient in pulses in the coming few years due to the concerted effort of scientists to get early maturing varieties of pulses and the increase in minimum support price. Going forward, the adoption of food safety and quality assurance mechanisms, such as Total Quality Management (TQM), including ISO 9000, ISO 22000, Hazard Analysis and Critical Control Points (HACCP), Good Manufacturing Practices (GMP), and Good Hygienic Practices (GHP) by the food processing industry will offer many benefits.

## REFERENCES


Batstone, D.J., Virdis, B. (2014). The role of anaerobic digestion in the emerging energy economy. *Curr. Opin. Biotechnol.* 27, 142–149. 10.1016/j.copbio.2014.01.013

Batstone, D.J., Hülsen, T., Mehta, C.M., Keller, J. (2015). Platforms for energy and nutrient recovery from domestic wastewater: a review. *Chemosphere*, 140 2–11. 10.1016/j.chemosphere.2014.10.021

Cai, T., Park, S.Y., Li, Y. (2013). Nutrient recovery from wastewater streams by microalgae: status and prospects. *Renew. Sustain. Energy Rev.* 19, 360–369. 10.1016/j.rser.2012.11.030

Davis, T.A., Volesky, B., Mucci, A. (2003). A review of the biochemistry of heavy metal biosorption by brown algae. *Wat. Res.*, 37 (18), 4311–4330.

House, T.W. (2012). National bioeconomy blueprint, April 2012. *Ind. Biotechnol.* 8 97–102. 10.1089/ind.2012.1524

http://bioenergycenter.org/

https://transportgeography.org

Hülsen, T., Batstone, D.J., Keller, J. (2014). Phototrophic bacteria for nutrient recovery from domestic wastewater. *Water Res.* 50, 18–26. 10.1016/j.watres.2013.10.051

*In: Agricultural Wastes* Editor: Camille N. Foster. Nova Science Publishers, Inc. ISBN: 978-1-63482-359-3 © 2015

Jenkins, D., Wanner, J. (2014). *Activated Sludge–100 Years and Counting.* London: IWA Publishing.

Jetten, M.S.M., Horn, S.J., Van Loosdrecht, M.C.M. (1997). Toward a more sustainable municipal wastewater treatment system. *Water Sci. Technol.* 35, 171–180.

Kleerebezem, R., Van Loosdrecht, M.C.M. (2007). Mixed culture biotechnology for bioenergy production. *Curr. Opin. Biotechnol.* 18, 207–212. 10.1016/j.copbio.2007.05.001

Loehr. (1978). Best management practices for agriculture and silviculture. In *Proceedings of the Cornell Agricultural Waste Management Conference.* Ann Arbor Science Publishers.

Lovins, L.H. (2008). Rethinking production. *State World*, 2008, 34.

Mahvi, A.H., Nouri, J., Babaei, A.A., Nabizadeh, R. (2005). Agricultural activities impact on groundwater nitrate pollution. *Int. J. Environ. Sci. Tech.*, 2 (1), 41–47.

Mohan, D., Sarswat, A., Ok, Y.S., Pittman, C.U. (2014). Organic and inorganic contaminants removal from water with biochar, a renewable, low cost and sustainable adsorbent: A critical review. *Bioresource Technol.* 160, 191–202.

Mondor, M., Masse, L., Ippersiel, D., Lamarche, F., Massé, D.I. (2008). Use of electrodialysis and reverse osmosis for the recovery and concentration of ammonia from swine manure. *Bioresour. Technol.* 99 7363–7368. 10.1016/j.biortech.2006.12.039

Pontius, F.W. (1990). *Water Quality and Treatment*, McGraw-Hill Inc., New York.

Qu, X., Pedro, J.J., Alvarez, Q. (2013). Application of nanotechnology in water and wastewater treatment. *Water Res.* 47, 3931–3946.

Sam, S.B., Dulekgurgen, E. (2016). Characterization of exopolysaccharides from floccular and aerobic granular activated sludge as alginate-like-exoPS. *Desalination Water Treat.* 57, 2534–2545. 10.1080/19443994.2015.1052567

Sarmah, A.K. (2009). Agricultural wastes. In Chapter 1. *Potential Risk and Environmental Benefits of Waste Derived from Animal Agriculture*; Editors: G. S. Ashworth and P. Azevedo. Nova Publishers, pp. 1–17.

Tarley, C.R.T., Arruda, M.A.Z. (2004). Biosorption of heavy metals using rice milling by-products. Characterization and application for removal of metals from aqueous effluents. *Chemosphere*, 54, 987–995.

Tontti, T., Poutiainen, H., Heinonen-Tanski, H. (2016). Efficiently treated sewage sludge supplemented with nitrogen and potassium is a good fertilizer for cereals. *Land Degrad. Dev.* 10.1002/ldr.2528

Verstraete, W., Van De Caveye, P., Diamantis, V. (2009). Maximum use of resources present in domestic "used water". *Bioresour. Technol.* 100 5537–5545. 10.1016/j.biortech.2009.05.047

Wett, B., Omari, A., Podmirseg, S.M., Han M., Akintayo O., Gómez Brandón M., et al. (2013). Going for mainstream deammonification from bench to full scale for maximized resource efficiency. *Water Sci. Technol.* 68 283–289. 10.2166/wst.2013.150

# 10 Bionanotechnology in Biofuel and Bioenergy Production

*Sumit Kaushik, Amit Yadav, Monika Singh, Namrata Gupta, Gyanendra Singh and Piyush Gupta*

## CONTENTS



## 10.1 INTRODUCTION

The term "bionanotechnology" is the section or part of science in which study of biology takes place on bio-machines, and its utility is in developing the building blocks of biological classes to overcome engineering problems and open up new opportunities for technological advancement. Synthetic biology (SynBio) is used in the production of biofuels by developing highly efficient enzymes that break down solid biomass or solid bacteria that produce biofuels directly. In the production of biofuels, available technologies incorporate biochemical or thermochemical conversion processes to separate biomass feedstock into transportable substances. Biotechnology has been used to develop new and efficient biofuels that are as efficient as gasoline while also posing fewer difficulties than ethanol. Chemical feedstocks, degradation processes, and technological diversity are all part of biofuel biotechnology's scope (Malik and Sangwan, 2012).

The world's demand for energy is steadily increasing every day due to the development of new underused technologies, and the consumption of fossil fuels

DOI: 10.1201/9781003270959-10

exacerbates environmental problems. One of the most essential programs at the national and international levels could be to address the existing and future energy needs. Conventional energy sources provide around 90% of the globe's demands, i.e., mineral energy, but continued use of renewable resources portends a decline in mineral energy use of about 50% by 2040. To meet this market demand, there is a strong need for special renewable, environmentally friendly, low-energy, energy-saving, energy sources that are cost-competitive and environmentally beneficial (Srivastava and Prasad, 2000). The use of biomass assets in bioenergy and biofuel production is widely considered worldwide. Biomass has the potential for being the most renewable and environmentally beneficial renewable energy source available, as it creates little $CO_2$ (but a lot of sulphur) in the environment. Bioenergy and biofuel are nontoxic, biological, organic, and combustible energy sources made from biomass that can be used instead of petroleum products. Biofuel as a source of energy is dependent on research, the biomass environment, improved production conditions (temperature, reaction time, reagents), and the type of catalyst employed in an environmentally friendly and feasible manner. The effective application of new developments in the manufacture of big biofuels from biomass can help the energy industries (Pugh et al., 2011; Puri et al., 2012).

The conversion of biomass into biofuels can be done in two ways. Bioethanol, biogas, biodiesel, biooil, bio-syngas, and biohydrogen are techniques and thermochemical and biochemical compounds. Biofuels such as bioethanol, biodiesel, and biogas are important for long-term energy sources as they replace residual energy. It is based on renewable or biomass sources such as plants, wheat, beetroot, corn, grass, and wood used as bioethanol for feed production. Furthermore, a procedure known as transesterification is used to obtain biodiesel from vegetable oils and fats. Renewable biofuels can be delivered to a renewable bio cell through processes such as liquefaction, enzymatic hydrolysis, saccharification, aging, transesterification, anaerobic absorption, and blurred aging. Table 10.1 shows the conversion of biofuels from biomass. Based on the nature and characteristics of biomass, biofuels are separated into three generations: first generation biofuels are derived from starchy plants and food grains; second-generation biofuels are found in lignocellulosic biomass (residue obtained in agriculture as grass, sugarcane, and sorghum) and wood; the third generation, crude oil in edible foods and organic and industrial waste, are found in macro and microalgae. Pretreatment, hydrolysis, saccharification, fermentation, anaerobic digestion, and biomass transesterification are all ongoing research projects aimed at improving various biofuel production processes (Babu et al., 2013).

Different types of nanoparticles and nanomaterials are utilized as nanocatalysts in the catalytic reaction, the breakdown of lignocellulosic biomass into simple carbohydrates and fermented sugars for bioethanol and biodiesel production. Various nanoparticles, such as steel oxide nanoparticles, nanofibers, carbon nanotubes, and silica, are used for biofuel production (Adelere and Lateef, 2016).

Nanotechnology has an important ability to harvest low value and very efficient ways of producing biofuel and bioenergy. This even plays a major part in the design, integration, and extraction of a variety of innovative energy and catalysts that will be utilized to produce bioenergy and biofuel from renewable resources. It has the

**TABLE 10.1**
**Feedstock Conversion Technologies for Biofuel Production**

| Type of Biofuel | Feedstock/Substrate | Conversion Technologies |
|---|---|---|
| Bioethanol | Cellulosic biomass: Plants of starch sugars (cereals, sugarcane, corn, potatoes), Lignocellulosic biomass: Sources of forest waste (including poplars); agricultural residues (including maize, sorghum, oats, barley, wheat, soybeans, cotton, bagasse, and rice straw); and energy crops (hybrid sorghum, sugarcane, miscanthus, switchgrass, eucalyptus, and pine) | Liquefaction, hydrolysis, saccharification, and fermentation pretreatment. |
| Biodiesel | Rapeseed, oil palm, soy, canola, jatropha, castor, and other oil crops are some examples. Cooking and frying oil should be discarded. | Determination by cold and warm pressing, purification, and transesterification cultivation, harvesting, extraction and transesterification |
| Biogas | Cattle dung, agricultural residue, sewage sludge, municipal waste, green waste and crops | Anaerobic digestion |
| Biohydrogen | Lignocellulosic biomass, biowaste, algae, and food waste | Dark fermentation, photo fermentation, biophotolysis |

potential to offer clean, nontoxic, more affordable, profitable, raw, and reliable alternatives to renewable energy sources. Nanotechnology may be one of the strategies that may improve bioenergy and biofuel production efficiency and reduce the cost of transport and high stock conversion (Serrano et al., 2009; Sekhon, 2014).

The biofuel manufacturing and purification process has drastic turnround by nanomaterial use. The lastest studies investigated for strategic improvement of powerful nano catalysts for biomass conversion, operational conditions, and decreasing the general processing value of biofuel manufacturing at business scale. This evaluation highlights the numerous applications of nanomaterials and nanoparticles for biofuel and bioenergy manufacturing in an eco-friendly and sustainable manner.

## 10.2 THE USE OF NANOTECHNOLOGY IN THE PRODUCTION OF BIOENERGY AND BIOFUEL

Worldwide, biofuel has received interest as a substitute for gasoline, as its chemical makeup makes it almost 85% comparable to gasoline, and its use requires no engine modification. Additionally, biofuels have gained recognition because of their renewable nature, sustainability, environmental friendliness, and lower greenhouse fuel line emissions. Generally, biofuels are comprised of agriculture residue, animal manure, vegetable oils, animal fat, and sugar plants and are obtained through anaerobic digestion, fermentation, and transesterification. However, challenges to biofuel manufacturing, like cost, unproven conversion techniques, and other technical problems, have slowed its growth. It is hoped that nanotechnology will be the answer to the financial and technical/environmental issues of bioenergy and biofuel manufacturing (Rai et al., 2014; Duhan et al., 2017). In order to boost biofuel yield and performance in gasoline and diesel, nanotechnology was introduced using catalysts and nano components, including nano-magnets, nanocrystals, nanofibers, nanodroplets, and more (Nizami and Rehan, 2018).

Nanoparticles and nanomaterials have multi-functional residues that include large surface areas, catalytic overgrowth, crystallinity, stability, adsorbency, environment, durability, versatile packages and reusable capabilities, recycling, and recycling that can help create a common path (Chandran et al., 2006). Nanotechnology can be useful in all biofuel technologies including pre-biomass/feedstock treatment, anaerobic digestion, fermentation, transesterification, biogas and fatty ester production, bioethanol and biohydrogen production. Nanotechnology-based biofuel production technologies are efficient, economical, and environmentally friendly, but generally these are at the standard laboratory and testing scale. Nanotechnology is a promising solution for changing conventional systems on a commercial scale. Various nanomaterials, including $TiO_2$, $Fe_3O_4$, $SnO_2$, ZnO, carbon nanotubes (CNTs), graphene, fullerene, silica, and metallic oxide nanoparticles (iron, zinc, nickel, gold and silver), are used in mass production for bioenergy and biofuel production. Magnetic nanoparticles, on the other hand, have a big surface area, making biofuel synthesis viable. In addition, those nanoparticles have a few fine-grained structures that can be free of problems found in the catalytic reaction while being made in the magnetic field (Ahmed and Douek, 2013). Metal oxides, ceramics, magnetic materials, semiconductors, quantum dots, lipids, polymers (synthetic or natural), dendrimers, and emulsion are some of the products utilized to make nanoparticles. However, the use of nanotechnology and nanomaterials may be one of the most effective alternatives to biofuel production by reducing conventional production costs.

## 10.3 BIOETHANOL PRODUCTION

Pre-treatment, hydrolysis, fermentation, and separation are the four key phases of the bioethanol manufacturing process from lignocellulosic biomass. Bioethanol is one of the potential fossil-fuel alternatives for ensuring energy security and reducing pollution. Bioethanol is also safer, less toxic, biodegradable, and more environmentally friendly than traditional fuels (Goldemberg, 2006; Wang et al., 2016). Agricultural residues, sugarcane, lignocellulosic substances, starch, and sugar biomass are used to do this (John et al., 2011). Wheat grass, corn kernels, rice straw, bagasse, and grass are examples of lignocellulosic biomass that are both less and more expensive because their conversion to bioethanol is time-consuming and expensive (Kulkarni et al., 2015; Malik and Tokas, 2018). The manufacturing of bioethanol is a complicated process that changes depending on the type of biomass employed. The later processes, such as early treatment, saccharification, fermentation, and product acquisition, are all part of the manufacturing process. Early diagnosis is critical for getting fermented sugar started and ready for fermentation. Bioethanol is produced by fermentation with the help of converting sugar/feedstock into biofuel. Pretreatment, hydrolysis, saccharification procedures, and fermentation methods are all used in the generation of biofuel from lignocellulosic biomass. In this process, biomass is first extracted with the help of hydrolysis to deliver monosaccharides and then boils the yeast to produce ethanol. Carbohydrates, cellulose, hemicellulose, and lignin are abundant in lignocellulosic biomass (Antunes et al., 2014). Lignin serves as a protective barrier for the body and needs to be removed to make carbohydrates available in the same way as hydrolysis. Therefore, early treatment is an important way to use lignocellulosic biomass to achieve excess sugar. Saccharification of the cellulosic component of lignocellulosic biomass hydrolyzed with the help of enzymatic activation and glucose (monomeric) can be produced at low temperatures without the need for stress. Major barriers to industrial production of bioethanol from complex food lignocellulosic biomass include poor concentration of unripe substances, high levels of enzymes, and the absence of boiling yeast. Operational parameters for the manufacture of bioethanol from lignocellulosic biomass have not been defined as a result of the increased production value; nanotechnology can be solved for complex conditions within the strategy and fuel production in a sustainable system. Various nanoparticles and nanomaterials have been proposed for bioethanol production. In this context, advanced strategies have been developed that can provide for the recovery and digestion of enzymes by reducing the number of reproduction. In view of these facts, nanotechnology may be helpful in blocking cellulase and hemicellulase enzymes that may be used to produce bioethanol. Table 10.2 shows biofuel production from lignocellulosic biomass application of nanotechnology.

The enzyme was transformed into immobilized magnetic nanomaterials (MNPs) that provide enzyme healing with the help of the proper magnetic field as well as re-enzyme recreational cycles. Immobilized cellulase enzymes, with the help of magnetic nanomaterials, accelerate the production of bioethanol. The structure of nanoparticles can be altered in accordance with the hydrolysis process and the saccharification of bioethanol production (Alftren and Hobley, 2014). Large amounts of bioethanol synthesis have been converted into gains without the use of any synthetic ingredients or toxic mixtures/mixtures (Khoshnevisan et al., 2011). Commercial enzyme resistance (Celtic CTec3) in 5 nanomaterials has been analyzed with the help of body adsorption and integration methods. The enzyme that has been transformed into immobilized nanoparticles was eliminated through combination

**TABLE 10.2**
**Biofuel Production from Different Feed Stocks Using Nanotechnologies**

| Feedstock/Biomass | Biocatalyst | Product |
|---|---|---|
| Tea that has been dissolved (*Camellia sinensis*) | Cobalt (Co) nanoparticles | Bioethanol (40.8%), biodiesel (40.8%) (95.7%) |
| Sugarcane leaves and jackfruit waste | $MnO_2$ immobilized cellulose | Bioethanol as a renewable energy source (21.96 %) |
| *Sesbania aculeate* | Cellulose bound with magnetic nanoparticles | Bioethanol (5.31g/l) |

binding techniques and body-marketing techniques (Alftren, 2013). For the synthesis of ethanol from cellulosic materials, the immutable-glucosidase enzyme in the polymer magnetic nanofibers has been changed. The incorporation of the static glucosidase enzyme into magnetic nanofiber converted to a given strong enzyme and reuse of this enzyme, which was transformed and separated by magnetic forces (Kumar et al., 2018). The enzyme-glucosidase was utilized as a biocatalyst in the synthesis of bioethanol; however it transformed to fungi and was unable to transport magnetic nanoparticles. Adsorption techniques were used to convert the enzyme to fixed $TiO_2$ NPs, which were then used to develop it for the breakdown of lignocellulosic biomass to make bioethanol (Chen et al., 2015). In any other study, the cellulase enzyme changed away from the fungus (*Aspergillus fumigates*) and was unable to transport $MnO_2$ NPs through a binding process and utilized agricultural waste hydrolysis to make bioethanol. Nanomaterials (silica, $TiO_2$, polymeric NPs, fullerene, grapheme, and CNTs) were reported to be inactive for bioethanol-producing enzymes (Lee et al., 2012). Microbial cells (*Saccharomyces cerevisiae*) have evolved into static nanoparticles that are subjected to a steady fermentation process and are used for complete bioethanol production. The immaturity of enzymes and yeasts in single-type NPs has proven that nanomaterials and nanoparticles can be simple, secure, and can pay for the efficient production of bioethanol from lignocellulosic biomass without exertion (Ahmad and Sardar, 2015). Uses of these nanomaterials in a one-of-a-kind conversion approach of bioethanol manufacturing may offer a sustainable manner through decreasing the price of uncooked biomass processing, manufacturing and delivery prices as properly dangerous environmental impacts (Verma et al., 2013). Enzyme immobilization on nanoparticles strategies is beneficial for bioethanol manufacturing and it is probably lets in reuse of enzyme; offer huge floor place for enzymes loading compared to business enzymes. The short duration of the nanomaterials enables them to absorb most of the enzymes into the molecular components of lignocellulosic biomass, and for this reason, nanomaterials can easily react with lignocellulosic additives with low complexity to deliver large amounts of carbohydrates and converted to bio-converted to use. The immaturity of the various biocatalysts including cellulase, cellobiose, hemicellulase, β-glucosidase, laccase, xylanase enzymes could be activated in one of the nanomatadium species that increased catalytic efficiency and balance of enzyme making bioethanol. The advantages of nanomaterials and nanostructures own best traits that provide on this area because of their physicochemical houses including huge floor place for excessive enzyme loading, mass transfer resistance, better enzymatic balance, powerful enzyme loading, and opportunity of enzyme reusability, which might reduce the price of business scale production of biofuel. These nanocatalysts are beneficial for price-powerful and eco-friendly technique for biofuel manufacturing from lignocellulosic biomass. Other techniques include nano-encapsulation silaffin binding and adsorption used to prevent enzymatic activity to convert the lignocellulosic biomass into bioethanol. Total bioethanol production can be increase by using nanoparticle in different phases of production process

Compared to fossil fuels, a large amount of biomass is produced in many industries that are referred to as "unwanted waste" and have no market value; however, there is no shortage of stock. Due to the growing interest in developing nanotechnologies they can develop and create new energy sources to meet global needs. Nanotechnologies can be used to pretreat or modify feedstocks and biomass, as well as to produce more efficient biocatalysts. In addition, the use of bio-nanomaterial can improve performance. Enzymes such as lipase and cellulase are used to convert lignocellulosic biomass/feedstock into bioethanol, biodiesel, and other products. Although bionanotechnology is relatively recent, few studies on the making of bioenergy and biofuels by lignocellulose biomass have been published.

## 10.4 BIODIESEL PRODUCTION

Macro and microalgae are now commonly recognized as ideal feed stocks for biofuels in the third generation. Microalgae could be a vital renewable fuel source because of its rapid growth rate, high oil content, increased biofuel productivity, $CO_2$ fixation ability, ease of handling and growing in water, absence of impact with food and feed crops, and ability to be cultivated on nonagricultural land. However, biodiesel produced from microalgae was 10–20 times more expensive per unit area than biodiesel manufactured from vegetable oils when all of these considerations were taken into account (Masran et al., 2016). It has been proposed that algae biomass be used as a feed stock. Algal biomass has the potential to yield up to 50% oil, which can be used in a variety of applications. Bionanotechnology, nanomaterials, and nanocatalysts were extensively used for isolation, accumulation of lipid, extraction, and act as a catalyst for the transesterification process. Additionally, they assist in cultivation and harvesting of algae for biodiesel manufacturing that might be extra efficient, economically possible, with excellent product and yields. Various steel oxide nanocatalysts together with $TiO_2$, CaO, MgO, and SrO, have been used because of their greater catalytic response for biodiesel manufacturing. Carbon-primarily based totally nanocatalysts i.e., carbon nanotubes, carbon nanofibers, graphene oxide, and biochar were used for biodiesel manufacturing from distinctive kinds of feedstocks. An immobilized enzyme (cellulase and lipase) became used on magnetic and steel oxide particles of nano range used in biofuel production by sea microalgae *Chlorella salina* and maximum yield (93.56%) became produced beneath standardized conditions. However, biodiesel manufacturing from moist algal biomass could be a cost-effective, easy, environmentally friendly, and riskless system. Many studies efforts ought to be taken for biofuel manufacturing from microalgae the usage of nanoparticles; there are a few demanding situations in reactor designs for excessive manufacturing with performance and balance of nanoparticles, low cost, recycling, socio-financial and environmental issues that ought to be taken into consideration. Nano-catalyzed is used for transesterification system for biodiesel manufacturing that would get replaced using convectional catalyst Nanotechnology facilitates in layout and change of photo- bioreactor or fermenter, advanced harvesting, drying and extraction techniques of microalgae biomass the usage of nano biocatalyst that greater biodiesel manufacturing system. MNPs primarily based totally catalysts have proven capacity to split from response media and reused enzyme for response ends in extra economically possible biodiesel manufacturing at industrial scale.

The use of magnetic nanoparticles and Cs / Al / iron oxide ($Fe_3O_4$) as a precursor for sunflower oil trans-esterification to generate biodiesel is estimated to be 94.8 percent effective. Similarly, biodiesel production includes Pongamia oil, rapeseed oil and sunflower oil while trans- esterified using Mg-Al hydrotalcite as a catalyst. The microalgal boom converted to biofuel has been simplified thanks to components such as nano and nanomaterials, particles, nanotubes, nanosheets, nano-droplets, and different nanostructures systems (Hossain et al., 2019).

Nanoparticles have been additionally included with microalgae cultivation technique consisting of mobileular suspension, mobileular separation and mobileular harvesting and biofuel conversion technologies. Another promising benefit of nanotechnology in biofuel enterprise has been used as biocatalysts in immobilization of enzyme at some point of lipase-catalyzed biodiesel and cellulosic ethanol manufacturing processes. Nanoparticles are also used in organic catalysts as potential providers of lipase immobilization. A flexible biocatalyst for biodiesel production was therefore discovered in stable lipase in magnetic nanoparticles. For example; Nanotechnology has played a critical role within biodiesel production environments by using or using nano waste and nanomaterials as catalysts. This era has brought about broaden safe, efficient, economic, durable, non-poisonous and extra solid nanocatalysts that retaining the capacity to completed higher high-satisfactory of product and yields in stipulated time period (Thangaraj et al., 2019).

## 10.5 BIOGAS PRODUCTION

Biogas manufacturing is a well-hooked up conversion generation that obtains electricity from biomass that is an opportunity supply to fossil fuel. Anaerobic digestion (AD) is taken into consideration as one of maximum crucial approaches that convert natural wastes into renewable electricity merchandise withinside the shape of methane, carbon dioxide and different hint factors and appropriate for electricity manufacturing that update traditional fuels. Biogas is primarily created through the digesting of carbon and nitrogen-rich anaerobic waste, agriculture waste, farm animal faeces, and human waste, as well as the extraction of anaerobically dependent power at a C/N rate. Anaerobic treatment of organic matter to produce biogas is a complicated microbiological operation that necessitates the co-existence of numerous microbial firms, each with its own metabolic potential and growth requirements. It's critical to satisfy the expanding needs of all the microorganisms engaged in the process of producing solid and green biogas. Through nanocatalysts that execute some of the best strategies for transforming green bioconversion, high substrate degradation, and improved delivery of biogas, nanotechnology can pave the way for increased biogas production.

The four types of nanomaterials used in the AD system are valent-free NPs, metal oxide NPs, carbon-based nanoparticles, and nanoparticles compounds. I employ iron nano-oxide, nano-valent iron, nano fly ash, nano fossil ash, and nano oxide bioactive metal in a favorable way during anaerobic digestion to increase methane generation. Bio-stimulated bacteria create nutrients and improve their activity as nanoparticles. In the digestion of anaerobic waste, nanomaterials raised methane concentration by 234 percent and biogas output by 180 percent. Methane content and hydrogen production rise with bacterial growth and promote enzymatic action, as iron plays a major role in electron transport.

Tests were performed with the effect of iron oxide nanoparticles (70–80 ppm) on the biodegradation of food waste in cellular digesters. Biogas generation in national parks has also improved in quantity and quality. Iron oxide nanoparticles' involvement (IONPs) and multiwalled carbon nanotubes have been used to develop biogas. Accumulated methane yields were significantly higher for high-density carbon nanotube reactors compared to IONP after 96 hours of anaerobic digestion. Research on the production of biogas from new compounds used by the nanoparticles of Ni, iron oxide, Co, Fe and methane (78.53%) and (116.76%) (Emmanuel and Fundam, 2016). During anaerobic digestion of municipal mud, the effects of varied bimetallic concentrations of copper and iron nanoparticles (nZVI / Cu) were found on the methane content. Nanoparticles have been employed as a growth promoter for bacteria that break down long-lasting natural compounds. Using nanomatadium and modified nanoparticles, anaerobic digestion productivity can be increased by up to 200 percent. Nanoparticles dissolve in the anaerobic environment and are used as a catalyst for microorganisms responsible for biological decay.

The effect of four iron NPs (Fe, Ni, Co, and iron oxide) during decomposition of chicken waste was investigated using several types of nanoparticles (metals, metal oxides, CNTs, and nano-ash) utilized as additives to increase anaerobic digestion function in biogas production.

In comparison to the control, NP supplementation boosted methane generation. In another significant technique, anaerobic digestion to produce biogas from cow dung, the impact of metal oxide nanoparticles (MONP) such as ferric oxide ($Fe_2O_3$) and $TiO_2$ has been examined biofuels and bioenergy. By removing the high chemical demand for oxygen and enhancing microbial activity, nanoparticles (NPs), as an addition to biodegradation, can increase biogas generation compared to control. During anaerobic digestion, iron nanoparticles act as additives that are able to promote interactions between organisms and methanogen-producing biogas through a bio-stimulation process. The discovery of NP properties may also enhance the microbial activity of acetotrophs, hydrogenotrophs and methanogens during anaerobic digestion converting biomass into biogas and methane. The addition of modified iron nanoparticles can serve as a catalyst for improving the digestion of anaerobic biogas production. In anaerobic degradation, bacterial biodiversity and the sustainability of the microbial population structures are critical. The breakdown of Zn homeostasis causes toxicity of zinc oxide NPs nanoparticles in the microbial community's lysosomes and mitochondria, supernatural abilities (Bhattacharya and Rajinder, 2005; Chandersekar et al., 2015).

## 10.6 BIOHYDROGEN PRODUCTION (ARTIFICIAL PHOTOSYNTHESIS)

By darkening and photochemistry of organic substrates, hydrogen can be generated in a biomanufacturing process. Hydrogen is created under anaerobic settings when fermentative bacteria convert organic substrates to organic acids, and when photochemical bacteria convert organic acids to hydrogen and carbon dioxide. Dark fermentation, on the other hand, produces a small amount of biohydrogen. Some internal and external limits or alterations to the regular fermentation conditions are implemented to boost the amount and quality of the product. Nanotechnology has recently been introduced to speed up the darkening fermentation process (Patel et al., 2018). NPs have shown very significant photocatalytic activity and play an important role in improving biohydrogen production in dark and photobiotic organisms that feed on sugars and biological wastes. Nanoparticles (NPs), especially inorganic NPs such as Ag, Fe, Au, Cu, Ni, Pd, silica, $TiO_2$, CNT and composites have increased biohydrogen production (Kumar et al., 2019).

Photosynthesis is a means of generating solar energy in the form of fuel. Furthermore, photosynthetic organisms now collect all fossil fuel energy used in the production of sunshine (plants, microorganisms, algae, etc.). The natural or artificial photosynthesis required to produce the fuel consists of three main components at the nanoscale:

(i) A redox reaction centre complex that receives sunlight and converts significant amounts of energy to electrochemical energy.
(ii) The water oxidation complex catalyses the conversion of water into hydrogen ions and stored electrons by converting the redox potential transformed to oxygen after being kept as an equivalent reducing agent.
(iii) A second catalytic system that produces fuels such as carbohydrates, lipids, or hydrogen gas using equivalent reducing agents.

To produce clean fuels such as biohydrogen, solar energy must be the main contributor. To achieve this, it is necessary to improve the efficiency of solar water separators. For hydrogen to become the primary fuel of a future economy based on renewable energy, high-capacity hydrogen storage technologies and high-performance fuel cells must be developed.

Nanotechnology mainly involves the synthesis of nanoparticles and fabrication of nanostructures with controlled sizes, shapes, preparations, and dispersions and their potential benefits in the fields of drug design and development, water decontamination, bioenergy, and biofuel production. It also involves manipulating materials at the nanoscale, either by scaling from groups of single atoms, or by refining or reducing bulk materials. Though physical and chemical processes can manufacture pure and well-defined nanoparticles, they are both costly and possibly dangerous to the environment and individuals. For particle manufacturing, biological entities such as microbes, plant extracts, or plant biomass can be used as an alternative to chemical and physical processes. Low-cost nanotechnology that is also environmentally friendly. Other creatures have been utilized as sustainable predecessors to create particles (bacteria, actinomycetes, fungi, yeasts, viruses, and so on). Nanotechnology is a stable and functional material that is also clean, environmentally benign, and nontoxic. As a result, a more stable, dependable, and sustainable approach for the synthesis of economically viable, ecologically sustainable, and socially adaptive nanomaterials is urgently needed. Nanomaterials can be made utilizing local resources. Nanoparticles have been created using a variety of bacteria, fungi, and algae.

Nanotechnology has high energy conversion efficiency, produces nearly little pollution, and has the potential to be mass produced in massive quantities. The eventual goal will be to incorporate promising nanotechnological approaches in engineering innovations to develop sustainable energy supplies until commercial deployment, as well as their implementation as a means of lowering the cost of renewable energy while increasing production and consumption efficiency.

## 10.7 MECHANISMS OF NANOPARTICLES FOR ENHANCEMENT OF BIOENERGY AND BIOFUEL PRODUCTION

Nanomaterials have high surface area and volume ratio because of its nano range size; hence, it is very useful in the production of bioenergy and biofuels. Insoluble enzymes such as biocatalysts are considered promising ways to improve bioenergy and biofuel production. As carriers, nanomaterials, and nanobiocatalysts (NBCs) not only increase catalytic efficiency, but also improve enzyme stability. In addition, magnetism, NP, and NBC aid in the detection of enzymes and reuse.

Nanoparticles are highly efficient, and their nanostructures allow them to insert into the cell wall of biomass and react with biomolecules. It releases hydrates of carbon in biofuel production. Enzymes capable of transporting nanoparticle carriers allow for a big range of pH performance parameters, temperature, and temperature stability compared to natural enzymes. Recently, nano additives such as aluminum oxide, cerium oxide, aluminum, MNP, and carbon nanotubes (CNTs) have been identified as powerful additives for biodiesel production from microalgae. The nanoparticle-biomass interaction process is very complex and requires a better understanding of how they work together, the modification process, the reduction of negative effects, and the efficiency of positive effects. Systematic representation of the synthesis of nanoparticles by methanogens by the anaerobic conversion of natural waste and biomass into methane (Dutta and Saha, 2019).

## 10.8 SAFETY CHALLENGES

The bioenergy and biofuel production has been increased by the use of nanomaterials and nanoparticles. The process of nanoparticle synthesis releases a variety of toxic compounds, whose decomposing metabolites pose a threat to the ecosystem. In addition, nanoproducts are based on metal nanocatalysts, CNTs, and toxic carbon nanofibers. When a person comes in contact with nanoparticles, these nanoparticles enter the human body by the processes of ingestion, smell, and penetration due to their small size so that nanoparticles can easily enter the cells of humans and animals. It has been reported that nanomaterials and nanoparticles are types of active oxygen. DNA damage from nanoparticles is also caused with nanoparticle reaction (Kim et al., 2009). Extensive research needs to develop on the in vivo interactions of nanoparticles, especially those used in the production of fossil fuel and fossil energy. Care should be taken after the investigation. The bioenergy and biofuel industries are growing rapidly and play a major part in production of renewable energy and show resistance to climate change. Nanomaterials are often used as catalysts in biodiesel production and as biogas production additives to improve efficiency. Nanotechnology can be used as a potential tool in biofuel production processes such as early treatment, advanced fermentation, enzymatic hydrolysis, ethanol conversion and development, transesterification, and methane production. Nanomaterial plays a key role in increasing energy efficiency and conversion processes, or in improving the design and performance of devices. The construction and integration of new materials will have a profound effect on sustainable energy use. Sustainable energy needs to address resource shortages and environmental issues. While biofuels are renewable and environmentally friendly, converting and producing them remains a daunting process. The use of nanotechnology is an ideal choice for improving the efficient, cost-effective, environmentally friendly, and socio-economically viable bioenergy and biofuel production process. However, there are some safety parameters that should be properly considered in relation to health and the environment before applying nanotechnology. Further research needs to focus on the use of nanometal oxides being explored to produce hydrogen and methane. Thus, it is evident that nanotechnologies are key components that could significantly contribute to the reduction of environmental hazards and use of fossil fuels in the future. The production of bioenergy and biofuel by nanotechnology could be potential approaches that enhance fermentation, anaerobic digestion, converting and upgrading ethanol, transesterification and lipid extraction (Guan et al., 2012).

## 10.9 CONCLUSION

The energy production in the new age of world is the biggest threat in the near future. As we know, during the modern era, we have developed so many resources by which we enhance the production of energy, but those are the insufficient to mete out daily energy requirement of world population and the huge developments in the modern era of world. So, the new approach and alternative resource from fossils fuel is bioenergy production with the help of bio nanotechnology. By this means, we control various problems such as environment pollution and depletion of fossil fuel. This is the new approach to carry energy production to the next level, with removal of various short comings in the present methods of energy production.

## REFERENCES

Abdelsalam, E., Samer, M., Attia, Y.A., Abdel-Hadi, M.A., Hassan, H.E. and Badr, Y. (2016). Comparison of nanoparticles effects on biogas and methane production from anaerobic digestion of cattle dung slurry. *Renew. Energy.* 87, 592–598.

Abraham, R.E., Verma, M.L., Barrow, C.J. and Puri, M. (2014). Suitability of magnetic nanoparticle immobilized cellulases in enhancing enzymatic saccharification of pretreated hemp biomass. *Biotechnol Biofules.* 7, 90–99.

Adelere, I.A. and Lateef, A. (2016). A novel approach to the green synthesis of metallic nanoparticles: The use of agro-wastes, enzymes, and pigments. *Nanotechnol. Rev.* 5, 567–587.

Ahlawat, S., Gulia, S., Malik, K., Rani, S. and Chauhan, R. (2019). Persistence and decontamination studies of chlorantraniliprole in Capsicum annum using GC-MS/MS. *J. Food Sci. Technol.* 56, 2925–2931.

Ahmad, R. and Sardar, M. (2015). Enzyme immobilization: An overview on nanoparticles as immobilization matrix. *Anal. Biochem.* 4, 178–185.

Ahmed, F.M., Rahman, R.S. and Gomes, D.J. (2012). Saccharification of sugarcane bagasse by enzymatic treatment for bioethanol production. *Malays. J. Microbiol.* 8(2), 97–103.

Ahmed, M. and Douek, M. (2013). The role of magnetic nanoparticles in the localization and treatment of breast cancer. *Biomed. Res.* 5, 1–11.

Alftren, J. (2013). *Immobilization of cellulases on magnetic particles to enable enzyme recyclingduring hydrolysis of lignocellulose.* Ph.D. thesis submitted to Institute for Food, Technical University of Denmark, Lyngby, Denmark.

Alftren, J. and Hobley, T.J. (2014). Immobilization of cellulase mixtures on magnetic particles for hydrolysis of lignocellulose and ease of recycling. *Biomass Bioenergy.* 65, 72–78.

Alvira, P., Tomas-Pejo, E., Ballesteros, M. and Negro, M.J. (2010). Pretreatment technologies for an efficient bioethanol production process based on enzymatic hydrolysis: A review. *Bioresour. Technol.* 101, 4851–4861.

Ambuchi, J.J., Zhang, Z. and Feng, Y. (2016). Biogas enhancement using iron oxide nanoparticles and multi-wall carbon nano-tubes. *Int. J. Chem. Eng.* 10, 1305–1311.

Amen, T.W., Eljamal, O., Khalil, A.M., Sugihara, Y. and Matsunaga, N. (2018). Methane yield enhancement by the addition of new novel of iron and copper-iron bimetallic nanoparticles. *Chem. Eng. Process.* 130, 253–261.

Antunes, F.A.F., Chandel, A.K., Mmilessi, T.S.S., Santos, J.C., Rosa, C.A. and Da Silva, S.S. (2014). Bioethanol production from sugarcane bagasse by a novel Brazilian pentose fermenting yeast Scheffersomyces shehatae UFMG-HM 52:2: evaluation of fermentation medium. *Int. J. Chem. Eng.* 4, 1–8.

Antunes, F.A.F., Gaikwas, S. and Da Silva, S.S. (2017). Bioenergy and biofuels: Nanotechnological solutions for sustainable production. In *Nanotechnology for Bioenergy and Biofuel Production*, Switzerland: Springer International Publishing AG, Nature, 3–18.

Asharani, P.V., Lianwu, Y., Gong, Z. and Valiyaveettil, S. (2011). Comparison of the toxicity of silver, gold and platinum nanoparticles in developing zebrafish embryos. *Nanotoxicology.* 5, 43–54.

Avhad, M.R. and Marchetti, J.M. (2015). A review on recent advancement incatalytic materials for biodiesel production. *Renew. Sust. Energ. Rev.* 50, 696–718.

Babu, V., Thapliyal, A. and Patel, G.K. (2013). *Advances in Biofuel Production.* Hoboken, JWS, Inc.

Baskar, G., Kumar, R.N., Melvin, X.H., Aiswarya, R. and Soumya, S. (2016). Sesbania aculeate biomass hydrolysis using magnetic nano biocomposite of cellulose for bioethanol production. *Renew. Energy.* 98, 23–28.

Bhattacharya, D. and Rajinder, G. (2005). Nanotechnology and potential of microorganisms. *Crit. Rev. Biotechnol.* 25, 199–204.

Challa, S.S. and Kumar, R. (2006). Nanomaterials: Toxicity. *Environ. Health Perspect.* Wiley-VCH. 143.

Chandersekar, K., Lee, Y.J. and Lee, D.W. (2015). Biohydrogen production: Strategies to improve process efficiency through microbial routes. *Int. J. Mol. Sci.*, 16, 8266–8293.

Chandran, S.P., Chaudhary, M., Pasricha, R., Ahmad, A. and Sastry, M. (2006). Synthesis of gold nanotriangles and silver nanoparticles using Aloe vera plant extract. *Biotechnol. Prog.* 22(2), 577–583.

Chen, L., Yokel, R.A., Hennig, B. and Toborek, M. (2008). Manu- factured aluminum oxide nanoparticles decrease expression of tight junction proteins in brain vasculature. *J Neuroimmune Pharmacol.* 3, 286–295.

Chen, S., Li, L., Sun, H. and Lu, B. (2015). Nanomaterials for renewable energy. *J. Nanomater.* 7, 2–10.

Cherian, E., Dharmendirakumar, M. and Baskar, G. (2015). Immobilization of cellulase onto MnO2 nanoparticles for bioethanol production by enhanced hydrolysis of agricultural waste. *Chin. J. Catal.* 36(8), 1223–1229.

Chisti, Y. (2007). Biodiesel from algae. *Biotechnol. Adv.* 25, 294–306.

Cho, E.J., Jung, S., Kim, H.J., Lee, Y.G., Nam, K.C., Lee, H.J. and Bae, H.J. (2012). Co-immobilization of three cellulases on Au-doped magnetic silica nanoparticles for the degradation of cellulose. *Chem. Commun.* 48, 886–888.

Duhan, J.S., Kumar, R., Kumar, N., Kaur, P., Nehra, K. and Duhan, S. (2017). Nanotechnology: The new perspective in precision agriculture. *Biotechnol. Rep.* 15, 11–23.

Duraiarasan, S., Razack, S.A., Manickam, A., Munusamy, A., Syad, M.B., Ali, M.Y., Ahmed, G.M. and Mohiuddin, M.D.S. (2016). Direct conversion of lipids from marine microalgae C. salina to biodiesel with immobilized enzymes using magnetic nanoparticles. *J. Environ. Chem. Eng.* 4, 1393–1398.

Dutta, N. and Saha, M.K. (2019). Nanoparticle-induced enzyme pretreatment method for increased glucose production from lignocellulosic biomass under cold conditions. *Agric. Food Sci.* 99(2), 767–780.

Eleutherio, E.C.A., Boechat, F.C., Silva Magalhaes, R.S., Rona, G.B. and Brasil, A.A.(2019). Molecular mechanisms involved in yeast fitness for ethanol production. *AIBM.* 12(5), 105–114.

Elganzoury, M. and Allam, N.K. (2015). Impact of nanotechnology on biogas production: A minireview. *Renew. Sust. Energ. Rev.* 50, 1392–1404.

Emmanuel, A. (2016). Nanotechnology as a tool for enhanced renewable energy application in developing countries. *J. Fund. Renew. Energy Appl.* 6(6).

Erdely, A., Dahm, M., Schwegler-Berry, D., Leonard, H.D., McKinney, W., Frazer, D.G., Antonini, J.M., Porter, D.W., Castranova, V. and Schubauer-Berigan, M.K. (2013). Carbon nanotube dosimetry: From workplace exposure assessment to inhalation toxicology. *Toxicology.* 10(1), 53–65.

Faisal, S., Yusuf Hafeez, F., Zafar, Y., Majeed, S., Leng, X., Zhao, S. and Li, X. (2019). A review on nanoparticles as boon for biogas producers-nano fuels and biosensing monitoring. *Appl. Sci.* 9(1), 9–72.

Farghali, M., Andriamanohiarisoamanana, F.J., Ahmed, M.M., Kotb, S., Yamashiro, T., Iwasaki, M. and Umetsu, K. (2019). Impacts of iron oxide and titaniumdioxide nanoparticles on bio- gas production: Hydrogen sulfide mitigation, process stability, and prospective challenges. *J. Environ. Manage.* 240, 160–167.

Feng, Y., Zhang, Y., Quan, X. and Chen, S. (2014). Enhanced anaerobic digestion of waste activated sludge digestion by the addition of zero valent iron. *Water Res.* 52, 242–250.

Fernandes, F.A.A., Gaikwad, S., Ingle, A.P., Pandit, R., Santos, J.C., Rai, M. and Da Silva, S.S. (2016). *In Hand book of Nanotechnology for Bioenergy and Biofuel Production*, edited by M. Rai and S.S. Da Silva, Switzerland, Springer International Publishing AG, Nature. 3–18.

Gardy, J., Hassanpour, A., Lai, X., Ahmed, M.H. and Rehan, M. (2017). Biodiesel production from used cooking oil using a novel surface functionalized TiO2 nano-catalyst. *Appl. Catal. B.* 207, 297–310.

Goldemberg, J. (2006). The ethanol program in Brazil. *Environ. Res. Lett.* 3, 1–5.

Guan, Q., Li, Y., Chen, Y., Shi, Y., Gu, J., Li, B., Miao, R., Chen, Q. and Ning, P. (2017). Sulfonated multi-walled carbon nanotubes for biodiesel production through triglycerides transesterification. *RSC Adv.* 7, 7250–7258.

Guan, R., Kang, T., Lu, F., Zhang, Z., Shen, H. and Liu, M. (2012). Cytotoxicity, oxidative stress, and genotoxicity in human hepatocyte and embryonic kidney cells exposed to ZnO nanoparticles. *Nanoscale Res. Lett.* 7(1), 602.

Gupta, I., Duran, N. and Rai, M. (2012). *Nano-silver Toxicity: Emerging Concerns and Consequences in Human Health in Nano- Antimicrobials: Progress and Prospects*, edited by, N. Cioffi and M. Rai, Berlin, Springer. 525–548.

Gupta, I.R., Anderson. A.J. and Rai, M. (2015). Toxicity of fungal generated silver nanoparticles to soil inhabiting Pseudomonas putida KT2440, a rhizospheric bacterium responsible for plant protection and bioremediation. *J. Hazard. Mater.* 286, 48–54.

Harun, R. and Danquah, M.K. (2011). Influence of acid pre-treatment on micro algal biomass for bioethanol production. *Process Biochem.* 46, 304–309.

Holm-Nielsen, J.B., Al Seadi, T. and Oleskowicz-Popiel, P. (2009). The future of anaerobic digestion and biogas utilization. *Bioresour. Technol.* 100, 5478–5484.

Hossain, N., Mahlia, T.M.I. and Saidur, R. (2019). Latest development in microalgae-biofuel production with nanoadditives. *Biotechnol. Biofuels.* 12, 125–141.

Hsieh, P.H., Lai, Y.C., Chen, K.Y. and Hung, C.H. (2016). Explore the possible effect of TiO2 and magnetic hematite nanoparticle addition on biohydrogen production by Clostridium pasteurianum based on gene expression measurements. *Int. J. Hydrog. Energy.* 41, 21685–21691.

Huang, X.J., Chen, P.C., Huang, F., Ou, Y., Chen, M.R. and Xu, Z.K. (2011). Immobilization of Candida rugosalipase on electro-spun cellulose nanofiber membrane. *J. Mol. Catal. B Enzym.* 70, 95–100.

Hussain, M., Ahmad, R., Liu, Y., Liu, B., He, M. and He, N. (2017). Applications of nanomaterials and biological materials in bioenergy. *J. Nanosci. Nanotechnol.* 17, 8654–8666.

Hussein, A.K. 2015. Applications of nanotechnology in renewable in energies: A comprehensive overview and understanding. *Renew. Sust. Energ. Rev.* 42, 460–476.

Ingle, A., Paralikar, P., Silva, D.S.S. and Rai, M. (2018). Nanotechnology-based developments in biofuel production: Current trends and applications in Sustainable Biotechnology-Enzymatic Resources of Renewable Energy, *Springer Sci. Rev.* 289–305.

Ivanova, V., Petrova, P. and Hristov, J. (2011). Application in the ethanol fermentation of immobilizedyeast cells in matrix of alginate/magnetic nanoparticles, on chitosan-magnetite microparticles and cellulose-coated magnetic nanoparticles. *Int. J. Chem. Eng* 3(2), 289–299.

Jebali, A., Ramezani, F. and Kazemi, B. (2011). Biosynthesis of silver nanoparticles by Geotrichum sp. *J. Clust. Sci.* 22, 225–232.

John, R.P., Anisha, G.S., Nampoothiri, K.M. and Pamdey, A, (2011). Micro and macro algal biomass: A renewable source for bioethanol. *Bioresour. Technol.* 102, 186–193.

Johnson, B.F. (2003). Nanoparticles in catalysis. *Catalysis.* 24, 47–159.

Joo, S.H., Delicio, L., Muniz, J. and Baek, S. (2018). Perspective: Catalytic increase of biogas production in an anaerobic codigestion system. *Int. J. Nanopart. Nanotechnol.* 4(1), 16.

Jordan, J., Kumar, C.S.S. and Theegala, C. (2011). Preparation and characterization of cellulase-boundmagnetite nanoparticles. *J. Mol. Catal. B Enzym. B: Enzymatic.* 68, 139–146.

Karellas, S.B. (2010). Development of an investment decision tool for biogas production from agricultural waste. *Renew. Sust. Energ. Rev.* 14, 1273–1282.

Kato, H. (2011). In vitro assays: Tracking nanoparticles inside cells. *Nat. Nanotechnol.* 6, 139–140.

Khoshnevisan, K., Bordbar, A.K., Zare, D., Davoodi, D., Noruzi, M., Barkhi, M. and Tabatabaei, M. (2011). Immobilization of cellulase enzyme on super paramagnetic nanoparticles and determination of its activity and stability. *Int. J. Chem. Eng.* 171, 669–673.

Kim, K.H., Lee, O.K. and Lee, E.Y. (2018). Nano-immobilized biocatalysts for biodiesel production from renewable and sustainable resources. *Catalysts.* 8, 68–89.

Kim, Y.J., Choi, H.S., Song, M.K., Youk, D.Y., Kim, J.H. and Ryu, J.C. (2009). Genotoxicity of aluminum oxide (A12O3) nanoparticle in mammalian cell lines. *Mol. Cell. Toxicol.* 5172–178.

Kootstra, A.M.J., Mosier, N.S., Scott, E.L., Beeftink, H.H. and Sanders, J.P.M. (2009). Differential effects of mineral and organic acids on the kinetics of arabinose degradation under lignocelluloses pretreatment conditions. *Biochem. Eng.* 43, 92–97.

Kulkarni, S.J., Shinde, N.L. and Goswami, A.K. (2015). A review on ethanol production from agricultural waste raw material. *Int. J. Sci. Eng.*1(4), 231–233.

Kumar, A., Singh, S.S. and Nai, L. (2018). Magnetic nanoparticle immobilized cellulase enzyme for saccharification of paddy straw. *Int. J. Curr. Microbiol.* 7(4), 881–893.

Kumar, G.K., Mathimani, T., Rene, E.R. and Pugazendni, A. (2019). Application of nanotechnology in dark fermentation for enhanced biohydrogen production using inorganic nanoparticles. *Int. J. Hydrog. Energy.* 44, 13106–13113.

Lee, W.S., Chen, I.C., Chang, C.H. and Yang, S.S. (2012). Bioethanol production from sweet potato by co immobilization of saccharolytic molds and Saccharomyces cerevisiae. *Renew. Energy.* 39, 216–222.

Li, X., Xu, H., Chen, Z.S. and Chen, G. (2011). Biosynthesis of nanoparticles by microorganisms and their applications. *J. Nanomater.* 2, 11.

Lin, Y. and Tanaka, S. (2006). Ethanol fermentation from biomass resources: Current state and prospects. *Appl. Microbiol. Biotechnol.* 69, 627–642.

Liu, X., He, H., Wang, Y. and Zhu, S. (2007). Transesterification of soybean oil to biodiesel using SrO as a solid base catalyst. *Catal. Commun.* 8, 1107–1111.

Liu, X., He, H., Wang, Y., Zhu, S. and Piao, X. (2008). Transesterification of soybean oil to biodiesel using CaO as a solid base catalyst. *Fuel.* 87, 216–221.

Lupoi, J.S. and Smith, E.A. (2011). Evaluation of nanoparticle immobilized cellulase for improved ethanol yield in simultaneous saccharification and fermentation reactions. *Biotechnol. Bioeng.* 108, 2835–2843.

Ma, W., Xin, H., Zhong, D., Qian, F., Han, H. and Yuan, Y. (2015). Effects of different states of Fe on anaerobic digestion: A review. *J. Harbin Inst. Technol.* 22, 69–75.

Mahmood, T. and Hussain, S.T., (2010). Nanobiotechnology for the production of biofuels from spent tea. *Afr. J. Biotechnol.* 9, 858–886.

Mahto, T.K., Jain, R., Chandra, S., Roy, D., Mahto, V. and Sahu, S.K. (2016). Single step synthesis of sulfonic group bear- ing graphene oxide: a promising carbonano material for biodiesel production. *J. Environ. Chem. Eng.* 4, 2933–2940.

Malik, K. and Tokas, J. (2018). Eco-friendly process of ethanol production from paddy straw. *J. Agrometeorol.* 20, 287–29.

Malik, P. and Sangwan, A. (2012). Nanotechnology: A tool for improving efficiency of bioenergy., *J. Eng. Appl. Sci.* 1, 1–5.

Mao, S.S., Shen, S. and Guo, L. (2012). Nanomaterials for renewable hydrogen production, storage and utilization. *Progress Nat. Sci.: Met. Mater.* 22, 522–534.

Marchetti, J.M. (2012). A summary of the available technologies for biodiesel production based on a comparison of different feedstock's properties. *Process Saf. Environ. Prot.* 90, 157–163.

Masran, R., Zanirun, Z., Bahrin, E.K.B., Ibrahim, M.F., Yee, P.L. and Abd-Aziz, S. (2016). Harnessing the potential of lignolytic enzymes for lignocellulosic biomass pretreatment. *Appl. Microbiol. Biotechnol.* 100, 5231–5246.

Mohanpuria, P., Rana, N.K. and Yadav, S.K. (2008). Biosynthesis of nanoparticles: Technological concepts and future applications. *J Nanopart Res.* 10, 507–517.

Mostafa, F., Hssankhani, A. and Rafie, H.R.(2013). Preparation and characterization of Cs/Al/Fe3 O4 nanocatalysts for biodiesel production. *Energy Convers. Manag.* 71, 62–68.

Nizami, A. and Rehan, M. (2018). Towards nanotechnology-based biofuel industry. *Biofuel Res. J.* 18, 798–799.

Obadiah, A., Kannan, R., Ravichandiran, P. and Kumar, S.V. (2012). Nano hydrotalcite as a novel catalyst for biodiesel conversion. *Dig. J. Nanomater. Biostruct.* 7(1), 321–327.

Palaniappan, K. (2017). An overview of applications of nanotechnology in biofuel production. *World Appl. Sci. J.* 35, 1305–1311.

Patel, S.K.S., Lee, J.K. and Kalia, V.P. (2018). Nanoparticles in biological hydrogen production: An overview. *Indian J. Microbiol.* 58, 8–18.

Pattanayak, M. and Nayak, P.L. (2013). Green synthesis and characterization of zero valent iron nanoparticles from the leaf extract of Azadirachta indica (Neem). *World J. Nano Sci. Eng.* 2, 6–9.

Pavlidis, I.V., Vorhaben, T., Gournis, D., Papadopoulos, G.K., Bornscheuer, U.T. and Stamatis, H. (2012). Regulation of catalytic behaviour of hydrolases through interactions with functionalized carbon based nanomaterials. *J Nanopart Res.*14, 842–850.

Pourmand, A. and Abdollahi, M. (2012). Current opinion on nanotoxicology. *DARU J. Pharm. Sci.* 20, 5–102.

Pugh, S., McKenna, R., Moolick, R. and Nielsen, D.R. (2011). Advances and opportunities at the interface between microbial bioenergy and nanotechnology. *Can J Chem Eng.* 89, 2–12.

Puri, M., Abraham, R.E. and Barrow, C.J. (2012). Biofuel production: Prospects, challenges and feedstock in Australia. *Renew. Sust. Energ. Rev.* 16, 6022–6031.

Rai, M., Santos, J.C.S., Soler, M.F., Marcelino, P.R.F., Brumano, L.P., Ingale, A.P., Gaikwad, S., Gade, A. and Da Silva, S.S. (2014). Strategic role of nanotechnology for production of bioethanol and biodiesel. *Nanotechnol. Rev.* 5, 3–17.

Rai, M., Avinash, P., Ingle, A.P., Paralikar, P., Kumar, J. and Da Silva, S.S. (2019). Emerging role of nano biocatalysts in hydrolysis of lignocellulosic biomass leading to sustainable bioethanol production. *Catal Rev Sci Eng.* 61, 1–26.

Rashid, N., Rehman, M.S.U., Sadiq, M., Mahmood, T. and Han, J.I., (2014). Current status, issues and developments in microalgae derived biodiesel production. *Renew. Sust. Energ. Rev.*, 40, 760–778.

Safarik, I., Prochazkova, G., Pospiskova, K. and Branyik, T. (2016). Magnetically modified microalgae and their applications. *Crit. Rev. Biotechnol.* 36, 931–941.

Saritha, M., Arora, A. and Lata (2011). Biological pretreatment of lignocellulosic substrates for enhanced delignification and enzymatic digestibility. *Indian J. Microbiol.* 52, 1–9.

Sekhon, B.S. (2014). Nanotechnology in agri-food production: An overview. *Nanotechnol Sci Appl.* 7, 31–53.

Serrano, E., Rus, G. and Garcia-Martinez, J. (2009). Nanotechnology for sustainable energy. *Renew. Sust. Energ. Rev.* 13, 2373–2384.

Simon-Deckers, A., Gouget, B., Mayne-Lhermite, M., Herlin- Boime, N., Reynaud,C. and Carriere, M. (2008). In vitro investigation of oxide nanoparticle and carbon nanotube toxicity and intracellular accumulation in A549 human pneumocytes. *Toxicology.* 253(1–3), 137–146.

Sreekanth, K.M. and Sahu, D. (2015). Effect of iron oxide nanoparticle in bio digestion of a portable food-waste digester. *J. Chem. Pharm. Res.* 7(9), 353–359.

Srivastava, A. and Prasad, R. (2000). Triglycerides-based diesel fuels. *Renew. Sust. Energ. Rev.*4, 111–133.

Srivastava, V. and Mika, S. (2018). Recent advancement in biodiesel production methodologies using various feedstocks: A review. *Renew. Sust. Energ. Rev.* 90, 356–369.

Stellwagen, D.R., Vander Klis, F., Van Es, D.S., de Jong, K.P. and Bitter, J.H. (2013). Functionalized carbon nanofibers as solid-acid catalysts for transesterification. *Chemsus Chem.* 6, 1668–1672.

Sujeeta, Malik, K., Mehta, S. and Sihag, K. (2018). Optimization of conditions for bioethanol production from potato peels waste. *Int. J. Chem. Stud.*, 6, 2021–2024.

Talukdar, D., Verma, D.K., Malik, K., Mohapatra, B. and Yulianto, R. (2017). Sugarcane as a potential biofuel crop in sugarcane biotechnology; challenges and rospects, edited by, Chakravarthi Mohan. *Springer Sci. Rev.* 1, 123–137.

Tang, J., Xiong, L., Wang, S., Wang, J., Liu, L., Li, J., Yuan, F. and Xi, T. (2009). Distribution, translocation and accumulation of silver nanoparticles in rats. *J. Nanosci. Nanotechnol.* 9, 4924–4932.

Thangaraj, B., Solomon, P.R., Muniyandi, B., Ranganathan, S. and Lin, L. (2019). Catalysis in biodiesel production. *Renew. Sust. Energ. Rev.* 3, 2–23.

Verma, M.L., Barrow, C.J. and Puri, M. (2013). Nanobiotechnology as a novel paradigm for enzyme immobilization and stabilization with potential applications in biodiesel production. *Appl. Microbiol. Biotechnol.* 97, 23–39.

Verma, M.L., Puri, M. and Barrow, C.J. (2016). Recent trends in nanomaterials immobilized enzymes for biofuels production. *Crit. Rev. Biotechnol.* 36, 108–119.

Verziu, M., Cojocaru, B., Hu, J., Richards, R., Ciuculescu, C., Filip, P. and Parvulescu, V.I. (2008). Sunflower and rapeseed oil transesterification to biodiesel over different nanocrystalline MgO catalysts. *Curr. Green Chem.* 10, 373–381.

Vishwakarma, V., Samal, S.S. and Manoharan, N. (2010). Safety and risk associated with nanoparticles: A review. *JMMCE.* 9(5), 455–459.

Wang, T., Zhang, D., Dai, L., Chen, Y. and Dai, X, (2016). Effects of metal nanoparticles on methane production from waste-activated sludge and microorganism community shift in anaerobic granular sludge. *Sci. Rep.* 6, 25857.

Waqas, M., Aburiazaiza, A.S., Minadad, R., Rehan, M., Barakat, M.A. and Nizami, A.S. (2018). Development of biochar as fuel and catalyst in energy recovery technologies., *J. Clean. Prod.* 188, 477–488.

Weiland, P. (2010). Biogas production: Current state and perspectives. *Appl. Microbiol. Biotechnol.* 85(4), 849–860.

Xia, T. (2008). Comparison of the mechanism of toxicity of zinc oxide and cerium oxide nanoparticles based on dissolution and oxidative stress properties. *ACS Nano*, 2, 2121–2134.

Zhang, Z., Ohara, I.M., Mundree, S., Gao, B., Ball, A.S., Zhu, N., Bai Z., Jin, B. (2016). Biofuels from food processing wastes. *Curr. Opin. Biotechnol.* 38, 97–105.

# 11 Recent Advancements in Bionanotechnology and IoT-Based Sustainable Air Quality Management

## *Surveillance and Monitoring Using Machine Learning Techniques*

*Ram Kumar Lakshminarayana and*
*Mohd. Zafar, Naveen Dwivedi*

**CONTENTS**



DOI: 10.1201/9781003270959-11

## 11.1 INTRODUCTION

Air pollution affecting human health is a serious issue around the globe, contributing to the deterioration of the atmosphere and environment. A variety of environment-related problems have emerged as a result of the rapid growth and expansion of large metropolitan cities. The air pollutants present in the air are the cause of air pollution. Common air pollutants affecting the air are nitrogen dioxide ($NO_2$), carbon monoxide ($CO_2$), ozone ($O_3$), and sulphur dioxide ($SO_2$), and the parameters that influence the air quality are location, density of population, wind speed, and distribution of pollutants (Almetwally et al., 2020). Normally, monitoring stations are placed at fixed locations, which poses the challenges of huge cost, big buildings, and large area and mobility. IoT technologies can address the challenges of fixed monitoring station costs and possibly of deploying "mini" stations in different locations. It also provides the advantages of accuracy, scalability, and ease of deployment.

The information related to past and current air quality trends reveal the mitigation future of air quality. It can play a decisive role in the planning of sustainable air quality management (AQM). In the last decades, several approaches in terms of forecasting and prediction models have been developed to document the different characteristics of air quality under the influence of climatic variations. The first approach toward development of numerical models of air pollutant dispersion involves the Box model, Gaussian model, and Euler model for the modeling and simulation of physical and chemical variables pertaining to atmospheric air pollution. Second approaches include the statistical models which exploit multivariate linear regression and correlation, kernel regression, and fuzzy time series analysis to forecast atmospheric pollutant variables. These approaches have limitations and cannot solve the problems associated with complex nonlinear systems. The advent of machine learning has overcome these problems and developed the third most popular artificial intelligence approaches. The artificial intelligence-based approaches uses artificial neural network (ANN), genetic algorithm (GA), hybrid GA-ANN, and neuro-fuzzy models for modeling and simulation of nonlinear complex behavior of atmospheric variables for AQM. With the recent advancements in the field of artificial intelligence, deep learning is growing faster, and several models, such as Auto encoder (AE), Deep Belief Networks (DBN), Deep Boltzmann Machines (DBM), Convolutional Neural Networks (CNN), and Recurrent Neural Networks (RNN) have been developed and used in air pollutant distribution and prediction of air quality.

This chapter focuses on recent information related to the air quality monitoring systems and associated challenges. The application of advanced available machine learning-based tools and techniques for efficient air quality monitoring and management are discussed in detail. At the end, the case studies on global applications of IoT-based air quality management devices are discussed.

## 11.2 BACKGROUND OF AIR QUALITY MONITORING SYSTEMS

Every country has defined their own regulations and policies considering the environmental situation in their region and have created infrastructure to monitor air pollution and inform their citizens. To protect people and the environment from toxic gases, air pollution regulations are strictly enforced throughout the world. The common approach is to place the monitoring stations across cities in high altitude locations. Air quality monitors have a role to play in supplying knowledge on environmental quality concentrations. A population exposure assessment and an evaluation of health effects are then conducted. Environmental threats are too high when pollution levels are too high, therefore it is imperative to reduce pollution and protect the atmosphere. In order to measure air quality accurately, the Air Quality Index value is used.

### 11.2.1 Air Quality Index

The air quality index (AQI) is a system for describing air quality and communicating risks. Information about the quality of the air in their surroundings is important to the public, especially for vulnerable groups like children, the elderly, and people with respiratory and cardiovascular diseases. AQI is used for deciding outdoor activities, in general, and especially by individuals (Grace and Manju, 2019).

### 11.2.2 Common Air Pollutants

Ozone($O_3$), carbon monoxide (CO), sulfur dioxide ($SO_2$), nitrogen dioxide ($NO_2$), carbon dioxide ($CO_2$), methane ($CH_4$), volatile organic compounds (VOCs)-Benzene, fine particulate matter (PM2.5), particulate matter (PM10), lead (Pb), and black carbon (BC) are common air pollutants. These pollutants are released directly from an identifiable source to the atmosphere and stay the same in the original form. Table 11.1 outlines the two categories of air pollutants – primary and secondary.

The common information required related to the pollutants of interest are type, detection limits, range to expect, and level. Type informs whether the pollutant is primary, which is directly emitted or secondarily formed by chemical reactions in the atmosphere. The *detection limit* of the pollutant is the lowest concentration in the

**TABLE 11.1**
**Types of Air Pollutants**

| Primary Pollutants | Secondary Pollutants |
|---|---|
| Sulfur and nitrogen oxides, carbon monoxide, hydrocarbons, lead, particulate matter, and volatile organic compounds (VOCs) | • Chemical interaction between primary pollutants<br>• Between primary pollutants and atmospheric constituents by oxidation and or hydrolysis<br>• Tetrasulphate acid, nitric acid |

environment. The sensor detection limits are: μg/m$^3$ (microgram per cubic meter), ppm (parts per million), ppb (parts per billion), one hour averaging time period (1 hr), one eight-hour averaging time period (8 hr), one 24-hour averaging time period (24 hr), one three-month averaging time period (3 mo), and one-year averaging time period (1 yr). The *range to expect* is the concentration ranges within countries' specific locations and proximity to the manor power plant or roadway, for instance. *Level* is the pollutant concentration that occurs for a period of time.

The key pollutants that increase urban air pollution are particulate matter (PM), sulphur dioxide ($SO_2$), nitrogen oxides (NOx), carbon monoxide (CO), lead (Pb) and ozone ($O_3$), and volatile organic compounds (VOCs). These pollutants are emitted from a variety of natural and anthropogenic sources such as automobiles, energy production (coal, biomass), construction and industrial activities, open burning, and dust. It is noteworthy that exposure to these pollutants may lead to substantial socio-economic impacts in the form of mortality, degradation in health, and reduced productivity. If there are limited air quality monitoring stations with uneven distribution in the city, interpolation is used to solve the problem. To protect humans from harm, the valuable information is provided by the prediction. The main relevant factors to the variation in the air quality are identified by feature analysis.

## 11.3 CHALLENGES OF AIR QUALITY MONITORING AND MANAGEMENT

The challenges with air quality monitoring and management are related to the classical problems, such as sparse monitoring data, emergent issues of siloed stakeholder engagement, limited science-driven mitigation measures, and sporadic use of burgeoning information and communication technology (ICT) options. The major critical issues are absence of real-time hyper-local air quality data and information related to traffic, meteorology, and energy use. Besides these, the following challenges are associated with air quality monitoring and management:

- Due to the high cost in the constructing the air monitoring station, there are limited number of stations in cities.
- Labeled training samples with fine-grained air quality is expensive.
- Air quality generated data are incomplete.
- With the advancement of data acquisition technologies, it is difficult to define the key features for prediction and interpolation.
- AQIs can vary in their approach for determining pollutant concentrations (Plaia and Ruggieri, 2011).
- High cost is associated with AQI stations, and they require significant resources to be routinely maintained and calibrated (Chong and Kumar, 2003).
- In a large urban area, AQM stations would offer an extraordinary global view, but they can't detect pollution hotspots inside or around the city center (Kumar et al., 2015).

## 11.4 APPLICATIONS OF ADVANCED AVAILABLE TOOLS AND TECHNIQUES FOR AIR QUALITY MONITORING AND MANAGEMENT

Recent studies have highlighted the emergence of many advanced technologies including IoT (low-cost sensing (LCS) wearable devices), social media associated computation, crowdsourcing, artificial intelligence, big data, satellites, multi-sector modeling, cloud computing, and GIS.

### 11.4.1 Satellites in Air Quality Monitoring

The (Sentinel-5p, 2021) satellite is the Copernicus mission dedicated to monitoring the atmosphere. It carries the state-of-the-art TROPO spheric Monitoring Instrument (TROPOMI) designed to map a multitude of trace gases such as nitrogen dioxide, ozone, formaldehyde, sulphur dioxide, methane, carbon monoxide, and aerosols, all of which affect the air we breathe, affecting our health and climate.

Sentinel-5p is one of the most important sources of data for the Copernicus Atmosphere Monitoring Service (CAMS), one of the Copernicus earth observation programs that works on the planet and the environment.

CAMS is implemented by the European Center for Medium Range Weather Forecasts. It provides consistent and quality-controlled information related to air pollution, solar energy, greenhouse gases, and climate. CAMS is fundamental, because just like for weather forecasting, satellite observations need to be processed with advanced numerical models to deliver readily useful information such as surface concentrations for forecasts. Satellite data are essential for monitoring air pollution and protecting our health, and Copernicus Sentinel-5p, merged with other sensors that are involved in collecting air quality data.

The satellite data available for aerosol inversion are:

- Moderate Resolution Imaginary Spectroradiometer (MODIS).
- Multi-Angle Implementation of Atmospheric Correction (MAIAC).
- Multi-angle Imaging Spectroradiometer (MISR).
- Landsat 8 operational land imager (OLI).

With the inversion method of linear and nonlinear regression, the ground level air-pollutant concentration is estimated from satellite-derived Aerosol Optical Depth (AOD). There is a possibility of inherent retrieval errors due software versioning and the retrieval algorithms.

### 11.4.2 Machine Learning Approaches – Monitoring, Data Mining, and Calibration for Air Pollution

In recent years, many researchers have been working on predicting air quality using IoT-equipped sensors, predicting accurate values using machine learning and deep

**TABLE 11.2**
**Traditional Approach for Air Quality Forecasts**

| Chemical Transport Model Simulations | Statistical Methods |
|---|---|
| Community Multiscale Air Quality (CMAQ) | Auto Regression Moving Average (ARMA) model |
| Weather Research and Forecasting model coupled with chemistry (WRF-Chem) | Auto Regression Integrated Moving Average (ARIMA) model |
| Nested Air Quality Prediction Modeling (NAQMS) | Multiple Linear Regression (MLR) |
| | Geographically Weighted Regression (GWR) |

learning algorithms. Table 11.2 summarizes the traditional approach for air quality forecasts.

Some commonly used machine learning and deep learning methods in air quality forecasts in Edge Computing are: feedback radial basis function (RBF), multilayer perceptron neural networks, geographical correlation into the DBN model, multi-target regression neural network, random forest regressions, recurrent neural network, long short-term memory (LSTM), Lag Layer–LSTM-Fully-Connected Network (Lag-FLSTN), GRU model, SAE model, convolutional neural networks-long short-term memory (CNN-LSTM), and graph neural networks-gated recurrent units (GNN-GRU).

#### 11.4.2.1 k-Nearest Neighbors

The k-Nearest Neighbors (kNN) algorithm is one of the most popular approaches to classifying data. Normally, the common way is to start with a dataset with known categories, a principal components analysis (PCA) is used to reduce redundancy information and data dimensionality. The data that are used for initial clustering are called training data. The next step is to add new data with an unknown category to the PCA plot. Following that, it is possible to classify the new data into the "nearest neighbors." If the "k" in the k-nearest neighbor is equal to 1, then we only use the nearest neighbor to define the category. In another case, if the new data are between 2 or more categories, based on the most votes of the new data, the category is decided (Dragomir, 2010).

#### 11.4.2.2 k-NN by Euclidean Distance

Let A and B are represented by the feature vectors $A=(x_1,x_2,....,x_m)$ and $B=(y_1, y_2,...y_m)$, where m is the dimensionality of the feature space. To calculate the distance between A and B, the normalized Euclidean metric is used by formula:

$$\operatorname{dist}\left(A \cdot B\right)=\sqrt{\frac{\sum_{i=1}^{m}(x_i-y_i)^2}{m}} \tag{11.1}$$

#### 11.4.2.3 kNN by Dynamic Time Warping Distance

In the k-Nearest Neighbor, in order to calculate the proximity, Euclidean is replaced by the dynamic time warping (DTW) algorithm.

If the data are of different length and size, DTW is capable of calculating the proximity. To calculate, the matrix *i* x *j* is plotted with time series data. The column of the matrix is represented by the size of the dataset and is denoted by *j*. The row of the matrix is represented by the size of the data model and is denoted by *i*. Initially, DTW is calculated with the first column and first row. After that, another value in the cell is calculated, and the value in the cell i, j is returned.

#### 11.4.2.4 Recurrent Neural Network

The recurrent neural network (RNN) is a form of feed forward neural networks (FNN). In FNN, signals travel in one direction from input to output. It does not have looping functionality between the input and the output, and it is straightforward (Baby and Alexander, 2018). Based on FNN, self-connection of neuron cyclic structures is introduced by RNN. Through self-connected neurons, the sequence of data can influence network outputs with input data memorized. With the advantage of the RNN's memory characteristics, it performs better than FNN in many applications. When the training time is too long, it faces the problem of vanishing and exploding gradients as it may fail to capture long-time dependencies in input data (Hochreiter and Schmidhuber, 1997).

#### 11.4.2.5 Long Short-Term Memory

Recurrent Neural Network allows the neural network to keep hold of the context. With long short-term memory (LSTM), it is also allowed to forget the context that is no longer required. LSTM is a type of neural network. In this approach, the node receives some input and computes it and provides it as the output; in the next cycle of input, the output of the early computation will also be fed as input to the node. LSTM contains input gate, forget gate, and output gate. And with the Mean Absolute Error, they predicted the measure of difference between predicted and observed values. Zhou et al. (2019), using deep multi-output LSTM with the air quality factors $PM_{2.5}$, $PM_{10}$, $O_3$, NOx, $NO_2$, $SO_2$, CO and five meteorological factors – rainfall, temperature, wind speed, wind direction, and relative humidity predicted the air quality with improved spatio-temporal stability and accuracy.

#### 11.4.2.6 Internet of Things and Edge Computing

Ordinary household appliances as well as industrial tools are now connected with embedded sensors for exchanging data over the internet, and it is called the Internet of Things (IoT). Cloud computing is mainly used to store the data as well as to process the huge volume of data over large geographic distances. Challenges such as transferring large volumes of data to the cloud and the associated loss of data, as well as maintaining the large number of sensors, have led to the introduction of edge computing.

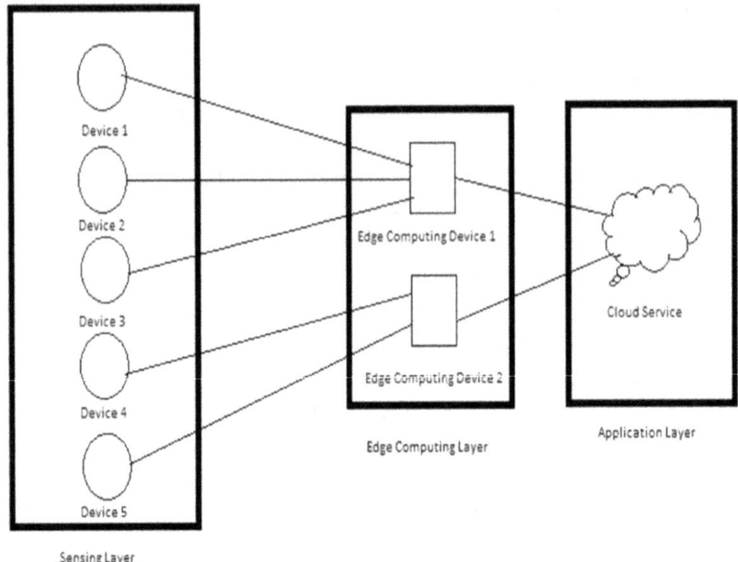


**FIGURE 11.1** Air pollution-monitoring edge computing architecture.

In the edge computing approach, processing in spite of happening in the cloud, will happen in the edge device. In this case, the IoT devices with the capability of processing the data will be deployed, and the subset of the data is sent to the cloud. By this approach, the network device with reduced network latency will yield data in a shorter response time.

In general, for air pollution monitoring systems, IoT architecture is balanced with a sensing layer, an edge computing layer, and an application layer. The sensing layer is deployed over the wide area network to sense the air quality using air quality sensors. The role of the edge layer is to establish consistent communication between the sensing and application layers. It also performs the processing of the data, applying different machine learning algorithms. The collaborative service to the consumers is provided by the application layer (Figure 11.1).

## 11.5 RECENT APPLICATIONS OF DEEP LEARNING FOR AIR QUALITY MONITORING AND PREDICTION

The knowledge and information of current and past air quality trends can play important roles in the mitigation of air quality. In the previous decade, several mathematical and deep learning-based approaches have been developed for air quality prediction, monitoring, and improvement in air quality trend characteristics. Table 11.3 represents the recent applications of deep learning technologies for air quality monitoring and prediction.

**TABLE 11.3**
**Recent Application of Deep Learning on Air Quality Monitoring and Prediction**

| S. No. | Approach | Study | Conclusion | References |
|---|---|---|---|---|
| 1. | Artificial neural network (ANN) model | Evaluation of air quality using ANN and multiple linear regression (MLR) models and data collected from a real urban air quality-monitoring network | • Five regulated air pollutants (nitrogen dioxide, nitrogen monoxide, ozone, carbon monoxide, and sulphur dioxide) were compared using a set of correlation and difference statistical measures and residuals distribution.<br>• ANN was found in most cases to be significantly superior, especially where the air quality network density is limited, leading to a decreased degree of spatial correlations among the monitoring sites. | Alimissis et al. (2018) |
| 2. | ANN model for the forecasting of annual PM10 (larger particulate) emissions | Optimized using a genetic algorithm and the ANN was trained using the following variables | • The model was trained and validated with the data for 26 EU countries from 1999 to 2006. PM10 emission data, collected through the Convention on Long-range Transboundary Air Pollution (CLRTAP) and the European Monitoring and Evaluation Programme (EMEP) or as emission estimations by the Regional Air Pollution Information and Simulation (RAINS) model, were obtained from Eurostat.<br>• The ANN model has shown very good performance and demonstrated that the forecast of PM10 emission up to two years can be made successfully and accurately. The mean absolute error for two-year PM10 emission prediction was only 10%, which is more than three times better than the predictions obtained from the conventional MLR and principal component regression models that were trained and tested using the same datasets and input variables | Antanasijevic (2013) |

*(Continued)*

**TABLE 11.3 (CONTINUED)**
**Recent Application of Deep Learning on Air Quality Monitoring and Prediction**

| S. No. | Approach | Study | Conclusion | References |
|---|---|---|---|---|
| 3. | Combination of principal components analysis (PCA) and ANN | Pattern recognition of Malaysian air quality based on the data obtained from the Malaysian Department of Environment (DOE) | • The combination of PCA and ANN was developed to determine its predictive ability for the air pollutant index (API). The PCA has identified that CH4, NmHC, THC, O3, and PM10 are the most significant parameters. The PCAANN showed better predictive ability in the determination of API with fewer variables<br>• The work has demonstrated the importance of historical data in sampling plan strategies to achieve desired research objectives, as well as to highlight the possibility of determining the optimum number of sampling parameters, which in turn will reduce costs and time of sampling | Azid et al. (2014) |
| 4. | Hybrid ARIMA and ANN model | Forecasting particulate matter in urban areas | • Box–Jenkins Time Series (ARIMA) and MLR models were applied to air quality forecasting in urban areas<br>• The hybrid ARIMA and ANN model could improve forecast accuracy for an area with limited air quality and meteorological data<br>• The hybrid model was able to capture 100% and 80% of alert and pre-emergency episodes, respectively | Díaz-Robles et al. (2008) |
| 5. | Support vector regression (SVR) based on Quantum-behaved particle swarm optimization (QPSO) algorithm | Improvement in forecasting accuracy of atmospheric pollutant concentration using a prediction model based on SVR | • The online direct prediction method is the most accurate in the three prediction methods. In comparison with particle swarm optimization (PSO) algorithm, QPSO algorithm is tested more efficiently for the improvement of global search ability and robustness during the procedure of parameter selection | Li et al. (2018) |

*(Continued)*

**TABLE 11.3 (CONTINUED)**
**Recent Application of Deep Learning on Air Quality Monitoring and Prediction**

| S. No. | Approach | Study | Conclusion | References |
|---|---|---|---|---|
| 6. | Neuro-fuzzy (NF) model | Air quality forecasting based on artificial intelligence-based Neuro-Fuzzy (NF) model | • The combination of ANN and fuzzy logic was used for air quality forecasting<br>• AERMOD could improve the forecasting ability of models<br>• Statistical analysis reflects that NF is performing better than others | Mishra and Goyal (2016) |
| 7. | Long short-term memory (LSTM) recurrent artificial neural networks | Prediction of air quality with deep learning LSTM | • A method for forecasting CO, $NO_2$, $O_3$, $PM_{10}$, $SO_2$, and pollen concentrations was developed using LSTM in recurrent neural networks. A nonparametric hypothesis test was used to back up the decision to select the most accurate and robust configuration. The results showed increases in accuracy when compared to two traditional approaches (as a benchmark) | Navares and Aznarte (2020) |
| 8. | Shallow multi-output long short-term memory (SM-LSTM) based ANN model | Exploration a deep learning multi-output neural network for regional multi-step-ahead air quality forecasts | • The proposed SM-LSTM model was evaluated by three time series of $PM_{2.5}$, $PM_{10,}$ and $NO_x$ at five air quality monitoring stations<br>• The results demonstrated that the proposed SM-LSTM model incorporated with three deep learning algorithms could significantly improve the spatio-temporal stability and accuracy of regional multi-step-ahead air quality forecasts | Zhou et al. (2019) |

*(Continued)*

**TABLE 11.3 (CONTINUED)**
**Recent Application of Deep Learning on Air Quality Monitoring and Prediction**

| S. No. | Approach | Study | Conclusion | References |
|---|---|---|---|---|
| 9. | Deep spatial-temporal ensemble (STE) model | Prediction of air quality based on deep STE model | • A deep STE model which comprises three components was developed<br>• The model evaluated data from 35 monitoring stations, and experiments showed that each component of our model makes a contribution to improvement in prediction accuracy, and the model is superior to baselines | Wang and Song (2018) |
| 10. | Wavelet based recurrent neural network model | Prediction of ambient air pollutant concentration | • The present paper proposes a wavelet based recurrent neural network (RNN) model to forecast concentrations of ambient CO, $NO_2$, NO, $O_3$, $SO_2$, and $PM_{2.5}$ which are the most prevalent air pollutants in urban atmospheres<br>• The model results demonstrated that a judicious selection of wavelet network design may be employed successfully for air quality forecasting | Prakash et al. (2011) |

**TABLE 11.4**
**Recent Applications of Nanotechnology for Air Quality Monitoring and Management**

| S. No. | Nanomaterials Used | Findings | References |
|---|---|---|---|
| 1. | Carbon nanotubes (CNTs) | • Multi-walled CNTs were fabricated using solution of (3-aminopropyl) triethoxysilane for $CO_2$ removal from air<br>• Significant adsorption of $CO_2$was reported by the amine functionalized CNTs | Gui et al. (2013) |
| 2. | Titanium dioxide-based nanofilter | • $TiO_2$-reduced-GO composite was fabricated and used for removal of methanol from air<br>• The active filter media was considered a promising candidate for improving environment quality through removal of air pollutants | Roso et al. (2015) |
| 3. | ZnO nanoparticles (NPs) | • Ultasonic-assisted precipitation method was used to synthesize ZnO NPs which are used for adsorptive removal of $H_2S$ gas<br>• It is reported that UZnO showed an increased hydrogen sulphide adsorption ability of 29.50 mg/g, compared to adsorption ability of 3.60 mg/g using material synthesized without ultrasonic treatment | Huy et al. (2019) |
| 4. | Graphitic carbon nitride/ titania composite-based nanocatalysts | • Graphitic carbon nitride/titanium dioxide composite photocatalysts with various graphitic carbon nitride/ titania was synthesized for removal of NOx pollutants<br>• The synthesized composite with a ratio of 1:4 showed increased photocatalytic capability in NO oxidation | Giannakopoulou et al. (2017) |
| 5. | Titanium dioxide-based nanocatalysts | • It is reported that $CO_2$ interacts strongly with $TiO_2$ disks as well as rods and efficiently removed on the titanium dioxide nanoparticle surface | Tumuluri et al. (2017) |
| 6. | Carbon nanotubes (CNTs, filters) | • CNTs were packed in a form of filters for air pollutant removal<br>• Efficient removal of different gaseous pollutants such as HC, NOx, $CO_2$, and CO was demonstrated in the range from 5.0% to 60.0% | Romero-Guzm´an et al. (2017) |

*(Continued)*

**TABLE 11.4 (CONTINUED)**
**Recent Applications of Nanotechnology for Air Quality Monitoring and Management**

| S. No. | Nanomaterials Used | Findings | References |
|---|---|---|---|
| 7. | Carbon nanotube filters | • MOF:UiO-66-$NH_2$ and CNT-modified filters were prepared and showed superior capture proficiency for ultrafine dust and $SO_2$ was reported | Feng et al. (2018) |
| 8. | Ag and Zn nanoparticles | • Ultraviolet light and silver/zinc oxide coating-based NPs were prepared and used for removal of airborne microbes<br>• It is reported that the silver/zinc oxide-coated air filter was efficiently utilized for removal of the germ contaminants from indoor air | Pokhum et al. (2018) |

## 11.6 RECENT ADVANCES IN NANOTECHNOLOGY FOR AIR QUALITY MONITORING AND MANAGEMENT

The various nanomaterials in the form of nano-adsorbents, nanocatalysts, nanofilters, and nanosensors are extensively utilized for the remediation of air pollutants (Saleem et al., 2022). Several studies have been conducted and examine the applicability of these nanoparticles extensively (Gui et al., 2013; Roso et al., 2015; Tumuluri et al., 2017; Saleem et al., 2022). Different nano-sensors have been analyzed for their efficiency in detecting toxic gases, such as nitrogen dioxide ($NO_2$), sulphur dioxide ($SO_2$), and hydrogen sulfide ($H_2S$). Similarly, issues related to poor air quality could be resolved using nanoadsorbents through significant adsorption of different contaminants existing in the air. Nanocatalysts are also utilized extensively for control of air contaminations and as semiconducting materials for photocatalytic remediation (Theerthagiri et al., 2019). Besides this, nanostructured membranes in the form of nanofilters have been used extensively for control of air contamination (Rubina et al., 2019; Samantara et al., 2019). The recent advancements in sensor technology facilitated the development of nanomaterial-enabled sensors which could detect ecological pollutants at the nanomolar and sub-picomolar levels (Saleem et al., 2022). Table 11.4 presents some recent applications of nanotechnology for air quality monitoring and management.

## 11.7 CONCLUSION

This chapter summarized the recent advancements in deep learning approaches for air pollution monitoring and prediction toward improvement in the quality of air under the perspective of urban development. In the 21st century, air pollution is a widespread global environmental issue apace with human-induced activities such as urbanization, industrial production, transportation, and high energy consumption. The brownfield cities (i.e., IoT and smart technology installations in an already established city) are facing challenges for effective management of air quality as urban environments are highly impacted by traffic and existing industrial technologies. Concise information related to deep learning approaches and recent IoT-based applications will be helpful in the selection of suitable tools and techniques for air quality monitoring and prediction.

## REFERENCES

Alimissis, A., Philippopoulos, K., Tzanis, C., Deligiorgi, D. (2018). Spatial estimation of urban air pollution with the use of artificial neural network models. *Atmos. Environ.* 191, 205–213.

Almetwally, A.A., Bin-Jumah, M., Allam, A.A. (2020). Ambient air pollution and its influence on human health and welfare: an overview. *Environ. Sc. Poll. Res.* 27(20), 24815–24830.

Antanasijevic (2013). PM10 emission forecasting using artificial neural networks and genetic algorithm input variable optimization. *Sci. Total Environ.* 443, 511–519.

Azid, A., Juahir, H., Toriman, M.E., Kamarudin, M.K.A., Saudi, A.S.M., Hasnam, C.N.C., Aziz, N.A.A., Azaman, F., Latif, M.T., Zainuddin, S.F.M. (2014). Prediction of the level of air pollution using principal component analysis and artificial neural network techniques: A case study in Malaysia. *Water Air Soil Pollut.* 225(8), 2063.

Baby, A., Alexander, A.A. (2018). A review on various techniques used in predicting pollutants. *IOP Conference Series: Materials Science and Engineering* (Vol. 396, No. 1, p. 012016). IOP Publishing.

Chong, C.Y., Kumar, S.P. (2003). Sensor networks: evolution, opportunities, and challenges. *Proc. IEEE.* 91(8), 1247–1256.

Díaz-Robles, L.A., Ortega, J.C., Fu, J.S., Reed, G.D., Chow, J.C., Watson, J.G., Moncada-Herrera, J.A. (2008). A hybrid ARIMA and artificial neural networks model to forecast particulate matter in urban areas: The case of Temuco, Chile. *Atmos. Environ.* 42(35), 8331–8340.

Dragomir, E.G. (2010). Air quality index prediction using K-nearest neighbor technique. *Bull. PG Unive. Ploiesti, Ser. Math. Inf. Phys. LXII.* 1(2010), 103–108.

Feng, S., Li, X., Zhao, S., Hu, Y., Zhong, Z., Xing, W., Wang, H. (2018). Multifunctional metal organic framework and carbon nanotube-modified filter for combined ultrafine dust capture and SO 2 dynamic adsorption. *Environ. Sci.: Nano.* 5(12), 3023–3031.

Giannakopoulou, T., Papailias, I., Todorova, N., Boukos, N., Liu, Y., Yu, J., Trapalis, C. (2017). Tailoring the energy band gap and edges' potentials of g-C3N4/$TiO_2$ composite photocatalysts for NOx removal. *Chem. Eng. J.* 310, 571–580.

Grace, R.K., Manju, S. (2019). A comprehensive review of wireless sensor networks based air pollution monitoring systems. *Wireless Personal Commun.* 108(4), 2499–2515.

Gui, M.M., Yap, Y.X., Chai, S.P., Mohamed, A.R. (2013). Multi-walled carbon nanotubes modified with (3-aminopropyl) triethoxysilane for effective carbon dioxide adsorption. *Int. J. Greenh. Gas Control.* 14, 65–73.

Hochreiter, S., Schmidhuber, J. (1997). Long short-term memory. *Neural Comput.* 9(8), 1735–1780.

Huy, N., Vo T.T.T., Hung T.N., Nguyen T.T., Tran T.K., Van Thanh, D. (2019). Facile one-step synthesis of zinc oxide nanoparticles by ultrasonic-assisted precipitation method and its application for $H_2S$ adsorption in air. *J. Phys. Chem. Solid.* 132, 99–103.

Kumar, P., Morawska, L., Martani, C., Biskos, G., Neophytou, M., Di Sabatino, S., Bell, M., Norford, L., Britter, R. (2015). The rise of low-cost sensing for managing air pollution in cities. *Environ. Int.* 75, 199–205.

Li, X., Luo, A., Li, J., Li, Y. (2018). Air pollutant concentration forecast based on support vector regression and quantum-behaved particle swarm optimization. *Environ. Model. Assess.* 1–18.

Mishra, D., Goyal, P. (2016). Neuro-fuzzy approach to forecast NO2 pollutants addressed to air quality dispersion model over Delhi, India. *Aerosol Air Qual. Res.* 16(1), 166–174.

Navares, R., Aznarte, J.L. (2020). Predicting air quality with deep learning LSTM: Toward comprehensive models. *Ecol. Inform.* 55, 101019.

Plaia, A. and Ruggieri, M., (2011). Air quality indices: a review. *Reviews in Environ. Sc. Bio/ Technol.* 10(2), pp.165–179.

Pokhum, C., Intasanta, V., Yaipimai, W., Subjalearndee, N., Srisitthiratkul, C., Pongsorrarith, V., Phanomkate, N., Chawengkijwanich, C. (2018). A facile and cost-effective method for removal of indoor airborne psychrotrophic bacterial and fungal flora based on silver and zinc oxide nanoparticles decorated on fibrous air filter. *Atmos. Poll. Res.* 9(1), 172–177.

Prakash, A., Kumar, U., Kumar, K., Jain, V. (2011). A wavelet-based neural network model to predict ambient air pollutants' concentration. *Environ. Model. Assess.* 16(5), 503–517.

Romero-Guzm´an, L., Reyes-Guti´errez, L.R., Romero-Guzm´an, E.T., Savedra-Labastida, E. (2017). Carbon nanotube filters for removal of air pollutants from mobile sources. *J. Miner. Mater. Char. Eng.* 6(1), 105–118.

Roso, M., Lorenzetti, A., Boaretti, C., Hrelja, D., Modesti, M. (2015). Graphene/TiO2 based photo-catalysts on nanostructured membranes as a potential active filter media for methanol gas-phase degradation. *Appl. Catal. B Environ.* 176, 225–232.

Rubina, A., Rubinov´a, O., Blasinski, P. (2019). Application of nanofilters for ventilation. *Int. Rev. Appl. Sci. Eng.* 10(1), 51–56.

Saleem, H., Zaidi, S.J., Ismail, A.F., Goh, P.S. (2022). Advances of nanomaterials for air pollution remediation and their impacts on the environment. *Chemosphere*, 287, 132083.

Samantara, A.K., Ratha, S., Raj, S. (2019. Functionalized graphene nanocomposites in air filtration applications. In *Functionalized Graphene Nanocomposites and Their Derivatives.* Elsevier, pp. 65–89.

Theerthagiri, J., Duraimurugan, K., Kim, H.S., Madhavan, J. (2019). Graphitic carbon nitride-based nanostructured materials for photocatalytic applications. In *Photocatalytic Functional Materials for Environmental Remediation*, pp. 291–307.

Tumuluri, U., Howe, J.D., Mounfield III, W.P., Li, M., Chi, M., Hood, Z.D. (2017). Effect of surface structure of $TiO_2$ nanoparticles on $CO_2$ adsorption and $SO_2$ resistance. *ACS Sustain. Chem. Eng.* 5 (10), 9295–9306.

Wang, J., Song, G., (2018). A deep spatial-temporal ensemble model for air quality prediction. *Neurocomputing.* 314, 198–206.

Zhou, Y., Chang, F.-J., Chang, L.-C., Kao, I.-F., Wang, Y.-S. (2019). Explore a deep learning multi-output neural network for regional multi-step-ahead air quality forecasts. *J. Cleaner Prod.* 209, 134–145.

# 12 Bionanotechnology
## *A Recommended Solution for Food Security, Climate Change, and Wastewater Treatment*

*Salman Khan, Sanjukta Vidyant, and Animesh Chatterjee*

## CONTENTS



## 12.1 INTRODUCTION

Food insecurity, hunger, and malnutrition must all be addressed as a worldwide concern. The present global food, nutrition, and agriculture governance architecture has failed to appropriately handle the system's current difficulties and ensure progress toward food security. Green biotechnology, biofortification, and nanotechnology are examples of new high-impact technologies that can help enhance agricultural productivity while also improving food quality and nutritional value. Plant varieties that can survive, and even thrive, in a variety of quickly changing and harsh environmental circumstances are required. Crops that are resistant to heat, drought, and other environmental challenges, have received a lot of attention. To say the least, the

DOI: 10.1201/9781003270959-12

techniques by which scientists are addressing this problem are innovative and environmentally friendly. Plant design, for example, can be altered to help them withstand harsh weather conditions. In times of extreme drought, the shape, distribution, and consistency of plant roots and leaves can be engineered to better catch and retain water. Roots can be manipulated to grow shallowly so that they stay close to the surface, collecting dew and runoff from precipitation. Likewise, by tightly regulating stomata, leaves can be adjusted to prevent evaporation through pores (Bhatnagar et al., 2008; Tester and Langridge, 2010). Simultaneously, work is being done to develop new types of staple crops that are high in micronutrients (biofortification). New high-impact technologies, such as nanotechnology and its applications, may allow individuals to eat foods without absorbing dangerous allergens or cholesterol, while also altering the taste and nutritional content of the food. However, early in the application process, research efforts should be directed to carefully assessing both benefits and hazards of such technologies. Nanotechnologies, genomics, and electronics can also be used to improve disease diagnoses; pesticide, fertilizer, and water supply; and soil quality monitoring and management.

Prosperity is dependent on economic growth. There is also a significant drive right now to guarantee that actions targeted at addressing global challenges are long-term rather than short-term. Through the deployment of green biotechnological solutions, biotechnology has the ability to boost economies while also addressing environmental issues. Sustainable development goals, on the other hand, are oriented to link the economy to environmental/social goals; as a result, "green growth" refers to the economic growth generated by industrial sectors that apply green biotechnology (the bioeconomy) (Ramos et al., 2017). At this point, it is worth clarifying the difference between green growth and sustainable development based on the difference between qualitative and quantitative growth. Green growth leads directly to a loss of the traditional jobs, but it does contribute to social development indirectly through green employment. We link green biotechnology to long-term development in this chapter.

## 12.2 BIONANOTECHNOLOGY TO COMBAT CLIMATE CHANGE

Food insecurity is exacerbated by global climate change. We address how green biotechnology could help to mitigate the link between climate change and food insecurity in this review. Rising temperatures and more frequent and extreme weather events, such as droughts, storms, and floods, are all repercussions of climate change (Gartland and Gartland, 2016). Agricultural and aquacultural productivity (including food crop and livestock yields), as well as forestry and fisheries, are all negatively impacted by these occurrences. Climate change may have an impact on the production of some foods, yield losses, and disease prevention efficacies. It can also have an impact on nutritional qualities, such as mineral and vitamin levels. Quantitative predictions of climate change consequences include a 4°C increase in mean world temperature by 2060, which will have a significant impact on global agricultural yields, such as wheat, rice, soya, and maize (Figure 12.1) (Lake et al., 2012). In other words, because climate change affects food availability, access to food, nutrient utilization,

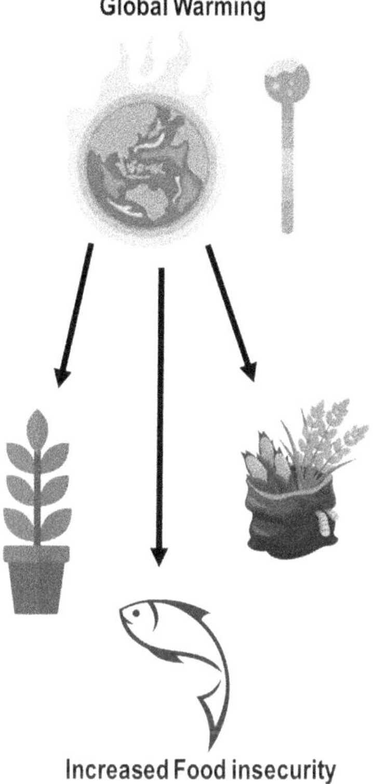


FIGURE 12.1 The effects of global warming on food sources.

and food system stability, it will exacerbate food insecurity. All of these data show that creative solutions to food insecurity are urgently needed. Green biotechnology has already been touted as a possible tool for combating global food insecurity. The following section delves deeper into biotechnology, specifically green biotechnology. Agriculture is always an important and stable sector because it produces and provides raw materials for food and feed industries. This alteration will be vital for achieving many goals in the coming years (Johnston and Mellor, 1961; Yunlong and Smit, 1994; Mukhopadhyay, 2014). Farming nutrient balances fluctuate noticeably with economic expansion, and as a result, the improvement of soil fertility in emerging countries is extremely important (Campbell et al., 2014). Agricultural development is necessary for eradicating poverty and hunger, which must be eradicated from the current condition. As a result, we should take one big step forward in agricultural growth. On this planet, the majority of people live in poverty, and they are concentrated in rural areas where agricultural expansion has been ineffective. The most important obsession nowadays is to create, which is accompanied by agricultural poverty and the nutritional method of obtaining food. As a result, new technology must be adopted that is solely focused on improving agricultural output (Yunlong and Smit, 1994).

## 12.3 BIONANOTECHNOLOGY AND FOOD SECURITY

Food insecurity is on the rise around the world. Hundreds of millions of people endure severe food shortages or famine-like disasters, which have been exacerbated by the COVID-19 epidemic and its catastrophic economic impacts, as well as climate extremes and violence. According to UN experts, the crises will worsen, putting decades of development gains and efforts to achieve the 2030 Sustainable Development Goals, which include ending hunger and malnutrition, in jeopardy. According to the most recent update of the 2021 Global Report on Food Crises, an estimated 161 million people were in crisis or worse (Integrated Food Security Phase Classification [IPC] Phase 3 or above) in 43 countries, up from 135 million in 2020, and needed immediate assistance to save their lives.

Food security exists when everyone has physical and financial access to enough safe and nutritious food to suit their dietary needs and food choices in order to live a healthy and active life (Mathews et al., 2018). More than 925 million people, including an average of 16% of the population in developing countries, were undernourished in 2020. Given that agriculture provides part or all of the income for 40% of the world's population, climate change is likely the greatest danger to global food security (Yashveer et al., 2014). Gradual temperature increases and extreme weather events caused by climate change would result in lower crop yields, increased soil degradation, and pollution from nitrogen runoff produced by the increased use of chemical fertilizers to boost food production (Pretty et al., 2011). Wheat yields have already started to fall globally (Menhas et al., 2016). Indeed, climate change is anticipated to reduce crop yields in Sub-Saharan Africa by 22% for wheat, 14% for rice, and 5% for maize by 2050, risking an alarming level of food insecurity (Montagu, 2020).

Adapting effectively to the progressive effects of climate change, improving our management of global warming-related agricultural risks, switching to crops that are better suited to altered environments, intensifying agriculture, and reducing deforestation for agricultural purposes are all opportunities to mitigate this decline in food security (Pretty and Bharucha, 2014; Vermeulen et al., 2012). Although it may seem contradictory, cutting deforestation by 10% may save 500 million tons of CO2-equivalent emissions over the next five years while also freeing up more area for food production (Smith et al., 2008). Sustainable intensification and climate-smart agriculture are two ways that are being employed around the world to address climate change and ensure food security. Sustainable intensification aims to boost crop yields from agricultural lands in order to enhance food production from a smaller land area (Pretty et al., 2011). It will take ecological, genetic, and market intensification to achieve this. Climate-smart agriculture strives to boost the sustainability and resilience of food production systems, reduce greenhouse gas emissions, and make it easier to achieve national food security and developmental goals by using breeding, technology, and policy instruments (Debnath, 2013). Green biotechnology plays a critical role in the efforts to mitigate the detrimental effects of climate change. It has made contributions in everything from traditional breeding and market-aided selection to genetic modification and the application of genomics in agriculture (Nasser et. al., 2021). Nanotechnology applications in food production can help meet the UN millennium objective of ensuring food security and safety (Sabourin and Ayande, 2015). Nanotechnology-based agrochemicals have greatly enhanced many crop management strategies. Nanotechnology aids in the increased efficacy of insecticides, herbicides, and fertilizers by allowing for controlled release while remaining environmentally benign. Chitosan and sodium alginate were utilized to encapsulate imidacloprid in a previous study, which improved its efficacy in soil applications (Guan et al., 2010). Nanomaterials are utilized in the processing, storage, and marketing of meat products (Gallocchio et al., 2015). Nano-feed, which is made up of nanosized additives and nanoclay, is used to provide better feed to animals by removing pathogens and toxins from processed foods in order to improve disease resistance, encourage the activation of the animal's own self-healing forces, improve bone growth and phosphate utilization, and lower mortality rates (Sekhon, 2014). Nanotechnology has the potential to increase animal feeding effectiveness and nutrition, improve plant pathogen detection, reduce animal losses due to disease, and improve agricultural, fishery, and animal production conservation and management (Singh, 2016; Pramanik and Pramanik, 2016). By creating healthy seeds and enhancing the effectiveness of fertilizers and pesticides, this technology can be utilized to increase crop productivity (Teng et al., 2011; Seabra et al., 2013). It is also utilized to improve crop output through soil and water cleansing, remediation, genetic engineering, and molecular-based crop breeding (Parisi et al., 2015; Wani and Kothari, 2018). Nanocarriers are utilized to deliver plant growth regulators, herbicides, and pesticides in a regulated manner. Treatment of mustard plants (*Brassica juncea*) with atrazine nanocapsules (carrier poly-epsilon-caprolactone) exhibited better herbicidal action, lower photosynthetic rates and stomatal conductance, increased oxidative stressors, weight loss, and growth reduction in a recent study (Oliveira et al., 2015). Some researchers have employed nanoparticle-mediated DNA or gene transfer in plants to generate insect-resistant plant types (Khot et al., 2012; Sekhon, 2014). Nanotechnology can help increase production by controlling nutrients (Gruere, 2012) as well as participate in the monitoring of water quality and pesticides for agriculture's long-term sustainability (Prasad et al., 2014). Nanomaterials contain so many different assets and functions that providing a broad assessment of their environmental and health impacts is impossible (Prasad et al., 2014). Chemical composition, form, surface structure, surface charge, behavior, level of agglomerates (clumping) or disaggregation, and other properties of NPs that determine toxicity may be associated with manufactured NPs (Ion et al., 2010). As a result, even nanomaterials with the same chemical makeup but various sizes or shapes might have variable levels of toxicity. The application of nanotechnology research in agriculture has become a vital and perhaps critical aspect for long-term development. Fullerenes, nanotubes, biosensors, controlled delivery systems, nanofiltration, and other agri-food applications have been observed (Ion et al., 2010; Sabir et al., 2014). This method has been shown to be effective in field based resource management, drug delivery systems in plants, and soil fertility maintenance. Furthermore, it is continually assessed in the utilization of biomass and agricultural waste as well as in food processing and packaging systems and vulnerability assessments (Floros et al., 2010). Nanosensors have already been applied to crops due to their advantages and speed in pollution management of land and sediments (Ion et al., 2010). Biosensors based on nano-detection technologies will be the major tools for identifying heavy metals at the lowest levels (Ion et al., 2010).

## 12.4 IMPACT OF BIONANOTECHNOLOGY ON ANIMALS AND HUMANS

The connection of sustainable development, climate change, and green biotechnology was discussed previously, with the main goal of green biotechnology being food security. Biotechnology, on the other hand, has various drawbacks, including socioeconomic and medical repercussions (Figure 12.2). To tackle these challenges using biotechnology, more research is required. Studies on genetic modification to improve weed management could lead to higher farm profits and less time spent weeding by female farmers, allowing them to spend more time caring for their children and improving child nutrition. Furthermore, biotechnology may provide cost-effective micronutrient deficiency treatments, such as vitamin A- and iron-rich crops (Domingo, 2007). Agricultural biotechnology could help maintain biodiversity and

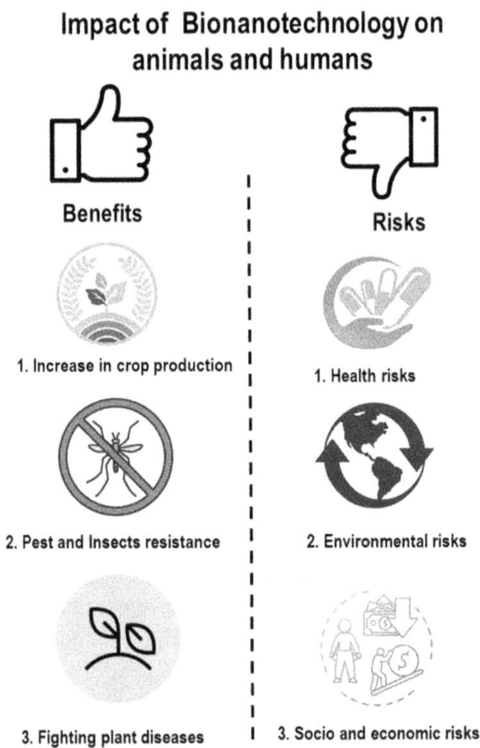


**FIGURE 12.2** The effects of green biotechnology on animals and humans.

protect vulnerable ecosystems by increasing productivity during food production and reducing the need to cultivate new lands (Mfutso-Bengo and Muula, 2007). The health effects of a glycaemic index (GI) food are determined by its composition. Iron-rich GI foods are likely to be useful to iron-deficient individuals. However, the transfer of genes from one species to another may also result in the transfer of allergic traits. In rats, for example, it was discovered that GI potatoes have specific effects on the digestive tract (Van de Wiel et al., 2016). The gene that was put into the potatoes was linked to a snowdrop flower lectin, which is known to be toxic to mammals. Animal lectin yields are often modest when compared to plant lectin yields, such as legume lectins. Lectins have a variety of functions, including anticancer, immunomodulatory, antifungal, HIV-1 reverse transcriptase inhibitory, and anti-insect properties, all of which could be useful. Antibacterial and antinematode properties are demonstrated by a modest number of plant lectins (Lam and Ng, 2011).

## 12.5 BIONANOTECHNOLOGY FOR WATER TREATMENT

The long-term evolution of the worldwide water issue is inextricably linked to the global population expansion and global warming. Continuous population growth, which is expected to approximately double from 3.4 billion in 2009 to 6.3 billion in 2050, is accompanied by an anticipated needed increase in agricultural productivity of 70% by 2050 (Bruinsma, 2009). Water is the most precious resource of human civilization, and a reliable supply of drinkable water is a basic human requirement. Nevertheless, we are still far from satisfying global expectations, and this issue will only get worse over time (Hillie and Hlophe, 2007). Growth in population, global warming, and deteriorating water quality all drive up demand (Ali and Aboul-Enein, 2004; Nemerow and Dasgupta, 1991; Tchobanoglous and Franklin, 1991). Fresh water (FW) is captured in only 2.5% of the world's oceans and seas (salt content of less than 1 g/L). Seventy percent (70%) of fresh water, on the other hand, is frozen as permanent ice. Only 1% of FW is suitable for consumption. More than 700 million people throughout the world do not even have access to clean drinking water This issue is particularly acute in developing states in Sub-Saharan Africa. As a result, treated wastewater in these impacted areas is required. Treatment technologies are approaching their limits in terms of providing adequate quality water to suit human demands (Qu et al., 2013). Organic, inorganic, and biological pollutants can all be found in water. Many pollutants are harmful and dangerous to persons and environments (Ali and Aboul-Enein, 2006;; Laws, 2000; Ali, 2012). Heavy metals are water contaminants with a high toxicity level. Arsenic is among the most poisonous elements, and it has been recognized as such since ancient times (Figure 12.3).

### 12.5.1 SIGNIFICANCE OF BIONANOTECHNOLOGY IN WATER PURIFICATION

Nanotechnology has produced revolutionary water treatment methods. Nanomaterials with high aspect ratio, reactivity, and customizable pore volume, as well as electrostatic, hydrophilic, and hydrophobic interactions, are used in adsorption, catalysis,

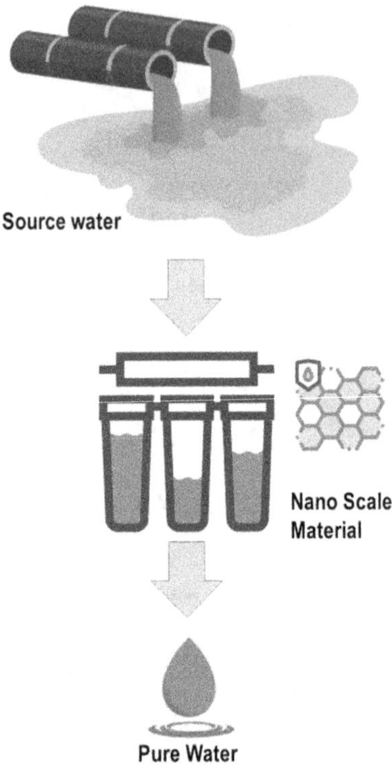


**FIGURE 12.3** The use of bionanotechnology in wastewater treatment.

sensing, and optoelectronics (Das et al., 2014). Nanotechnology-enabled processes are extremely efficient, adaptable, and multipurpose, and they deliver high-quality, low-cost water and wastewater treatment solutions. Nano object-based materials are long-lasting and have a high specific surface area (SBET). In other words, a large surface-to-volume ratio regulates how contaminants and bacteria interact (Qu et al., 2013b). Nanotechnology-enabled water treatment technologies pose significant challenges to current methods. Nanotechnology can also be used to purify and use atypical water sources in a cost-effective manner. The use of nanoparticles to treat industrial effluent is equally essential and ubiquitous. The current remediation solutions are successful, but they are expensive and time consuming. Nanotechnologies are useful in wastewater treatment because they remove impurities and aid in the recycling process, resulting in clean water (Figure 12.4). As a result, industry saves money, time, and resources while addressing a variety of environmental challenges (Kanchi, 2014). It should be highlighted that nanomaterials used to clean drinking water must be harmless and environmentally friendly. When unsafe particles come into contact with the human body, they can cause serious injury to important organs. Nano-objects may translocate to other organs due to their dimensional properties, increasing the risk of biological injury. As a result, toxicity performance tests must be strictly included in safety data sheets, SOPs, and other associated normative

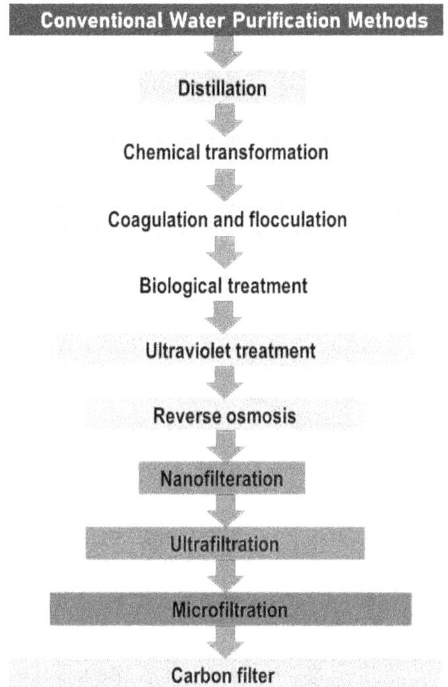


**FIGURE 12.4** Different water purification methods.

documents before being introduced to the industry (Figure 12.5). (Theron et al., 2008; Qu et al., 2013a; Faust and Aly, 1983; Hashim et al., 2012; Kumar et al., 2014; Ramakrishna et al., 2006).

### 12.5.2 Advantages of Bionanotechnology

Our findings bring us to the conclusion that we should not rely on biotechnology to alleviate the problem of food insecurity without further social, economic, and clinical research into the impact of green biotechnology applications. Such research is required to ensure that biotechnology can deliver food security while posing tolerable negative effects. As a result, we'll look at some recent examples of green biotechnology in agriculture to show its broad range of benefits. Meanwhile we also contribute some points to the disadvantages of green biotechnology. It provides useful techniques (for example, mutation breeding, improved conventional breeding, transgenic modifications, DNA insertion, gene transfer, gene editing, and somatic hybridization (Mazur, 2001; Yan and Kerr, 2002) for modifying plants to make them more adaptable to biotic and abiotic stresses such as extreme pH, salinity, drought, heat, and flooding.

Green biotechnology has numerous advantages in terms of boosting food security and mitigating the harmful effects of climate change. Enhancing global food production and agricultural yields, lowering pesticide residues on foods, protecting

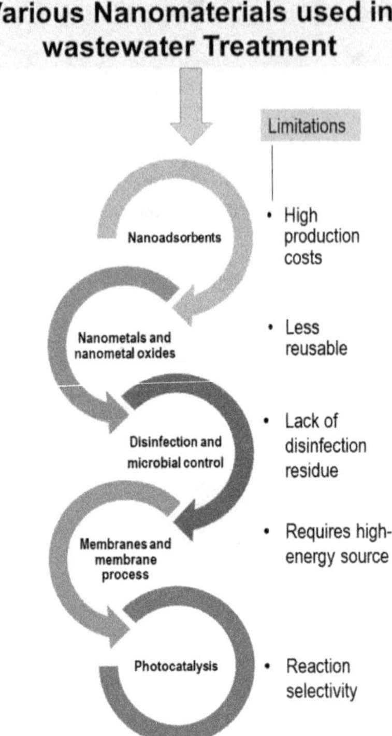


**FIGURE 12.5** Application of various nanomaterials to wastewater treatment and their limitations.

farmlands, creating jobs, and increasing worker productivity are just a few of the benefits (Brookes and Barfoot, 2014).

The following are some specific instances of the many benefits of green biotechnology for sustainable development.

1. Increased crop yields from the same amount of land (up to a 30% increase). For example, in a study in Africa, to address drought tolerance, the adoption of genetically modified (GM) maize with improved drought tolerance resulted in a 25% increase in yield in dry conditions, reducing the need to convert important natural resources like rainforests into farmland (Brookes and Barfoot, 2014).
2. Efficient protection against insect-induced crop yield losses. The utilization of insect-resistant maize resulted in a reduction of the amount of insecticide applied to 7.6 thousand tons of maize in 2012.
3. Tractors emit fewer greenhouse gases globally. The use of genetically modified crops reduced the need for spraying and irrigation, resulting in a 2.1 million tons reduction in $CO_2$ emissions from tractors in 2013 (Brookes and Barfoot, 2015, 2016).

4. Greater water-use efficiency, which aids in making crops more drought-resistant. Drought-tolerant maize hybrids have shown gains in water-use efficiency of up to 30% (Hao et al., 2015).
5. Protection from erosion and compaction for soils. Because GM crops require less soil tillage, the result is the sequestration of an additional 25.9 billion kg of carbon in the soil in 2013
6. Job creation in the green sector. For example, the adoption of transgenic crops in Argentina resulted in the creation of nearly one million new jobs over a ten-year period (Kamle et al., 2017).

## 12.6 DISADVANTAGES OF GREEN BIOTECHNOLOGY

Green biotechnology has certain adverse consequences for the ecology as well as human health and well-being. We'll take a closer look at the disadvantages of green biotechnology. Various concerns have been raised regarding the potential detrimental effects of green biotechnology on human health and well-being (Engel et al., 2002).

### 12.6.1 ALLERGENS AND TOXINS

An allergen is a type of antigen that produces an abnormally vigorous immune response in which the immune system fights off a perceived threat that would otherwise be harmless to the body. It is a pathogenic immune reaction that occurs when an antigen is transferred to a specific food component. The ingestion of the principal allergens (e.g., alimentary proteins) causes several allergic reactions in the cutaneous, respiratory, and circulatory systems, and these reactions can lead to serious health problems such anaphylactic shock. (Huang, 2017; Kramkowska et al., 2013). It is reported that approximately 2% of all adults and 6% of all children worldwide suffer from allergic reactions to food components (Bernstein et al., 2003; Huang, 2017; Kramkowska et al., 2013; Ladics et al., 2011).

Three potential negative outcomes of the novel proteins in genetically modified food have heightened public concern about their use in crops: they can act as allergens or toxins. That they can alter metabolic pathways in plants, resulting in the development of new allergies or poisons; or that they can reduce the nutritional quality of food. (Agnolo, 2005).

Food allergies have been linked to the consumption of foods derived from GM crops in numerous cases. The procedure of transferring genetic material from *Bacillus thuringiensis* to traditional maize cells to make StarLink maize resulted in increased production of the protein CRY9C, making this maize allergenic. The Environmental Protection Agency (EPA) has approved StarLink for use in animal feed. After the award of this licence, more concerns about StarLink maize surfaced as a result of its identification in human food products and reports. Another case of food allergy linked to the ingestion of a GM food is a GM soybean with enhanced methionine, an amino acid generated by a gene. This amino acid is the cause of hazardous side effects of ingesting the GM soybean. Furthermore, there is an increasing risk of potential morbidity as a result of GM food use due to tumor formation. For example, studies published in 2002 found that drinking milk from GM cows raises IGF-1 levels in consumers, resulting in a link between GM milk consumption and the development of lung, breast, and colon malignancies.

Furthermore, pesticide-resistant plants have been linked to an increased incidence of lymphoma development in animals and humans who eat food made from such plants. A decision tree approach for evaluating GM food allergenicity was created in the aforementioned publications (Batista and Oliveira, 2009; Huang, 2017; Key et al., 2008; Kramkowska et al., 2013), and it was later refined by the World Health Organization (WHO) (Batista and Oliveira, 2009; Huang, 2017; Key et al., 2008; Kramkowska et al., 2013).

As a result, a 90-day toxicity study in mice is required, as is allergenicity testing, which involves comparing the possible allergens of the GM crop to those of its conventional counterpart. As a result, it was highly advised that GM foods containing the CaMV 35S promoter not be commercially distributed (Agnolo, 2005).

### 12.6.2 Antibiotic Resistance Genes

During genetic modifications, antibiotic resistance genes are employed to discover features of interest. The prospect of transmitting antibiotic-resistant genes to GM plants or animals is a key issue for opponents of genetic engineering technology. The adoption of these gene markers, according to opponents of genetic modification, could lead to the emergence of new antibiotic-resistant bacteria strains. This in turn could result in the transfer of modified genes to bacteria in human and animal microfarad. This would encourage disease-causing microorganisms to develop antibiotic resistance, resulting in reduced antibiotic efficacy (Huang, 2017; Metcalfe, 2003). Some natural poisons or allergies could be transferred from donor microorganisms to genetically modified plants (Uzogara, 2000).

Furthermore, recent research has focused on the possibility of antibiotic-resistant genes being transferred to harmful bacteria in the animal or human gut.

Antimicrobial therapy efficacy would be reduced as a result. Recent research has also looked into the possibility of antibiotic-resistant genes being transferred to harmful bacteria in the animal or human gut. Antimicrobial therapy efficacy would suffer as a result (Richards et al., 2003). Moreover, some ecologists have claimed that releasing genetically modified crops into the environment could have unfavorable consequences for animals or humans who come into close contact with them. There are also concerns that pesticide-resistant transgenic crops may be dangerous to humans or animals who ingest them (Alberts et al., 2016).

## 12.7 CONCLUSION

We have explained the problem of food insecurity caused by climate change as well as recommended solutions to this problem throughout this systematic study. Green biotechnology is a new and crucial potential solution to food insecurity that is currently being evaluated. Green biotechnology techniques can be used to boost crop yields and nutritional value, as well as raise insect resistance and crop tolerance to biotic and abiotic challenges. As a result, from an economic standpoint, it has a good impact on long-term development. Yet, there are concerns about the prospective risks of green biotechnology to humans, animals, and ecosystem health, due to fear of the unknown and a lack of research on the long-term effects of green biotechnology on human and animal health. Due to the existence of such dangers, some components of green biotechnology will be unable to meet the UN Sustainable Development Goals relating to health unless appropriate safeguards are taken. These concerns suggest that additional research and examination into the long-term effects of green biotechnology is required if human health is to be protected. Sustainable development goals must be satisfied, as well as environmental goals.

## ACKNOWLEDGMENTS

The authors are grateful to the book editors for the opportunity to contribute.

## REFERENCES

Alberts, J.F., van Zyl, W.H., Gelderblom, W.C.A. (2016). Biologically based methods for control of fumonisin-producing *Fusarium* species and reduction of the fumonisins. *Front Microbiol* 7, 548. https:// doi.org/10.3389/fmicb.2016.00548

Ali, I. (2012). New generation adsorbents for water treatment. *Chem Rev* 112, 5073–5091.

Ali, I., Aboul-Enein, H.Y. (2004). *Chiral Pollutants: Distribution, Toxicity andAnalysis by Chromatography and Capillary Electrophoresis*. John Wiley & Sons, Chichester, UK.

Ali, I., Aboul-Enein, H.Y. (2006). *Instrumental Methods in Metal Ions Speciation: Chromatography, Capillary Electrophoresis and Electrochemistry*. Taylor & Francis Ltd, New York.

Ali, I., Aboul-Enein, H.Y., Gupta, V.K. (2009). *Nano Chromatography and Capillary Electrophoresis: Pharmaceutical and Environmental Analyses*. John Wiley & Sons, Hoboken, NJ.

Batista, R., Oliveira, MM. (2009). Facts and fiction of genetically engineered food. *Trends Biotechnol.* 27, 277–286. https://doi. org/10.1016/j.tibtech.2009.01.005

Bernstein, J.A., Bernstein, I.L., Bucchinil. (2003). Clinical and laboratory investigation of allergy to genetically modifed foods. *Environ. Health Perspect.* 111, 1114–1121. https:// doi.org/10.1289/ ehp.5811

Bhatnagar-Mathur, P., Vadez, V., Sharma, K.K. (2008). Transgenic approaches for abiotic stress tolerance in plants: Retrospect and prospects. *Plant Cell Rep.* 27, 411–424.

Brookes, G., Barfoot, P. (2014). Economic impact of GM crops: The global income and production effects. *GM Crops Food.* 5, 65–75. https://doi.org/10.4161/gmcr.28098

Brookes, G., Barfoot, P. (2015). Environmental impacts of genetically modifed (GM) crop use 1996–2013: Impacts on pesticide use and carbon emissions. *GM Crops Food.* 6, 103–133. https://doi. org/10.1080/21645698.2015.1025193

Bruinsma, J. (2009). By how much do land, water and crop yields need to increase by 2050? The resource outlook to 2050. Presented at the Food and Agriculture Organization of the United Nations Expert Meeting entitled "How to Feed the World in 2050", Rome, Italy, June 24–26.

Campbell, B.M., Thornton, P., Zougmoré, R., van Asten, P., Lipper, L. (2014). Sustainable intensification: What is its role in climate smart agriculture? *Curr. Opin. Environ. Sustain.* 8, 39–43. https://doi.org/10.1016/j.cosust.2014.07.002

D'Agnolo, G. (2005). GMO: Human health risk assessment. *Vet. Res. Commun.* 29(Supplement 2), 7–11. https://doi.org/10.1007/s11259-005-0003-7

Das, R., Ali, M.E., Hamid, S.B.A., Ramakrishna, S., Chowdhury, Z.Z. (2014). Carbon nanotube membranes for water purification: A bright future in water desalination. *Desalination.* 336, 97–109.

Debnath, K. (2013). *Book Review: Conway, Gordon, (2012). One BillionHungry: Can We Feed the World?* Cornell University Press, ISBN 0-8014-7802-2, pp. 456. J Agric Econ 64, 738–740. https://doi.org/10.1111/1477-9552.12029

Domingo, J.L. (2007). Toxicity studies of genetically modifed plants: A review of the published literature. *Crit. Rev. Food Sci. Nutr.* 47, 721–733. https://doi.org/10.1080/10408390601177670

Engel, K.H., Frenzel T., Miller, A. (2002). Current and future benefts from the use of GM technology in food production. *Toxicol. Lett.* 127(1–3), 329–336. https://doi.org/10.1016/S0378 -4274(01)00516-1

Faust, S.D., Aly, O.M., 1983. *Chemistry of Water Treatment.* Butterworth, Stoneham, MA.

Floros, J.D., Newsome, R., Fisher, W., Barbosa-Cánovas, G.V., Chen, H., Dunne,C.P. (2010). Feeding the world today and tomorrow: The importanceof food science and technology. *Compr. Rev. Food Sci. Food Saf.* 9, 572–599. https://doi.org/10.1111/j.1541-4337.2010.00127.x

Gallocchio, F., Belluco, S., Ricci, A. (2015). Nanotechnology and food: Brief overview of the current scenario. *Procedia Food Sci.* 5, 85e88.

Gartland, K.M.A., Gartland, J.S. (2016). Green biotechnology for foodsecurity in climate change. *Ref Module Food Sci.* 1–7

Gruère, G.P. (2012). Implications of nanotechnology growth in food andagriculture in OECD countries. *Food Policy.* 37, 191–198. https://doi.org/10.1016/j.jhazmat.2014.05.079

Guan, H.A., Chi, D.F., Yu, J., Li, H., (2010). Dynamics of residues from a novel nanoimidacloprid formulation in soyabean fields. *Crop Prot.* 29(9), 942e946.

Hao, B., Xue, Q., Marek, T.H. (2015). Soil water extraction, water use, and grain yield by drought-tolerant maize on the Texas High Plains. *Agric. Water Manag.* 155, 11–21. https://doi.org/10.1016/j.agwat.2015.03.007

Hashim, D.P., Narayanan, N.T., Romo-Herrera, J.M., Cullen, D.A., Hahm, M.G., Lezzi, P., Suttle, J.R., Kelkhoff, D., Munoz-Sandoval, E., Ganguli, S., Roy, A.K., Smith, D.J., Vajtai, R., Sumpter, B.G., Meunier, V., Terrones, H., Terrones, M., Ajayan, P.M., 2012. Covalently bonded three-dimensional carbon nanotube solids via boron induced nanojunctions. *Sci. Rep.* 2, 363.

Hillie, T., Hlophe, M. (2007). Nanotechnology and the challenge of clean water. *Nat. Nano.* 2, 663–664.

Huang, K. (2017). *Safety Assessment of Genetically Modified Foods.* Springer, Singapore. https://doi.org/10.1007/978-981-10-3488-6 i.2011.09.003

Ion, A.C., Ion, I., Culetu, A. (2010). Carbon-based nanomaterials:Environmental applications. *Univ. Politehn. Bucharest.* 38, 129–132.

Johnston, B.F., Mellor, J.W. (1961). The role of agriculture in economic development. *Am. Econ. Rev.* 51, 566–593.

Kamle, M., Kumar, P., Patra, J.K., Bajpai, V.K. (2017). Current perspectives on genetically modifed crops and detection methods. *3 Biotech* 7, 219. https://doi.org/10.1007/s13205-017-0809-3

Kanchi, S. (2014). Nanotechnology for water treatment. *J. Environ. Anal. Chem.* 1, 102.

Key, S., Ma, J.K.C., Drake, P.M.W. (2008). Geneticallymodifed plants and human health. *J. R. Soc. Med.* 101, 290–298. https://doi. org/10.1258/jrsm.2008.070372

Khot, L.R., Sankaran, S., Maja, J.M., Ehsani, R., Schuster, E.W. (2012). Applications of nanomaterials in agricultural production and crop protection: A review. *Crop Prot.* 35, 64e70.

Kramkowska, M., Grzelak, T., Czyzewska, K. (2013). Benefts and risks associated with genetically modifed food products. *Ann. Agric. Environ. Med.* 20(3), 413–419.

Kumar, S., Ahlawat, W., Bhanjana, G., Heydarifard, S., Nazhad, M.M., Dilbaghi, N. (2014). Nanotechnology-based water treatment strategies. *J. Nanosci. Nanotechnol.* 14, 1838–1858.

Ladics, G.S., Cressman, R.F., Herouet-Guicheney, C. (2011). Bioinformatics and the allergy assessment of agricultural biotechnology products: Industry practices and recommendations. *Regul. Toxicol. Pharmacol.* 60, 46–53. https://doi.org/10.1016/j.yrtph.2011.02.004

Lake, I.R., Hooper, L., Abdelhamid, A. (2012). Climate change andfood security: Health impacts in developed countries. *Environ. Health Perspect.* 120, 1520–1526. https ://doi.org/10.1289/ehp.11044 24

Lam, S.K., Ng, T.B. (2011). Lectins: Production and practical applications. *Appl. Microbiol. Biotechnol.* 89, 45–55. https ://doi.org/10.1007/s0025 3-010-2892-9

Laws, E.A. (2000). *Aquatic Pollution: An Introductory Text*, third ed. John Wiley & Sons, New York.

Mathews, J.A., Kruger, L., Wentink, G.J. (2018). Climate-smart agriculture for sustainable agricultural sectors: The case of Mooifontein. *Jamba: J. Disaster Risk Stud.* 10, 492. https ://doi.org/10.4102/jamba.v10i1.492

Mazur, B.J. (2001). Developing transgenic grains with improved oils, proteins and carbohydrates. *Novartis Found. Symp.* 236, 233–241. https://doi.org/10.1002/9780470515778.ch17

Menhas, R., Umer, S., Shabbir, G. (2016). Climate change and its impact on food and nutrition security in Pakistan. *Iran J. Public Health.* 45, 549–550.

Metcalfe, D.D. (2003) Introduction: What are the issues in addressing the allergenic potential of genetically modified foods? *Environ. Health Perspect.* 111, 1110–1113. https://doi.org/10.1289/ehp.5810

Mfutso-Bengo, J.M., Muula, A.S. (2007). Potential benefits and harm ofbiotechnology in developing countries: The ethics and socialdimensions. *Afr. J. Med. Med. Sci.* 36, 63–67.

Mukhopadhyay, S.S. (2014). Nanotechnology in agriculture: Prospects and constraints. *Nanotechnol. Sci. Appl.* 7, 63–71. https://doi.org/10.2147/NSA.S39409

Nasser, H.A., Mahmoud, M., Tolba, M.M. (2021). Pros and cons of using green biotechnology to solve food insecurity and achieve sustainable development goals. *Euro-Mediterr. J. Environ. Integr.* 6, 29. https://doi.org/10.1007/s41207-020-00240-5

Nemerow, N., Dasgupta, A., (1991). *Industrial and Hazardous Waste Treatment.* VanNostrand Reinhold, New York.

Oliveira, H.C., Moreira, R.S., Martinez, C.B.R., Grillo, R., Jesus, M.B., Fraceto, L.F., (2015). Nanoencapsulation enhances the post-emergence herbicidal activity of atrazine against mustard plants. *PloS One* 10, 1e12.

Parisi, C., Vigani, M., Cerezo, E.R., (2015). Agricultural nanotechnologies: What are the current possibilities? *Nano Today* 10, 124e127.

Pramanik, S., Pramanik, G., (2016). Nanotechnology for sustainable agriculture in India. In: Rajan, S., Dasgupta, N., Lichtfouse, E. (Eds.), *Nanoscience in Food and Agriculture 3. Sustainable Agriculture Reviews*, vol. 23. Springer, Cham, pp. 243e280.

Prasad, R., Kumar, V., and Prasad, K.S. (2014). Nanotechnology in sustainableagriculture: Present concerns and future aspects. *Afr. J. Biotechnol.* 13, 705–713. https://doi.org/10.5897/AJBX2013.13554

Pretty, J., Bharucha, Z.P. (2014). Sustainable intensification in agriculturalsystems. *Ann. Bot.* 114, 1571–1596. https ://doi.org/10.1093/aob/mcu205

Pretty, J., Toulmin, C., Williams, S. (2011). Sustainable intensificationin African agriculture. *Int J Agric Sustain* 9, 5–24. https ://doi.org/10.3763/ijas.2010.0583

Qu, X., Alvarez, P.J.J., Li, Q. (2013a). Applications of nanotechnology in water and wastewater treatment. *Water Res.* 47, 3931–3946.

Qu, X., Brame, J., Li, Q., Alvarez, P.J.J. (2013b). Nanotechnology for a safe and sustainable water supply: Enabling integrated water treatment and reuse. *Acc. Chem. Res.* 46, 834–843.

Ramakrishna, S., Fujihara, K., Teo, W.-E., Yong, T., Ma, Z., Ramaseshan, R. (2006). Electrospun nanofibers: Solving global issues. *Mater. Today.* 9, 40–50.

Ramos, J.L., García-Lorente, F., Valdivia, M., Duque, E. (2017). Green biofuelsandbioproducts: Bases for sustainability analysis. *Microb. Biotechnol.* 10, 1111–1113.

Richards, H.A., Han, C.T., Hopkins, R.G. (2003). Safety assessment of recombinant green fuorescent protein orally administered to weaned rats. *J. Nutr.* 133, 1909–1912. https://doi.org/10.1093/ jn/133.6.1909

Sabir, S., Arshad, M., and Chaudhari, S.K. (2014). Zinc oxide nanoparticles forrevolutionizing agriculture: Synthesis and applications. *Sci. World J.* 2014, 8. https://doi.org/10.1155/2014/925494

Sabourin, V., Ayande, A., (2015). Commercial opportunities and market demand for nanotechnologies in agribusiness sector. *J. Technol. Manage. Innov.* 10(1), 40e51.

Schellekens, H. (1999). Health risks of genetically modifed foods. *Lancet* 354, 70–71. https://doi.org/10.1016/S0140-6736(05)75334-4

Seabra, A.B., Rai, M., Duran, N., (2013). Nano carriers for nitric oxide delivery and its potential applications in plant physiological process: A mini review. *J. Plant Biochem. Biotech.* 23(1), 1e10.

Sekhon, B.S., (2014). Nanotechnology in agri-food production: An overview. *Nanotechnology, Science and Applications* 7(2), 31e53.

Singh, N.A., (2016). Nanotechnology definitions, research, industry and property rights. In: Ranjan, S., Dasgupta, N., Lichtfouse, E. (Eds.), *Nanoscience in Food and Agriculture 1. Sustainable Agriculture Reviews*, vol. 20. Springer, Cham, pp. 43e64.

Smith, P., Martino, D., Cai, Z. (2008). Greenhouse gas mitigation inagriculture. *Phil. Trans. R Soc B Biol. Sci.* 363, 789–813. https://doi.org/10.1098/rstb.2007.2184

Tchobanoglous, G., Franklin, L.B., (1991). *Wastewater Engineering: Treatment, Disposal and Reuse.* McGraw Hill Inc., New York.

Teng, B.S., Wang, C.D., Yang, H.J., Wu, J.S., Zhang, D., Zheng, M., Fan, Z.H., Pan, D., Zhou, P. (2011). A protein tyrosine phosphatase 1B activity inhibitor from the fruiting bodies of Ganodermalucidum (Fr.) Karst and its hypoglycemic potency on streptozotocininduced type 2 diabetes mice. *J Agr Food Chem.* 59, 6492e6500.

Tester, M., and Langridge, P. (2010). Breeding technologies to increase crop production in a changing world. *Science* 327, 818–822.

Theron, J., Walker, J.A., Cloete, T.E., (2008). Nanotechnology and water treatment: Applications and emerging opportunities. *Crit. Rev. Microbiol.* 34, 43–69.

Uzogara, S.G. (2000). The impact of genetic modifcation of human foods in the 21st century: A review. *Biotechnol. Adv.* 18, 179–206. https://doi.org/10.1016/S0734-9750(00)00033-1.

Van de Wiel, C.C.M., van der Linden, C.G., Scholten, O.E. (2016). Improving phosphorus use efficiency in agriculture: Opportunities for breeding. *Euphytica* 207, 1–22. https://doi.org/10.1007/s10681-015-1572-3.

Van Montagu, M. (2020). The future of plant biotechnology in a globalized and environmentally endangered world. *Genet. Mol. Biol.* 43(Supplement 2), e20190040. https://doi.org/10.1590/1678-4685-gmb-2019-0040.

Vermeulen, S.J., Aggarwal, P.K., Ainslie, A. (2012). Options for supportto agriculture and food security under climate change. *Environ. Sci. Policy.* 15, 136–144. https://doi.org/10.1016/j.envsc.

Wager, R. (2009). Comment on "The future of agriculture." *EMBO Rep.* 10, 104–105. https://doi.org/10.1038/embor.2008.250.

Wani, K.A., Kothari, R. (2018). Agricultural nanotechnology: Applications and challenges. *Ann. Plant Sci.* 7, 2146e2148.

Yan, L., Kerr, P.S. (2002). Genetically engineered crops: Their potential use for improvement of human nutrition. *Nutr. Rev.* 60, 135–141. https://doi.org/10.1301/00296640260093797.

Yashveer, S., Singh, V., Kaswan, V. (2014). Green biotechnology, nanotechnology and bio-fortification: Perspectives on novel environment-friendly crop improvement strategies. *Biotechnol. Genet. Eng. Rev.* 30, 113–126. https://doi.org/10.1080/02648725.2014.99262 2.

Yunlong, C., Smit, B. (1994). Sustainability in agriculture: A general review. *Agric. Ecosyst. Environ.* 49, 299–307. https://doi.org/10.1016/0167-8809(94)90059-0.

# Index

For Product Safety Concerns and Information please contact our EU representative GPSR@taylorandfrancis.com
Taylor & Francis Verlag GmbH, Kaufingerstraße 24, 80331 München, Germany

www.ingramcontent.com/pod-product-compliance
Lightning Source LLC
LaVergne TN
LVHW010544110826
845149LV00003B/553

* 9 7 8 1 0 3 2 2 2 0 3 9 0 *